OFFSHORE STRUCTURE
MODELING

ADVANCED SERIES ON OCEAN ENGINEERING

Series Editor-in-Chief
Philip L- F Liu (*Cornell University*)

Advanced Series on Ocean Engineering – Volume 9

OFFSHORE STRUCTURE MODELING

SUBRATA KUMAR CHAKRABARTI
Chicago Bridge & Iron Technical Services Co.
Plainfield, Illinois
USA

World Scientific
Singapore • New Jersey • London • Hong Kong

Published by

World Scientific Publishing Co. Pte. Ltd.

P O Box 128, Farrer Road, Singapore 9128

USA office: Suite 1B, 1060 Main Street, River Edge, NJ 07661

UK office: 73 Lynton Mead, Totteridge, London N20 8DH

OFFSHORE STRUCTURE MODELING

ISBN 981-02-1512-6
      981-02-1513-4 (pbk)

Printed in Singapore by JBW Printers & Binders Pte. Ltd.

DEDICATION

This third book by the author is dedicated with pride to the following people that had the greatest contribution in shaping his professional life.

- *His professor and Ph.D. thesis adviser, Dr. William L. Wainwright who gave him the first lessons in writing a technical paper which became the first publication by the author.*

- *His first supervisor, Mr. William A. Tam, then Director of Marine Research, who had faith in the author's ability and gave him an opportunity to work on hydrodynamic related subjects.*

- *Dr. Basil W. Wilson, then consultant of Chicago Bridge and Iron Co. who taught the author many of the basics of ocean engineering and mooring systems. The author had the good fortune then in cooperating with Dr. Wilson on several research projects.*

ACKNOWLEDGEMENTS

Since joining CBI twenty-five years ago I have been involved in model testing of offshore and marine structures. Many colleagues of mine have helped me understand tricks of model testing over the years. In particular, Erik Brogren provided guidance on model construction details and Alan Libby explained intricacies of instrumentation design.

Many experts reviewed the chapters of this book. Dr. Devinder Sodhi and Prof. Tom Dawson reviewed Chapter 2. Prof. Dawson also checked sections of Chapter 7. Dr. E.R. Funke reviewed Chapter 4. Prof. Christian Aage took the time to review Chapters 4, 7 and 10 and provided valuable comments. Chapter 5 was reviewed by Prof. Bob Hudspeth, Dr. E. Mansard, and Dr. Andrew Cornett. Dr. Erling Huse commented on Chapter 6. Chapter 7 was read by Prof. Li and Dr. Ove Gudmestad. Prof. S. Bhattacharyya improved on Chapter 8. Chapter 9 was reviewed by Dr. O. Nwogu. Keith Melin again reviewed the entire book and provided many editorial and other comments which improved its quality. I am grateful to these individuals. I am, however, responsible for any shortcomings in this book.

CBI Technical Services provided the secretarial help which made this book possible. Many individuals helped in putting the manuscript together, of which the most noteworthy is Ms. Danielle Cantu who finalized the manuscript in its printed form by retyping and reformatting it many times. Finally, I acknowledge the patience of my wife Prakriti (Nature) for providing me the time at home over the last 3 years to complete the book.

PREFACE

The offshore industry has matured over the years through innovation, initiatives and experience. The industry has advanced a long way to its present stage from its first installation of an exploration structure in coastal waters of the Gulf of Mexico in the 1940's. The early structures were fixed to the ocean floor and looked much like the electrical transmission towers common on land. Today the shape, size and type of offshore structures vary depending on the required applications. Offshore structures are abundant in all parts of the world. In addition to structures fixed at their base, moored floating structures and vertically tethered structures have been installed for exploration, production, storage and offshore processing of crude oil. While offshore exploration has been relatively trouble free, there have been a few catastrophic failures. Therefore, design of these structures for accident-free operation under anticipated conditions is vital for the continued success and growth of the offshore industry. One of the means of verifying the design of a structure is the testing of scale model of the structure in a simulated ocean environment during its design phase.

Model testing has been performed through the history of mankind. Systematic hydraulic scale model testing goes back to the nineteenth century. However, even today modeling is partly an art as well as a science. In modeling, certain laws of similarity are followed. Several text books are available that deal with these similarity laws. In many cases, these laws can not all be satisfied in a model test. In these cases, it is necessary to selectively distort some of these scaling laws to perform the model tests. Although this distortion is somewhat of a compromise, valid modeling results can still be expected.

While the technical literature dealing with offshore structures discusses model testing, a comprehensive book in this area discussing design, construction, instrumentation, testing and analysis of physical model is lacking. It is desirable that a single text contain the theoretical and practical aspects of physical modeling. Such a book should be valuable to engineers dealing with the design, construction, installation and operation of offshore structures. This requirement inspired me to write this book on modeling. This book provides reasonably detailed coverage of the technology of model testing. As such, it has applications throughout the entire field of engineering, reaching far beyond its focus on offshore structures. It should be equally appealing to engineers and scientists involved with the design and construction of unique structures.

The Introduction discusses the general need for model testing. A brief history of testing has been given in this section. Some of the general structural areas where model testing has been required have also been mentioned. Chapter 2 describes the modeling laws. It begins with the general requirements for similarity. A general discussion of the famous Buckingham pi theorem and an application of the pi theorem has been included. A few specific examples of modeling are discussed here including structural modeling, testing in uniform flows and modeling distortion. These are considered unique cases that may be applicable to an offshore structure. In hydraulic testing, however, Froude model law is discussed in detail in this section. A few textbooks that deal with the similarity laws and model testing have been discussed and referenced here. The methods of model construction are considered in Chapter 3. The physical requirements of the model necessary for scale testing are explained. Generally, the construction technique varies based on these requirements. For example, the fixed structures are generally used in the measurement of loads and stresses imposed by the environment. In this case, the dynamic properties of the structure are not a concern. This, however, is not true for compliant or floating structures. The compliant structure, in addition, must satisfy additional scaling laws. The static and dynamic properties that must be satisfied by these structures are discussed in detail. The techniques used in verifying these properties before testing can take place are illustrated.

The testing of offshore structures requires specialized facilities. Many of the small facilities that exist at the universities and other educational institutions are used as teaching tools in discussing the needs and methods of model testing. However, many larger commercial facilities are in existence in various parts of the world. Many of these facilities are described in detail in Chapter 4 including their capabilities and limitations. This section will be useful to a design engineer in choosing a suitable facility for his particular test requirements. The important components of these testing facilities, such as the wave generators, the current generators, the towing carriages and beaches are described. The most important feature of these testing facilities is the wave generation capability. A few theories of wave generation and beach reflection are presented for those interested in designing wavemakers. These facilities are used in duplicating the ocean environment. The modeling of the ocean environment is the subject of Chapter 5. Simulation of various types of waves, such as random two-dimensional and three-dimensional waves, wave groups, and higher harmonic waves, are discussed in detail. The wind and current generation and co-generation of current and waves are also explained in this section. Another important requirement of any model testing is the measurement of the responses of a structure model. This includes the inputs to structures from external sources and the corresponding outputs. In hydraulic testing, this measurement is further complicated by the presence of fluids. Various types of measuring instruments and measuring techniques are introduced in Chapter 6. Design methods of a few specialized instruments are described and methods of the waterproofing these instruments are discussed. The calibration procedure for special instruments is shown. Typical calibration curves of a few of these are also included. The important considerations in the recording of data output from these instruments are given here.

The actual modeling of various offshore structures is described in Chapters 7-9. Various areas covered in these chapters are outlined in the following:

FIXED STRUCTURES	OFFSHORE OPERATIONS	FLOATING STRUCTURES
Gravity Platform	Transport of Jackets	Buoys
Storage Structures	Towing of Structures	Single Point Moorings
Piled Jackets	Launching of Structures	Moored Tankers
Subsea Pipelines	Submergence of Structures	Tension Leg Platforms
	Pipelaying	Compliant Structures
		Semisubmersibles

Chapter 7 deals with the fixed structures. The methods of installing load measuring devices on fixed structures and associated problems are described in detail. Various examples are included. Actual recorded data are given to illustrate the validity as well as inadequacies of the techniques. Offshore operations include special techniques in delivering the completed structure to the offshore sites. These require launching, transportation, submergence and installation of these structures. These various stages are generally model tested to insure a proper installation procedure and to identify potential unforeseen problems. This is the subject of Chapter 8. Chapter 9 covers the area of seakeeping tests. These tests include floating structures moored to the ocean floor by mooring lines, articulated columns, vertically tethered structures, e.g., tension leg platforms and compliant structures. Special care is needed to insure proper duplication of model response without introduction of additional effects through setup. Example cases are discussed and illustrated.

In all these tests, data are recorded by the various instruments installed on the model. These data require special routines to reduce raw recorded data to a usable form for application in a design and analysis. Special care is taken to avoid spurious data entering into the test data. The data analysis techniques that may be used in reducing test results are described in Chapter 10. Examples are taken from specific tests to illustrate these techniques.

It should be noted here that the dimensional units in describing quantities in the book are somewhat mixed. This reflects the slow transition in this country from the English to the metric system. Wherever possible dual units have been provided. In general metric system has been followed. However, with conversion factor noted, some examples have been left in the English units. It is hoped that readers will have appreciation for both systems in following the material in the book.

A list of symbols has been included at the end of the book. Attempt has been made to maintain consistency throughout the book. A few variables have been used with dual meanings in two different parts of the book with little confusion. Some local variables have not been included in the list to limit its size. They have been defined locally where they appear. Subscripts and superscripts have been defined in the list. A list of all the abbreciations appearing in the book has also been included for convenience.

As is evident from its contents, this book should be a valuable addition to the library of all offshore engineers and naval architects whether they are involved in the research, design, construction or offshore operations. All hydraulic and ocean engineering curriculums in universities offer a course in modeling. This book should be a very appropriate reference for such a course. It is written such that it may be used as a text for a junior or senior level course in a four year engineering curriculum. Since the book deals with many subjects in modeling that go beyond the specifics for offshore structures, it should be found useful by all engineers and scientists interested in structural or hydraulic testing. Finally, it should also be valuable as a reference to many model testing facilities as a complement to their expertise in the area.

TABLE OF CONTENTS

OFFSHORE STRUCTURE MODELING

"I contend that unless the reliability of small scale experiments is emphatically disproved, it is useless to spend vast sums of money on full scale trials which afterall may be misdirected unless the ground is thoroughly cleared beforehand by an exhaustive investigation on a small scale."

William Froude, 1886

CHAPTER 1

INTRODUCTION

1.1 MODEL TYPES

Engineers constantly deal with models. Models are fundamental to communication. We learn from models. These models may be physical, mathematical, graphic or semantic. Mathematical models and methods are invaluable in guiding experimental work. The converse is also true. Regardless of the mode of model construction, each of these models represents a set of facts and has an intrinsic value. Each allows us to confirm a new concept by demonstrating the suitability, workability, and constructability of the concept. Models are used as tools that verify a design.

Models may be classified in two major categories: display models and engineering models. Display models are generally used to explain a concept or sell a product. An example of such a model of an exploration drilling structure is shown in Fig. 1.1. These models provide visual aid for engineering drawings and may also have moving parts. This latter category of display models is called a working model. It may be properly scaled and generally shows all of the major components. However, some of the details may be omitted for simplicity. The working schemes of these models are designed to demonstrate or highlight any special feature of the product.

Engineering models, on the other hand, are used to collect data useful in the design of the product under consideration. These models may be divided into two major categories: constructability models and measurement models. The constructability models are built to scale to insure that the concept design is feasible for construction. Instrumentation is generally not used on these types of models. Often, a new process or a system may be verified with a model of this type. In this case, a portion of the system or an entire pilot plant is built and operated as a proof of the concept. For example, the habitability of a floating offshore hotel is the most important aspect of its operation which may be model tested at an early stage of development. Quite often, improvements to the original process are developed by this type of modeling. The results, however, are not directly scalable to the full size process or prototype, due to possible additional constraints.

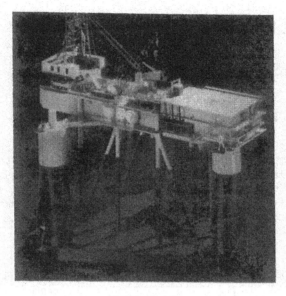

FIGURE 1.1
DISPLAY MODEL OF A DRILLING RIG

The measurement models, on the other hand, are mainly used to obtain engineering data from a scale model for direct use in the design or operation of the prototype. These model properties and their test environment follow certain similarity laws, often derived from a dimensional analysis. Important nondimensional quantities are taken into account in the modeling laws.

1.2   BRIEF HISTORY OF MODEL TESTING

From the beginning of recorded history, models have been used not only to visualize various structures (e.g., pyramids) but also as a working plan from which prototype structures have been constructed. In particular, ship models have enjoyed a long and useful history dating back at least to the time of the Pharaohs (as witnessed by ship models in the uncovered tombs).

Ship designers and shipwrights used essentially the same model techniques well into the 1600's. These models were, in essence, the precursor of analog computer, the basic precept being--as the model behaves, so will the prototype. Even without a strong analytical background, this method has been successfully used for centuries.

Stability models for ships have been (and in some cases still are) used to determine placement of cargo and ballast by the cargo officer on board ship.

Working mechanical models came into use during the industrial revolution, and several fine examples have been preserved in the British museum and the Smithsonian. In these cases, the design of the prototype structure was scaled directly from the working model. Many of these models were tested for considerable lengths of time prior to building the prototype so that areas of insufficient structural strength could be incrementally strengthened until such time as the fatigue limits were satisfactory with respect to the expected life cycle of the prototype.

The first major insight into the phenomena of modeling fluid mechanics was gained by Reynolds and Froude, wherein they developed criteria for both the viscous and inertial effects respectively. They were followed by Lamb, Stokes, Boussinesq and others until the present state of hydrodynamics was reached.

Although Osborne Reynolds (1842-1912) was a vigorous investigator of almost all physical phenomena, he was one of the first to elaborate upon scale effects in the areas of marine propulsions and hydrodynamic research. While exploring the apparently discordant results of experiments on the head loss during flow in parallel tubes by Poiseuille and Darcy, Reynolds argued that since viscous forces tend to produce stability while inertia forces tend to cause instability, the change of one type of motion to another could rely upon the ratio of these forces and would occur at some critical value of this ratio. The expression is now universally called the Reynolds number.

William Froude (1810-1879) proposed the idea of a bilge keel on the trial voyage of a ship to observe the rolling motion of the vessel. His first paper dealt with this rolling motion which he judged to have a greater need for study than the problems associated with hull form. In 1870 he began a series of experiments to study the resistance of ships using a towing tank. In 1876, the British Association set up a committee of which Froude was a member to study the propulsion stability and sea-going quality of ships.

These experiments led to his statement now known as Froude's law of comparison: "In fact, we are thus brought to the scale of comparison which was just now enunciated, that the entire resistance of a ship and a similar model are as the cube of their respective dimensions if their velocities are as the square roots of their dimension".

1.3   PURPOSE OF MODEL TESTING

Models of physical systems are, in essence, systems from which the behavior of the original physical system (i.e., prototype) may be predicted. The use of models is particularly advantageous when the analysis of the prototype is complicated or uncertain and when construction of the prototype would be uneconomical and risky without preliminary prediction of performance.

Model testing offers great savings when compared to full scale tests. However, it can still become an expensive undertaking. The larger the model, the better the test data, and the easier it is to scale up to the prototype values; but the cost of the model test can also increase substantially. Therefore, the model test should be planned carefully to reduce the time and cost of testing, while maintaining reliability.

While there are a number of peripheral reasons for model testing, the major justifications for model testing are as follows:

- Investigate a problem or situation which can not be addressed analytically.

- Obtain empirical coefficients required in analytical prediction equations.

- Substantiate an analytical technique by predicting model behavior and direct correlation between the predicted behavior and the actual behavior.

- Evaluate the effect of discarded higher order terms in a simplified analytical prediction model by correlating the discrepancy between the predicted model behavior and the actual model behavior.

Thus, model testing is an experimental procedure which is most generally used where the analytical techniques fail to predict the expected behavior of the prototype either within the tolerances required or within the confidence level required for good design.

Models may be, generally, classified in three groups. The first group consists of small-scale replicas of the prototypes, in which the behavior of the model is identical in nature with that of the prototype, but responses differ in magnitude with the chosen scale factor.

A second group is that of distorted models, in which a general resemblance exists between model and prototype; but some factors, such as distance (for example, water depth in an open channel flow), are distorted. The distortion of a model requires the use of prediction factors in translating model behavior to prototype behavior.

A third class of models is known as the analog. It consists of systems which are dissimilar in physical appearance to their prototypes but which are governed by the same class of characteristic equation, such as the Poisson or the Laplace equation. In such models, each element in the prototype has a corresponding item in the model. For example, a simple one-degree-of-freedom mechanical vibrating system may have, as an analog, an R-C-L electrical circuit where the inductance (L) of the model is proportional to the mass of the prototype, capacitance (C) to spring constant, resistance (R) to viscous damping, applied voltage to driving force, charge to displacement, current to velocity and so on. Sometimes, a mechanical simulator is applied to the model subjected to waves which simulate, for example, the wind load. Examples of mechanical simulators will be described in Chapter 4.

In hydrodynamics and in free-surface fluid flow problems (e.g., open oceans or rivers, etc.), the models are, generally, exact duplicates of the prototype [Kure (1981)]. Occasionally, there is a need for some distortion in certain directions due to the limited size of a testing facility and choice of the scale factor.

While this book is focused on physical modeling, there is another area of modeling that is gaining popularity and confidence quickly. This is the area of computer modeling which is essentially theoretical analysis of a problem using computational techniques. Numerical modeling, especially finite element modeling, is rapidly increasing in power and sophistication. At some point, engineers may be able to conduct full-scale modeling of a numerical prototype in a numerical ocean, with full accounting of inertial and viscous forces. Such numerical models will be created and tested with a possible savings of both time and money. Numerical sea state need not be limited by wave-making machines, and conditions leading to distortions in model geometry and properties need not be relaxed. Numerical models will obviate the need for separate environmental load models and seakeeping models.

Computers will become a more useful engineering tool as they continue to increase in speed and memory and decrease in cost. Numerical wind tunnels are gaining acceptance. There has been some discussion about the development of a numerical wave tank (Newman, MIT, during Weinblum lecture series in 1990). Digital computer models are replacing/supplanting some scale models, for example, in the analysis of airplanes and missiles. It is, however, too early to numerically model a complex offshore structure in a realistic sea state. Scale models constitute the most accurate and state of the art engineering tool available today. Given the physical nonlinearities and geometric complexities of many offshore structures, it is unlikely that virtual offshore modeling will become a routine procedure for years to come.

While computer models have been increasingly successful in simulating an everwidening range of engineering problems, it is nevertheless essential that advances in these models are validated and verified against experiment. Experimental measurements are themselves conditioned to the requirements of the computational models. Hence it is important that scientists working on experiments communicate with researchers developing computer codes as well as those carrying out measurements on prototypes.

Numerical experiments will probably never replace wave tank experiments completely, because many physical uncertainties will still prevail in a numerical model. It is expected that physical and numerical experiments will complement each other and guide the development of an efficient structure design.

1.4   MODELING CRITERIA

In building an engineering model, one tries to produce a facsimile of a particular product that will perform in a manner similar to the actual product being designed and constructed (called the prototype). The model may be full size, as is often the case in automotive and other industries where the product is relatively small and mass producible; or it may be small relative to the prototype as is often the case with very large structures of unique design as well as expensive design of limited products. In either case, the model must act in a manner physically similar to the prototype, and the similarity must be governed by rules that quantify the actions and allow them to be scaled up for use in predicting the action of the prototype. In some instances, the model may be larger than the prototype, e.g., in scaling small components or parts.

The rules for quantifying and scaling model responses are called the laws of similitude. For the study of fluid phenomena, there are three basic laws. The first is geometric similitude or similarity of form. Under this law, the flow field and boundary geometry of the model and of the prototype must have the same shape. Consequently, the ratios of all model lengths to their corresponding prototype lengths are equal.

The second basic law of similitude is kinematic similitude, or similarity of motion. According to this law, the ratios of corresponding velocities and accelerations must be the same between the model and the prototype. Thus, two different velocity components in the model must be scaled similarly with respect to the corresponding velocities of the prototype. Given geometric similitude, in order to maintain kinematic similitude, dynamic similitude or similarity of forces acting on the corresponding fluid must exist. Five forces that may affect the fluid structure interaction due to the flow field around the structure are the forces due to pressure, gravity, viscosity, surface tension and elasticity. Thus, the ratios of these forces between the model and the prototype must be the same.

For wave action, the surface tension is generally quite small and neglected. The elasticity is generally ignored for large offshore structures. The ratio of the inertia force to the viscous force is called the Reynolds number, while the ratio of the inertia force to the gravity force is the Froude number. While the Reynolds and Froude effects are generally present, the Froude number is considered the major scaling criterion in the water wave problems.

Geometric similarity ensures the identity of functional relations among different parameters. Thus, it is a necessary condition for the existence of dynamic similarity. Distortion is common in hydraulic modeling (e.g., estuary) which invariably provides deviation from geometric similarity. For example, the distribution of velocity of a flow in a cross-section depends very strongly on the cross-section geometry. Hence, by distorting the model and thus the shape of the cross-section, the velocity distribution and its effects are inevitably distorted.

Due to economic and practical reasons, in a conventional modeling technique, water is used as the model fluid. If the model operates with the prototype fluid (i.e. water) then the realization of dynamic similarity of a gravity dependent phenomenon in a small scale model is impossible in principle. It is well known that with water both Reynolds and Froude similitude cannot be achieved.

The importance of any Reynolds number, however, decreases as its value increases. This phenomenon will be illustrated further in discussing specific modeling problems. Hence, when modeling hydrodynamic phenomenon, the usual practice is to build the model as large as possible and, therefore, minimize the importance of the viscosity parameter. Sometimes, turbulence is artificially induced in the flow ahead of the model to obtain the effect of a large Reynolds number.

There are three major factors that influence the scale selection in waves: (1) model construction, (2) tank blockage, and (3) wave generation capability. For building models, generally a larger scale results in easier construction and material selection. In terms of tank blockage, the smaller the scale is, the better will be the quality of waves in the tank with less contamination due to reflection from the model. Finally, the wavemaker must be capable of producing the scaled wave heights and periods desired for the tests. Each of these criteria may lead to different "best scales". The wave generating properties in the tank, having fixed limitations, generally control the model scale.

The scale is chosen as a compromise between cost of the project and the technical requirements for similitude. It should be noted that the popularly held belief that the larger the scale model, the better it is, may not always be true. Several other considerations dictate the scale selection. Sometimes, more than one model is justified to study the different phenomena experienced by a structure. For example, the entire structure may be at one scale, whereas a small section of it may be studied using a larger scale model.

FIGURE 1.2
MOTIONS OF A RESIDENT TANKER AND MOORING TOWER
(SCALE = 1:48)

This is illustrated by an example of a buoyant tower. A permanently moored buoyant tower-tanker system is often used offshore for processing and storing crude oil. A 1:48 model of the tower and tanker moored with a yoke is shown in Fig. 1.2 being tested in a wave tank. This model allowed the study of the response of the overall system. The mooring system is often designed with a quick disconnect mechanism for emergency detachment. As a demonstration model for the design concept of the mechanical system, a 1:12 model of the quick disconnect assembly is shown in Fig. 1.3. The ship yoke and support system is mounted on a plywood barge. The operators could board the barge and manage the winch controls to simulate docking. This larger model is used for training personnel and give qualitative information on sea state limitations acceptable for such an operation.

FIGURE 1.3
MODEL OF RIGID YOKE ARM CONNECTING ARTICULATED TOWER AND TANKER UNDER TEST (SCALE = 1:12)

While it is not possible to recommend an optimum scale factor for a structure without investigating all parameters of importance, a common scale factor used in water wave effects in a tank is 1:50. A range of scales for a wave tank is typically between 1:10 and 1:100.

1.5 PLANNING A MODEL TEST

For a model test to be successful, proper planning is required. Care should be taken to consider all aspects of model testing with the ultimate goal in mind. It is often

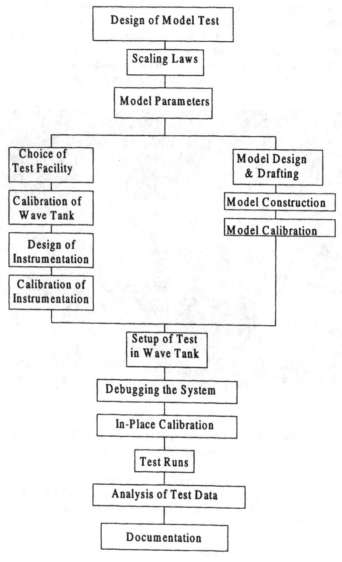

FIGURE 1.4
EXECUTION OF A MODEL TEST

a common mistake to expect too many results from one model test. Instead, the test plan should focus on a few most important aspects of modeling. This will help in choosing a proper scale factor and other parameters of the test.

A general procedure for a model test is outlined in Fig. 1.4. One of the most important criteria for successful model testing is the evaluation of the scaling law. The scaling parameters for the results sought in a model test must be established first before a scale factor is chosen. Once the scale factor is established, the input parameters may be computed. This may help in deciding on the best testing facility from those available. Concurrently, the model sizes may be determined, and the design of the model may continue. In the design of the model, proper attention must be given to the attachment and effect of instrumentation on the model. The model, instrumentation and the wave tank must be properly calibrated before the model is placed in the tank. For example, the random wave to be used in testing should be generated in the absence of the model in the tank. The in-place calibration of the model is very important. Valuable information may be obtained from the pre- and post-test calibration. A check of the calibration during the test runs insures accuracy of collected data. Also, testing should be properly documented for future reference. Any unusual observations must be noted by the test engineer in the lab notebook. These areas are explored in more details in the future chapters.

1.6   REFERENCES

1.   Kure, K., "Model Tests With Ocean Structures", Applied Ocean Research, Vol. 3, No. 4, 1981, pp. 171-176.

CHAPTER 2

MODELING LAWS

2.1   GENERAL DISCUSSIONS OF SCALING LAWS AND METHODS

Most physical systems can be investigated through small scale models whose behavior is related to that of the prototype in a prescribed manner [Soper (1967)]. The problem in scaling is to derive an appropriate scaling law that accurately describes this similarity. This requires a thorough understanding of the physical concepts involved in the system. One method of relating the model properties to the prototype properties is the parametric approach in which the Buckingham Pi Theorem is applied to all applicable variables to derive a group of meaningful dimensionless quantities. This method assumes that nothing is known about the governing equations of the system. If, however, these equations are known apriori, then the scaling law may be deduced from these equations by writing them in dimensionless form.

There are two generally accepted methods by which scaling laws relating two physical systems are developed. The two physical systems in this context are the prototype and the model. The first method is based on the inspectional analysis of the mathematical description of the physical system under investigation. Fluid mechanics problems deduce the condition of similitude from the Navier Stokes equation. A method of similarity analysis is adopted here (see references at the end of this chapter). The dynamics of the physical system are described by a system of differential equations. These equations are written in nondimensional terms. Since the simulated physical system duplicates the full-scale system, these nondimensional quantities in the differential equations must be equal for both. Then, the equality of the corresponding nondimensional parameters governs the scaling laws. This method assures similarity between the two systems but is dependent upon knowing explicitly the governing equations for both the prototype and the model.

A second method is based upon the well-known Buckingham Pi Theorem. In this approach, the important variables influencing the dynamics of the system are identified first. This is the most important step in the similarity analysis by this method. If a significant parameter is omitted, then the resulting scaling laws will be erroneous. On the other hand, if too many variables including those least significant are included, then the scaling laws become too complicated and, often, impossible to satisfy. It should be realized that a complete similitude cannot be obtained except at a one-to-one scale.

Therefore, the parameters of least significance are neglected. The scale is chosen as a compromise between cost, complexity and technical requirements for similitude.

Once the variables are identified, their physical dimensions are noted. Based on the Buckingham Pi Theorem (to be introduced shortly), an independent and convenient set of nondimensional parameters (pi terms) is constructed from these variables. The equality of the pi terms for the model and prototype systems yields the similitude requirements or scaling laws to be satisfied. The model and prototype structural systems are similar if the corresponding pi terms are equal.

In order to arrive at the prototype values from the model test results, similitude relationships are used to formulate the system equations and prediction equations. Let us define the characteristic equation of a system as

$$\pi = \phi(\pi_1, \pi_2,...\pi_n) \tag{2.1}$$

where the function ϕ is generally an unknown. The quantities $\pi_1,...\pi_n$ are the dimensionless group on which dimensionless π is considered dependent. The corresponding characteristic equation for the model is

$$\pi_m = \phi_m (\pi_{1m}, \pi_{2m},...\pi_{nm}) \tag{2.2}$$

where the subscript m stands for model. The two phenomena between the prototype and the model are identical if the functions $\phi = \phi_m$ which provides the system equations

$$\pi_i = \pi_{im} \qquad i = 1, 2,...n \tag{2.3}$$

In this case

$$\pi = \pi_m \tag{2.4}$$

This relationship provides the prototype values from the model test and is called the prediction equation of the system. Sometimes, due to the limitation in scaling, distortion modeling is used. Then

$$\pi_{im} = \varepsilon \, \pi_i \tag{2.5}$$

so that
$$\pi = \alpha\pi_m \tag{2.6}$$

In this case, the relationship between ε and α has to be established.

2.2   BUCKINGHAM PI THEOREM

The Buckingham Pi theorem may be stated as follows. Let the number of fundamental units, e.g., M,L,T (mass, length, time) needed to express all variables included in a problem be R. Let the number of variables employed to describe a phenomenon be N. It can be shown that the equation giving the relation between the variables will contain N-R dimensionless ratios which are independent. In other words, only   N-R   dimensionless quantities are required to establish the functional relationship.

The relationship among the variables is expressed by an exponential equation. Then the values of the exponents are solved for, assuming dimensional homogeneity. This is illustrated with the following example [Pao (1965)]. Consider a stationary sphere of diameter D immersed in an incompressible fluid flowing past the sphere in a steady flow.   The flow around the fixed sphere will introduce a drag force, F_D, whose magnitude will depend on several parameters: namely, the diameter of the sphere, D; the approaching velocity of flow, v; and the fluid properties, i.e., density, ρ, and viscosity, μ. Therefore, a functional relationship is expected between the drag force and these variables,

$$F_D = \phi(D, v, \rho, \mu) \tag{2.7}$$

An exponential form for this relation is

$$F_D = CD^a v^b \rho^c \mu^d \tag{2.8}$$

where C is an arbitrary dimensionless constant.

Converting this equation to their dimensional equivalent in an MLT system gives

$$\frac{ML}{T^2} = L^a \left(\frac{L}{T}\right)^b \left(\frac{M}{L^3}\right)^c \left(\frac{M}{LT}\right)^d \tag{2.9}$$

Equating the exponents of each dimension for dimensional homogeneity, we have

For M: $c + d = 1$
For L: $a + b - 3c - d = 1$
For T: $-b - d = -2$

This gives us 4 unknowns and 3 equations. Writing the equation in terms of one unknown,

$$F_D = C\,D^{2-d}\,v^{2-d}\,\rho^{1-d}\,\mu^d \tag{2.10}$$

Rearranging terms,

$$\frac{F_D}{D^2\rho v^2} = C\left(\frac{vD\rho}{\mu}\right)^{-d} \tag{2.11}$$

Note that the term within parenthesis is the definition of Reynolds number. Then, the general form of the relationship becomes

$$\overline{F_D} = C\phi(\mathrm{Re}) \tag{2.12}$$

where $\overline{F_D}$ is the nondimensional drag force on the sphere which is represented as a function of Reynolds number (Re). Note that Eq 2.7 involved 5 variables in an MLT system so that only (5-3 =) 2 nondimensional quantities are needed (Eq. 2.12) for a functional relationship. An experiment with a sphere in steady flow produces such a relationship.

Thus, the dimensionless form of a property of the system can be uniquely determined by a functional relationship in terms of N-3 nondimensional variables. Dynamic similarity means that all properties expressed as dimensionless forms are identical in model and prototype and the corresponding scale is equal to unity [Yalin (1982)]. This method is simple and attractive in that it does not depend on any mathematical relations which may not even be known for the particular system. It relies only on the parameters themselves. The best way to illustrate this is through an example.

Example 1

In fluid mechanics, the common dimensional variables are obtained in three categories

- Geometry of the structure boundary in the flow field, e.g., length, width, etc.

- Fluid properties, e.g., density, viscosity, etc.

- Properties of fluid motion, e.g., pressure, velocity, etc.

Consider this general problem of a structure immersed in a fluid flow. The parameters to be included are written as an arbitrary function of a dimensionless constant π,

$$\pi = \phi(p, v, l_1, l_2, \rho, \mu, \sigma, E, g) \tag{2.13}$$

Relating π in two different unit systems, we obtain,

$$\phi\left(\hat{p}, \hat{v}, \hat{l_1}, \hat{l_2}, \hat{\rho}, \hat{\mu}, \hat{\sigma}, \hat{E}, \hat{g}\right) = \phi(p, v, l_1, l_2, \rho, \mu, \sigma, E, g) \tag{2.14}$$

where

$\hat{p} = ML^{-1}T^{-2}p$	fluid pressure
$\hat{v} = LT^{-1}v$	fluid velocity
$\hat{l_1} = Ll_1$	longitudinal dimension
$\hat{l_2} = Ll_2$	transverse dimension
$\hat{\rho} = ML^{-3}\rho$	fluid mass density
$\hat{\mu} = ML^{-1}T^{-1}\mu$	fluid dynamic viscosity
$\hat{\sigma} = MT^{-2}\sigma$	surface tension
$\hat{E} = ML^{-1}T^{-2}E$	elasticity modulus
$\hat{g} = LT^{-2}g$	acceleration due to gravity

The basic independent dimensional units are L, M and T and there are 9 independent variables. The number of independent dimensionless quantities is 9 - 3 = 6.

The dimensionless quantities are chosen as

$$p/\rho v^2, l_2/l_1, \mu/l_1 \rho v, \sigma/\rho l_1 v^2, E/\rho v^2 \text{ and } gl_1/v^2$$

Note that these dimensionless quantities are not unique. For example, any two of these quantities may be combined to form a new one. However, the total number will remain at 6 in this example. Then, the general solution is given by

$$\pi = \phi(l_2/l_1, p/\rho v^2, \rho v l_1/\mu, v^2/g l_1, \rho v^2 l_1/\sigma, \rho v^2/E) \tag{2.15}$$

It is noted that the quantities $\rho v \ell /\mu$, $v^2/g\ell$, $\rho v^2 \ell /\sigma$ and $\rho v^2/E$ are known as Reynolds number, Froude number, Weber number and Cauchy number, respectively.

2.2.1 Dimensionality of Wave Motion

Let us consider the dimensional aspect of wave motion analysis. There are several parameters that are used in describing a two-dimensional progressive wave. Some of these parameters are the wave height (H), wave period (T), water depth (d), wave length (L), wave frequency (ω), wave number (k) and wave speed (c). Many of these parameters are interrelated. The independent quantities that are necessary and sufficient to characterize the wave motion are H, d, T and g [Sarpkaya and Isaacson (1981)]. All other quantities are related to these four independent variables in a manner prescribed by a particular wave theory. The two-dimensional coordinate system (x, y) and time (t) are also needed for a complete description of a spatial and time dependent variable. Consider the horizontal water particle velocity:

$$u = \phi(H,k,\omega,g,x,y,t) \tag{2.16}$$

Applying the pi theorem, there are 8 variables giving 6 dimensionless variables (in an L, T system) having the relationship

$$\frac{u}{H\omega} = \phi(ky, kH, \omega^2/gk, kx, \omega t) \tag{2.17}$$

For linear theory, the dependence on kH may be waived and the water particle velocity is given by,

$$\frac{2u}{H\omega} = \frac{\cosh ky}{\cosh kd}\cos(kx - \omega t) \tag{2.18}$$

along with the dispersion relationship

$$\frac{\omega^2}{gk} = \tanh kd \tag{2.19}$$

These two equations satisfy the functional relationship in Eq. 2.17.

2.3 NONDIMENSIONAL HYDRODYNAMIC FORCES

The principal types of forces encountered in a hydrodynamic model test are:

Gravity force:	$F_G = Mg$
Inertia force:	$F_I = M\dot{u}$
Viscous force:	$F_V = \mu A \, (du/dy)$
Drag force:	$F_D = 1/2 \; C_D \rho A u^2$

$$\text{Pressure force:} \quad F_p = pA$$
$$\text{Elastic force:} \quad F_e = EA$$

in which M = mass of the structure; u, $\dot{u}$ = velocity and acceleration of fluid   (or structure); y = vertical coordinate; A = area; and p = pressure of fluid.

TABLE 2.1

COMMON DIMENSIONLESS NUMBERS IN FLUID FLOW PROBLEMS

DIMENSIONELSS NUMBER	DEFINITION	REMARK
Froude Number, Fr	v^2/gD	Inertia/Gravity
Reynolds Number, Re	$\rho vD/\mu$	Inertia/Viscous
Strouhal Number, St	$f_e D/v$	Vortex Shedding Frequency
Keulegan-Carpenter Number, KC	vT/D	Period Parameter
Ursell Number, Ur	HL^2/d^3	Depth Parameter
Cauchy Number, Cy	$\rho v^2/E$	Elastic Parameter

Hydrodynamic scaling laws are determined from the ratio of these forces. The dynamic similitude between the model and the prototype is achieved from the satisfaction of these scaling laws. Several ratios may be involved in the scaling.   One of these may be more predominant than the others. In most cases, only one of these scaling laws is satisfied by the reduced-scale model of the prototype structure. Therefore, it is important   to   understand the   physical process experienced by the structure and to choose the most important scaling law which governs this process.

From the above forces, the following ratios may be defined:

Froude Number, Fr	Inertia Force/Gravity Force, F_I/F_G
Reynolds Number, Re	Inertia Force/Viscous Force, F_I/F_V
Iverson Modules, Iv	Inertia Force/Drag Force, F_I/F_D
Euler Number, Eu	Inertia Force/Pressure Force, F_I/F_p
Cauchy Number, Cy	Inertia Force/Elastic Force, F_I/F_e

There are a few other dimensionless numbers one experiences in fluid flow.   The common dimensionless numbers are listed in Table 2.1.

Of   these dimensionless scaling laws, the most common in the water wave problem is the Froude's law. While the Reynolds number plays an important role in many fluid flow problems, the Reynolds similitude does not practically exist in scale model technology.

2.4   FROUDE'S MODEL LAW

The Froude number considers the effect of gravity on the system in question. Thus, it contains the gravitational acceleration term. The Froude number is defined as the ratio of the inertia force to the gravitational force developed on an element of fluid in a medium. Let us consider an element of fluid as a block having dimensions dx, dy, and dz. The gravitational force on the block is given by:

$$W = \rho g \ dx \ dy \ dz \tag{2.20}$$

The inertia force is given by the product of mass and acceleration,

$$F_I = \rho \ dx \ dy \ dz (du / dt) \tag{2.21}$$

where u = fluid block velocity which may be defined as dy/dt. Then, the ratio of the inertia force to the gravitational force is obtained as

$$\frac{F_I}{W} = \frac{udu}{gdy} \tag{2.22}$$

Dimensionally then the Froude number is given by

$$Fr = \frac{F_I}{W} \rightarrow \frac{u^2}{gl} \tag{2.23}$$

Sometimes, the square root of the quantity on the right-hand side is defined as the Froude number, Fr.

In the case of water flow with a free surface, the gravitational effect predominates. The effect of other factors, such as viscosity, surface tension, roughness, etc., is generally small and can be neglected. In this case, Froude's law is most applicable. The Froude number for the model and the prototype in waves can be expressed by

$$Fr = \frac{u_p^2}{gl_p} = \frac{u_m^2}{gl_m} \tag{2.24}$$

where the subscripts p and m stand for prototype and model, respectively.

From geometric similarity,

$$l_p = \lambda l_m \tag{2.25}$$

where λ is the scale factor for the model. Then

$$u_p = \sqrt{\lambda} u_m \tag{2.26}$$

Similarly, force is given by $F = Mg$, where M is the displaced mass ($= \rho l^3$). Considering the same fluid density between the model and prototype,

$$M_p = \lambda^3 M_m \text{ and therefore, } F_p = \lambda^3 F_m \tag{2.27}$$

Consider the example of a moored ship as a spring mass system. The mass of the ship between the prototype and the model is related by Eq. 2.27. The (linear) spring constant K having the unit of force/length should be related by

$$K_p = \lambda^2 K_m \tag{2.28}$$

using Froude's law. The natural period of the system is given by

$$T_N = 2\pi \left(\frac{M}{K} \right)^{\frac{1}{2}} \tag{2.29}$$

Then, the ratio of the natural periods between the model and the prototype is given by

$$T_{Np} = \sqrt{\lambda} T_{Nm} \tag{2.30}$$

In Froude scalings, the acceleration in the model equals the acceleration in the prototype. For example, the acceleration of water particles under waves is given by the relationship

$$\dot{u}_p = \dot{u}_m \tag{2.31}$$

The advantages of the choice of Froude's law are not only that it directly scales the most important criteria of the mechanism, but also that there is a large background of experimental procedures and data reduction techniques available from the years of Froude scale modeling.

2.5 SCALING OF A FROUDE MODEL

A general assumption is made here that the model follows the Froude's law. The common variables found in the study of fluid mechanics are grouped under appropriate subheadings and are listed in Table 2.2. The units of these quantities are listed in the M-L-T (mass-length-time) system. If the variable is dimensionless, the "units" column includes the entry "NONE". Using Froude's law and the scale as λ, the suitable multiplier to be used to obtain the prototype value from the model data is shown. Where applicable, appropriate assumptions or short definitions are included under "Remarks". In the following sections, specific examples are considered from solid and fluid mechanics to show how the Froude models address the scaling criteria. From these examples, it should be clear that while Froude models do not scale all of the parameters, they satisfy the most important and predominant factor in scaling a system in wave mechanics, namely inertia.

2.5.1 Wave Mechanics Scaling

As shown in the table of variables (Table 2.2), the wave height, wave length and water depth scale linearly (as $1/\lambda$) in a Froude model. The time and wave period scale as $1/\sqrt{\lambda}$. The wave force and moment scale as $1/\lambda^3$ and $1/\lambda^4$, respectively.

In the study of wave mechanics (especially in the wave-structure interaction problem), three nondimensional numbers are most important. They are (in the order of importance): Froude number, Reynolds number and Strouhal number. Another dimensionless number, the Keulegan-Carpenter parameter, is also preferred in showing the dependence of the inertia and drag coefficient.

In many problems with waves, inertia is the most predominant force in the system. That is why Froude scaling is used more extensively than any other in the model study. The wave force on a structure whose members are one order of magnitude smaller than the length of the wave depends on the Reynolds number; and thus, in a Froude model, this force does not necessarily scale as λ^3. This will be illustrated in a subsequent section.

The total wave force f per unit length $d\ell$ on a small vertical cylindrical member of diameter D (Fig. 2.1) is obtained from inertia and drag effects added together [Morison, et al. (1950)]. It is written as

TABLE 2.2
MODEL TO PROTOTYPE MULTIPLIER FOR THE VARIABLES
COMMONLY USED IN MECHANICS UNDER FROUDE SCALING

VARIABLE	UNIT	SCALE FACTOR	REMARKS
GEOMETRY			
Length	L	λ	Any characteristic dimension of the object
Area	L^2	λ^2	Surface area or projected area on a plane
Volume	L^3	λ^3	For any portion of the object
Angle	None	1	e.g., between members or solid angle
Radius of Gyration	L	λ	Measured from a fixed point
Moment of Inertia Area	L^4	λ^4	
Moment of Inertia Mass	ML^2	λ^5	Taken about a fixed point
Center of Gravity	L	λ	Measured from a reference point
KINEMATICS & DYNAMICS			
Time	T	$\lambda^{1/2}$	Same reference point (e.g., starting time) is considered as zero time
Acceleration	LT^{-2}	1	Rate of change of velocity
Velocity	LT^{-1}	$\lambda^{1/2}$	Rate of change of displacement

TABLE 2.2 CONTD.

VARIABLE	UNIT	SCALE FACTOR	REMARKS
Displacement	L	λ	Position at rest is considered as zero
Angular Acceleration	T^{-2}	λ^{-1}	Rate of change of angular velocity
Angular Velocity	T^{-1}	$\lambda^{1/2}$	Rate of change of angular displacement
Angular Displacement	None	1	Zero degree is taken as reference
Spring Constant (Linear)	MT^{-2}	λ^2	Force per unit length of extension
Damping Coefficient	MT^{-1}	$\lambda^{5/2}$	Resistance (viscous) against oscillation
Damping Factor	None	1	Ratio of damping and critical damping coefficient
Natural Period	T	$\lambda^{1/2}$	Period at which inertia force = restoring force
Momentum	MLT^{-1}	$\lambda^{7/2}$	Mass times linear velocity
Angular Momentum	ML^2T^{-1}	$\lambda^{9/2}$	Mass moment of inertia times angular velocity
Torque	ML^2T^{-2}	λ^4	Tangential force times distance
Work	ML^2T^{-2}	λ^4	Force applied times distance moved
Power	ML^2T^{-3}	$\lambda^{7/2}$	Rate of work

TABLE 2.2 CONTD.

VARIABLE	UNIT	SCALE FACTOR	REMARKS
Impulse	MLT^{-1}	$\lambda^{7/2}$	Constant force times its short duration of time
Force, Thrust, Resistance	MLT^{-2}	λ^3	Action of one body on another to change or tend to change the state of motion of the body acted on
STATICS			
Stiffness	ML^3T^{-2}	λ^5	Modulus of elasticity times the moment of inertia, EI
Stress	$ML^{-1}T^{-2}$	λ	Force on an element per unit area
Moment	ML^2T^{-2}	λ^4	Applied force times its distance from a fixed point
Shear	MLT^{-2}	λ^3	Force per unit cross sectional area parallel to the force
Section Modulus	L^3	λ^3	Area moment of inertia divided by the distance from the neutral axis to the extreme fiber
HYDRAULICS			
Kinetic Energy	ML^2T^{-2}	λ^4	Capacity of a body for doing work due to its configuration
Pressure Energy	ML^2T^{-2}	λ^4	Energy due to pressure head
Potential Energy	ML^2T^{-2}	λ^4	Capacity of a body for doing work due to its configuration
Friction Loss	ML^2T^{-2}	λ^4	Loss of energy or work due to friction

TABLE 2.2 CONTD.

VARIABLE	UNIT	SCALE FACTOR	REMARKS
SCOUR			
Particle Diameter	L	λ	For same prototype material
Free Settling Velocity	LT^{-1}	$\sqrt{\lambda}$	Final velocity of a freely falling particle in a medium
Sediment Number	None	1	Nondimensional no. based on velocity and particle size
Shield's Number	None	1	Nondimensional no. based on velocity and particle size
WAVE MECHANICS			
Wave Height	L	λ	Consecutive crest to trough distance
Wave Period	T	$\sqrt{\lambda}$	Time between two successive crests passing a point
Wave Length	L	λ	Distance between two successive crests at a given time
Celerity	LT^{-1}	$\sqrt{\lambda}$	Velocity of wave (crest, for example)
Particle Velocity	LT^{-1}	$\sqrt{\lambda}$	Rate of change of movement of a water particle
Particle Acceleration	LT^{-2}	1	Rate of change of velocity of a water particle
Particle Orbits	L	λ	Path of a water particle (closed or open)

TABLE 2.2 CONTD.

VARIABLE	UNIT	SCALE FACTOR	REMARKS
Wave Elevation	L	λ	Form of wave (distance from still waterline)
Wave Pressure	$ML^{-1}T^{-2}$	λ	Force exerted by a water particle per unit area
Keulegan-Carpenter Parameter	None	1	Dependence of hydrodynamic coefficients on this parameter
STABILITY			
Displacement (Volume)	L^3	λ^3	Volume of water moved by a submerged object (or part thereof)
Righting & Overturning Moment (Hard Volume)	ML^2T^{-2}	λ^4	Moment about a fixed point of a displaced weight and dead weight, respectively
Natural Period	T	$\sqrt{\lambda}$	Period of free oscillation in still water due to an initial disturbance
Metacenter	L	λ	Instantaneous center of rotation
Center of Buoyancy	L	λ	Distance of C.G. of displaced volume from a fixed point
Soft Volume	L^3	λ^3	Volume of trapped air in a member
Buoyancy Pickup per Unit Angle	L^3	λ^3	Increase in displaced volume per unit tilt angle

TABLE 2.2 CONTD.

VARIABLE	UNIT	SCALE FACTOR	REMARKS
MATERIAL PROPERTIES			
Density	ML^{-3}	1	Mass per unit volume
Modulus of Elasticity	$ML^{-1}T^{-2}$	λ	Ratio of tensile or compressive stress to strain
Modulus of Rigidity	$ML^{-1}T^{-2}$	λ	Ratio of shearing stress to strain

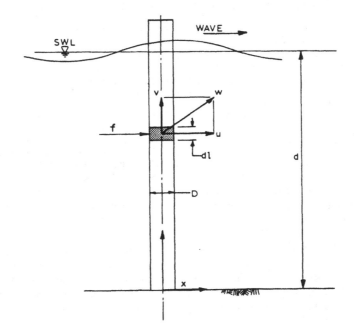

FIGURE 2.1
DEFINITION SKETCH OF WAVE FORCE ON SMALL CYLINDER

$$f = \rho C_M \frac{\pi}{4} D^2 \dot{u} + \frac{1}{2}\rho C_D D |u| u \qquad (2.32)$$

The first term on the right hand side depends on the inertia force which is proportional to the water particle acceleration, $\dot{u}$. The second term is the drag force proportional to the square of the water particle velocity, u. All the quantities on the right hand side follow Froude scaling with force except for the two quantities, C_M and C_D.

The hydrodynamic coefficients, C_M (inertia coefficient), and C_D (drag coefficient), are nondimensional. It has been found [Sarpkaya and Isaacson (1981)] that they are functions of the Keulegan-Carpenter parameter, KC, (defined as $u_0 T/D$) and the Reynolds number, Re,(defined as $u_0 D/\nu$) where the subscript o refers to maximum value and $\nu = \mu/\rho$. Therefore, it is important to understand how these parameters scale from the model to the prototype.

According to Froude's law, the velocity and wave period scale as the square root of the scale factor, while the linear dimensions scale linearly. Therefore,

$$(KC)_p = (KC)_m \tag{2.33}$$

whereas

$$(Re)_p = \lambda^{\frac{3}{2}} (Re)_m \tag{2.34}$$

Because the Keulegan-Carpenter number follows Froude's law, dependence on KC ensures that the model values are applicable to prototype. However, if the quantities strongly depend on Reynolds number, direct scaling is not possible. Moreover, for a small scale model in a wave tank, the prototype Reynolds number can not even be approached due to low fluid velocity in model (Eq. 2.34). Therefore, as C_M and C_D are strong functions of Re, the results from the model test are not directly applicable to the design.

Fluid flow past a small member of a structure creates a low pressure behind the member and causes vortices to shed from the surface of the member. This vortex formation behind the member is found to be a function of the Strouhal number. The eddy-shedding frequency, f_e, in the Strouhal number is dependent on Re. Therefore, the Strouhal numbers in a model and a prototype are different and do not follow Froude's law.

2.5.2   *Current Drag Scaling*

The drag force per unit length exerted on a bluff cylindrical member by a uniform current flow is proportional to the square of the mean current velocity U. The force is given as

$$f_D = \frac{1}{2}\rho D C_D U^2 \qquad\qquad (2.35)$$

The drag force acts parallel to the component of current that is normal to the member axis (i.e., current acts on the projected area normal to flow). The drag coefficient is based on data from Hoerner (1965).

Experiments have shown that the flow characteristics in the boundary layer are most likely to be laminar at Re $< 10^5$, whereas the boundary layer is turbulent for Re $> 10^6$ [Berkley (1968)]. Thus, most small model proportions and test conditions, scaled by Froude number, will result in laminar flow conditions while full-scale conditions are evidently turbulent. Thus, in reality, two different scaling laws (e.g., Froude and Reynolds) apply simultaneously to the model. Since both scaling factors cannot be satisfied concurrently during model tests, it is convenient to employ the Froude scaling process, and allowances are made for the variation in Reynolds number.

The dependence of the drag coefficient on the Reynolds number is quite strong in so far as the value of the Reynolds number tends to characterize the flow regime as laminar, transition or turbulent. Once the flow regime is turbulent, the drag coefficient is only weakly dependent on the Reynolds number. The turbulent flow may be verified by visual examination and pressure transducer data. It may be confirmed that the resistance is only weakly a function of the Reynolds number within the turbulent region, by testing two models with identical shapes but with different scales. By using a half size and full scale model, the Reynolds number will be doubled, and the dependence of the resistance and, therefore, the drag coefficient on the Reynolds number will be determined. When this relationship is confirmed, model data can be applied to prototype designs.

A practical answer to this problem in model is to deliberately "trip" the laminar flow by some kind of roughness near the bow of the structure. In testing tanker models, various methods have been employed including struts placed upstream of the vessel, wires attached at a point just aft of the bow or sandstrips, studs or pins attached directly to the hull. It has been shown that studs appear to be the most effective method of stimulating turbulent flow at lower velocity and over a broader region of the wetted surface area.

2.5.3 Wave Drag Scaling

For flow past a circular cylinder, the low pressure region present on the downstream side (the wake) accounts for a major part of the drag force. In steady flow, the drag coefficient as a function of Reynold's number is known for a smooth circular cylinder. However, the variation of C_D against Reynolds number in waves has not been established for high Re.   Moreover, the problem is complicated by the fact that the velocity is not constant, but rather changes in magnitude between 0 and a maximum value (depending on wave parameters) as well as in direction. Thus, how the Reynolds number should be defined is open to question. Usually, Re is given in terms of the horizontal water particle velocity for a fixed structure. For a moving structure, Re may be defined in terms of the maximum relative velocity between the structure and the fluid.

The   drag coefficient for steady flow has been observed to decrease with   the increase in Reynolds number except for a small region at the start of the supercritical zone. Drag coefficient is not well defined at very high Reynolds number even for a steady flow.   It is also a function of cylinder surface roughness, and other conditions.

However, assuming that the roughness coefficient between the prototype and model is about the same, and leaving out the region between subcritical and supercritical zone, it may be observed that the C_D in model is generally higher than that in the prototype. One way to circumvent this problem is to test the model at higher Reynolds number to try to reach the prototype Re so that a C_D value for the prototype may be established. However, since Re scales as $\lambda^{3/2}$, it is generally not possible to increase the velocity, u, within the range of the wave generating tank to obtain prototype Re. This could result in a serious modeling problem in special cases.

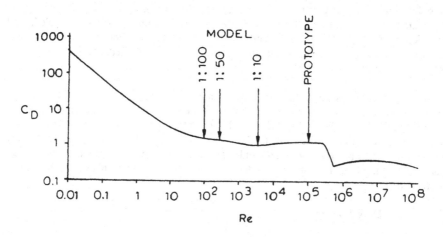

FIGURE 2.2
STEADY DRAG COEFFICIENTS FOR SMOOTH CIRCULAR CYLINDER

For structures whose dimensions compare to the wave length, wave forces are mainly inertial, and the drag forces are generally one order of magnitude smaller than the inertia forces. The velocity-dependent drag force on a structural member is 90° out of phase with the acceleration-dependent inertia force. The resultant maximum force (which is the sum of the inertia and drag force) is only slightly higher than the inertia force alone. However, for a member which is small compared to wave length, there is considerable influence of drag on the resultant maximum force. Since the drag force on the model (scaled according to the Froude's law) is greater than that on the prototype, the influence is expected to be higher in the model. This is illustrated by the following example [Chakrabarti (1989)]:

Example 2

A section of a cylinder of diameter 14.6m (48 ft) and of length 0.3m (1 ft) is at the still water level in a 146m (480 ft) water depth. The cylinder is to withstand a wave of 15 sec period and 30.5m (100 ft) height. A model is built using a scale factor of λ = 48. Calculate the phase shifts of the force in the model and prototype. Then, change the cylinder diameter to 1.5m (4.8 ft) and repeat the calculations above. Assume C_M = 2.0 for both and obtain C_D from the steady-state uniform flow curve (Fig. 2.2). Use Froude's law for modeling.

$$\text{Prototype Reynolds number, } Re_p = \frac{u_0 D}{\nu} \tag{2.36}$$

where D = 14.6m (48 ft) and ν = 0.15 x 10^{-5} m²/s (1.575 x 10^{-5} ft²/sec). The maximum water particle velocity is calculated using linear wave theory:

$$u_0 = \frac{gHk}{2\omega} = 6.5 m/s \ (21.2 \ ft/sec) \tag{2.37}$$

Then the prototype Re_p = 6.46 x 10^7.

The model has a diameter of 0.3m (1 ft) and a length of 6.3mm (1/4 in). Model water depth is 3m (10 ft), wave period is 2.16 sec and wave height 0.63m (2.08 ft). The kinematic viscosity of fresh water is 0.105 x 10^{-5} m²/s (1.126 x 10^{-5} ft²/sec). The water particle velocity amplitude,

$$u_0 = 0.94 \text{ m/s (3.06 ft/sec)} \tag{2.38}$$

The model Reynolds number, Re_m = 2.72 x 10^5. From the uniform flow curve (Fig. 2.2), the drag coefficients,

$$C_{D_p} = 0.6 \tag{2.39}$$

$$C_{D_m} = 1.1 \tag{2.40}$$

Maximum Inertia Force (Eq. 2.32):

$$\text{Prototype}: 2 \times 1.98 \times (\pi / 4 \times 48^2) \times 1 \times 8.88 \ = 63.6 kips$$
$$\text{Model}: \quad 2 \times 1.94 \times (\pi / 4 \times 1^2) \times 1 / 48 \times 8.88 \ = \underline{0.58 lbs}$$
$$\text{Scale} \ = \lambda^3$$

Maximum Drag Force (Eq. 2.32):

$$\text{Prototype}: 0.5 \times 0.6 \times 1.98 \times 48 \times 1 \times 21.2^2 = 12.8 kips$$
$$\text{Model}: \quad 0.5 \times 1.1 \times 1.94 \times 1 \times 1 / 48 \times 3.06^2 \ = \underline{0.21 lbs}$$
$$\text{Scale} \ = \lambda^{2.84}$$

Note that the prototype drag is only 20 percent of the inertia force, while the model drag is nearly 36 percent of its inertia force. Maximum total force on the prototype = 63.6 kips, phase shift = 90°. Maximum total force on the model = 0.58 lbs., phase shift = 90°.

In the second case, the member diameter is 1/10th so that the prototype Reynolds number is 6.46 x 10^6 while the model Reynolds number is 2.72 x 10^4. From the steady-state curve, then, the prototype C_{Dp} = 0.6 and the model C_{Dm} = 1.1. The calculated maximum forces on a 0.3m (1 ft) long prototype and a 6.3mm (1/4 in.) long models section are

	Inertia	Drag
Prototype:	0.636 kips	1.28 kips
Model:	0.0058 lbs	0.021 lbs
Scale =	λ^3	$\lambda^{2.84}$

Thus, in this case, the prototype drag force is 201 percent of the inertia, and the model drag force is 362 percent of the inertia force. Maximum total force on the prototype = 1.28 kips, phase shift = 30°. Maximum total force on the model = 0.021 lbs, phase shift = 16°.

Thus, the two examples presented show the two extreme cases of inertia dominance and drag dominance depending only on the member size. It also demonstrates that the model drag force is relatively higher than the prototype drag force compared to the inertia forces (which scale as λ^3). Thus, the phase shift of the maximum total force is higher in the model than that in the prototype.

2.6   HYDROELASTIC STRUCTURAL SCALING

In a hydrodynamic model study, it is a common practice to measure forces exerted on the model. Stresses are generally calculated from this measurement rather than measured directly. In order to measure the stresses, an elastic model is required. Thus, not only is the Froude similitude maintained, but the Cauchy similarity between the model and the prototype is generally required.

In a Froude model, the quantities, e.g., cross-sectional area, moment of inertia, section modulus, etc., will follow Froude scaling. However, if the same prototype material is used to build the model, the stiffness of the model, EI, where I is the moment of inertia, will not scale the prototype stiffness. Thus, the model behavior will not correctly predict the prototype performance, as is illustrated by the following examples.

Let us consider the fiber stress in a beam of a given section (Fig. 2.3). The moment of inertia of the section with respect to the neutral axis is I and the distance to the outer fiber from the neutral axis is represented by y. If F is the force acting on the section, then the shear stress is given by F/A where A is the cross-sectional area. If M_B is the bending moment on the section, then the bending stress is written as $M_B y/I$. Therefore, the total stress on the extreme fiber of the section becomes:

$$S = \frac{F}{A} + \frac{M_B y}{I} \tag{2.41}$$

Now, if the force and moment on the model are such that $F_p = \lambda^3 F_m$ and $M_{Bp} = \lambda^4 M_{Bm}$, the stress will scale as

$$S_p = \lambda S_m \tag{2.42}$$

Consider the section on a cantilever beam of length ℓ with F acting at the free end; the maximum deflection at the free end is obtained from

$$\delta_{max} = \frac{F\ell^3}{3EI} \tag{2.43}$$

For the deflection to scale linearly, $(\delta_{max})_p = \lambda(\delta_{max})_m$, we have

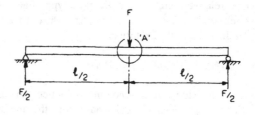

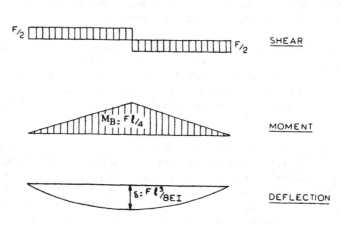

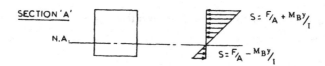

FIGURE 2.3
SCALING FOR SIMPLY SUPPORTED BEAM

$$(EI)_p = \lambda^5 (EI)_m \tag{2.44}$$

But

$$I_p = \lambda^4 I_m \tag{2.45}$$

so

$$E_p = \lambda E_m \tag{2.46}$$

Thus, to scale the stiffness of the structure, a suitable material should be chosen of the model so that the Young's modulus scales linearly with the scale factor (e.g., for λ =48, equivalent to prototype steel, the model E_m should be 625,000 psi).

Often, however, the same material as the prototype is used in model testing. In this case, the model is said to be distorted. Since $E_p = E_m$, the stiffness for a Froude model will scale as

$$(EI)_p = \lambda^4 (EI)_m \tag{2.47}$$

Thus, for the deflection to scale linearly ($\delta_p = \lambda \delta_m$), the model should be subjected to a (distorted) force and moment given by the relationship to the prototype as

$$F_p = \lambda^2 F_m \tag{2.48}$$
$$M_{Bp} = \lambda^3 M_{Bm} \tag{2.49}$$

In this case, the stress

$$S_p = S_m \tag{2.50}$$

Since the model in this case is stiffer, the force and moment applied to obtain the scaled down deflection are larger so that the stress level is the same as the prototype.

2.7   DISTORTED MODEL

Long models are defined as those that are large in one dimension compared to the other dimensions. This type of model may require distortion of scales in the short direction. In fluid mechanics, this distortion of the long model may be necessary to avoid the problem of viscous boundary layers at a rigid boundary or capillary effect.

A model is distorted when it has more than one geometric scale. For example, the vertical scale (β) in the model is different from the horizontal scale (λ). The ratio between β and λ is called the rate of distortion [LeMehaute (1976)]. In general, scale models used in the study of water waves are not distorted with the exception of a few special cases. Let us consider the wave velocity by linear theory which is given as

$$c = \frac{gT}{2\pi} \tanh \, kd \tag{2.51}$$

For the wave velocity to scale properly (as $\sqrt{\lambda}$) according to Froude's law, the depth parameter (d/L) must be the same between the model and the prototype. Since the ratio of wave lengths L_p/L_m is given by the horizontal scale λ, it is clear that the ratio of the water depth (d_p/d_m) should also be λ. Then the wave motion is in similitude.

In the case of long waves, however, distortion is possible. Since, in long waves, tanh kd may be approximated by kd, the wave velocity is given by

$$c = \sqrt{gd} \tag{2.52}$$

In this case, the water depth in the model can be distorted by using a scale β, different from the wave length scale λ. Then, the velocity scale is given by

$$\frac{c_p}{c_m} = \left(\frac{d_p}{d_m}\right)^{\frac{1}{2}} = \beta^{\frac{1}{2}} \tag{2.53}$$

The wave period (or time) scales as a combination of the two scale factors

$$\frac{T_p}{T_m} = \left(\frac{L_p}{L_m}\right)\left(\frac{d_m}{d_p}\right)^{\frac{1}{2}} = \frac{\lambda}{\sqrt{\beta}} \tag{2.54}$$

The distorted models are further discussed in Chapter 9.

Let us consider another example of structural modeling. In this case, it is desired to model an elastic underwater oil storage tank made of rubber-like material. In this case, the differential equations of motion have been derived by LeMehaute (1965) to describe the motion of two fluids separated by the membrane and subjected to surface waves. The similarity requirements deduced from these equations give

$$\left[gD\frac{\rho_w - \rho_0}{\rho_0}\right]_p = \lambda\left[gD\frac{\rho_w - \rho_0}{\rho_0}\right]_m \tag{2.55}$$

and

$$\left[\frac{eE}{\rho_0 R}\right]_p = \lambda \left[\frac{eE}{\rho_0 R}\right]_m \tag{2.56}$$

where ρ_w, ρ_0 = densities of water and oil, respectively, D = depth of oil in the tank, R = radius of the bag, e = membrane thickness and E = modulus of elasticity of membrane. From the first relationship,

$$D_p = \lambda D_m \tag{2.57}$$

and

$$\left[\frac{\rho_w - \rho_0}{\rho_0}\right]_p = \left[\frac{\rho_w - \rho_0}{\rho_0}\right]_m \tag{2.58}$$

This latter condition is satisfied by using lighter fluid for oil in the model, e.g., kerosene, alcohol or gasoline, to account for the difference between the sea water and fresh water. For the model of elasticity, we have

$$\left(eE\right)\big|_p = \lambda^2 \left(eE\right)\big|_m \tag{2.59}$$

Thus, the material and thickness of the model should be chosen such that their product satisfies the relationship in Eq. 2.56. For example, for a Firestone product 6.4 mm (1/4 in.) thick and at a scale of 1:30, a 0.4 mm (1/64 in.) thick rubber material will satisfy the scaling law.

The choice of lighter fluid in the model has the added advantage here to model the Reynolds number associated with the motion inside the tank. Thus,

$$v_p = \lambda^{\frac{3}{2}} v_m \tag{2.60}$$

where v = kinematic viscosity. For example, v for crude oil is 0.9 x $10^{-5} m^2/s$ ($10^{-4} ft^2$/sec); while for alcohol and gasoline, it is 0.15 x $10^{-5} m^2/s$ (1.6 x $10^{-5} ft^2$/sec) and 0.46 x $10^{-6} m^2/s$ (5 x $10^{-6} ft^2$/sec), respectively.

2.8 REFERENCES

1. Berkley, W.B., "The Application of Model Testing to Offshore Mobile Platform Design", Society of Petroleum Engineers of AIME, Paper No. SPE 2149, Houston, Texas, 1968.

2.   Bridgeman, P.W., Dimensional Analysis, Revised Edition, Yale University Press, New Haven, Connecticut, 1965.

3.   Buckingham, E., "On Physically Similar Systems; Illustrations of the Use of Dimensional Equations", Physics Review, Volume 4, 1914, pp. 345-376.

4.   Chakrabarti, S.K., "Modeling of Offshore Structures (Chapter), Application in Coastal Modeling, V.C Lakhan and A.S. Trenhaile (Editors), Elsevier Oceanography Series 49, 1989.

5.   Hansen, A.G., Similarity Analysis of Boundary Value Problems in Engineering, Prentice-Hall Inc., Englewood Cliffs, New Jersey, 1964.

6.   Haszpra, O., Modelling Hydroelastic Vibrations, Pitman Publishing, London, 1979.

7.   Hoerner, S.F., Fluid-Dynamic Drag, Published by the Author, Midland Park, New Jersey, 1965.

8.   Langhaar, H.L., Dimensional Analysis and Theory of Models, John Wiley & Sons, New York, New York, 1951.

9.   LeMehaute, B., "On Froude Cauchy Similitude", Proceedings on Specialty Conference on Coastal Engineering, Santa Barbara, California, ASCE, Oct., 1965.

10.   LeMehaute, B., An Introduction to Hydrodynamics and Water Waves, Springer Verlag, New York, New York, 1976.

11.   Morison, J.R., O'Brien, M.P., Johnson, J.W., and Shaaf, S.A., "The Forces Exerted by Surface Waves on Piles", Petroleum Transactions, AIME, Vol. 189, 1950, pp. 149-157.

12.   Murphy, G., Similitude in Engineering, Ronald Press Co., New York, New York, 1950.

13.   Pao, R.H.F., Fluid Mechanics, John Wiley and Sons, Inc., New York, New York, 1965.

14.   Sarpkaya, T. and Isaacson, M., Mechanics of Wave Forces on Offshore Structures, van Nostrand Reinhold Co., New York, New York, 1981.

16. Sedov, L., <u>Similarity and Dimensional Methods in Mechanics</u>, Academic Press, New York, New York, 1959.

17. Skoglund, V.J., <u>Similitude Theory and Applications</u>, International Textbook Co., Scranton, Pennsylvania, 1967.

18. Soper, W.G., "Scale Modeling", International Science and Technology, Feb., 1967, pp. 60-69.

19. Szucs, E., <u>Similarity and Models</u>, Elsevier, Amsterdam, The Netherlands, 1978.

20. Yalin, M.S. "Discussion on Hydraulic Modelling in Maritime", Institute of Civil Engineers, Conference Proceedings, Thomas Telford Ltd., London, 1982, pp. 9-12.

CHAPTER 3

MODEL CONSTRUCTION TECHNIQUES

3.1 GENERAL REQUIREMENTS FOR MODELS

The art of designing and building a working model that is both accurate in its properties and capable of withstanding the prolonged water environment is both difficult and time consuming. The models are usually equipped with a variety of instrumentation and proper provisions must be given for their attachment. Waterproofing of the instrumentation and their attachment to the model is critical to the success of the model test. Quite often, models must also possess surface finishes that are attractive for photographic requirements and visible in contrast under water.

Models are made of a variety of materials, such as, wood, metal, Fiberglass, plastics, concrete or composite materials. Quite often, models are built from multiple materials for different model components. Selection of the most suitable material to build a particular model depends on several factors. One of the most important of these is the size of the model. The other is the modeling criterion. Most offshore structure models are built following Froude's law, and geometric similitude is almost invariably maintained wherever possible. The size of the model is generally limited by the dimensions of the testing facility and its simulation capability of the environment. If standard parts needed for the model are available in the selected modeling material, then the computed scale factor may be changed somewhat to enable the use of these standard items, such as standard material thickness, pipes, angles, etc. The weight of the model is also an important consideration due to the limits in the available handling equipment. Of course, cost plays a major role in the choice of the scale factor. While the larger the model, the more expensive it may be to build, the small size of a model may sometimes make it expensive as well, as a result of the difficulty in fabricating and handling small, delicate components. Thus, cost should be considered in the decision of a model scale, λ. The instrumentation should also be considered along with model size. For example, force scales as λ^3 (Froude's law). Therefore, if loads on a structure are measured, the size of the model loads may be a controlling consideration in determining the limits of the model scale. If the model loads are too small or too large, they may be difficult to measure with acceptable accuracy.

3.2   MODEL TYPES

The model for an offshore structure may be classified in the following categories depending on the purpose of testing:

- Models for environmental load measurement;
- Seakeeping models;
- Elastic models; and
- Models of attachments (e.g., mooring or anchoring system)

Different techniques are used in designing and building these types of models. Examples are given here (as well as in Chapters 7, 8, and 9) to illustrate the model construction techniques in these areas. The scaling technique and building of an elastic model will be addressed in Chapter 9.

3.3   ENVIRONMENTAL LOAD MODELS

The models used to determine environmental loads are held fixed in waves, wind and/or current. For the measurement of loads, the model is supported on load cells. In order that the weight of the model may be supported on load cells, the structure must be strong enough to represent a rigid body. Sometimes, additional internal support is needed to develop the structural strength required by the model. The outer dimensions of the model are maintained; however, the thickness of the wall or the internal geometry is not necessarily scaled. The model is ballasted in the tank so that the load on the load cells is within the range of measurement. Additional ballast may sometimes allow the tension load cells to register both tensile and compressive (i.e. oscillatory) loads without going slack.

Lead (or steel) weight placed inside the model is a quite common method of ballasting a model used for load measurement. Sometimes, the model is flooded in place inside the wave tank. Ballasting with water, however, should be given careful consideration in order not to introduce any problem with internal sloshing. Communication of flow inside the model should always be avoided to insure the measurement of true external loads. Communications may introduce internal pressure which will alter the external loads registered.

The interior surfaces of these models are primed where possible. The exterior is generally primed and top-coated with high visibility epoxy paint. Often contrasting radial and circumferential striping is applied to the model so as to provide improved visibility and photographic definition. For surface piercing structures, it also provides qualitative data on the wave run-up at the model.

The problems associated with the construction of an environmental load model are best described through an actual example of the construction of a model. The model chosen provides most of the common design techniques necessary for an environmental load model. A shelf-mounted Ocean Thermal Energy Conversion (OTEC) platform model was tested in a wave tank to determine wave loads. The model construction involved several features described above.

3.3.1 OTEC Platform Model

A large shelf-mounted OTEC platform model was constructed for load determination tests. The platform concept included large power modules represented by submerged blocks which required verification for wave loads in the design calculation.

As shown in Fig. 3.1, the OTEC platform frame consisted of 4 vertical columns, a deck, 2 horizontal pontoons and 2 horizontal braces. This configuration was identified as the baseline structure. The columns emanated from the circular pods (as shown in Fig. 3.1) and were connected at the top by a stiffened deck and at the bottom in pairs by 2 horizontal pontoons. The prototype and the corresponding model dimensions for the members of the baseline platform are shown in Table 3.1.   The platform members by themselves did not provide sufficient rigidity.   Therefore, additional members were introduced without affecting the load measurement. Two smaller tubular braces were introduced to connect the 4 pods.   These members connected the pods on opposite faces where there are no pontoons.

TABLE 3.1
BASIC OTEC PLATFORM MEMBER DIMENSIONS
MODEL SCALE = 1:25.75

DESCRIPTION	QUANTITY	DIMENSIONS (FT)	
		PROTOTYPE	MODEL
Vertical Columns	4	30ϕ x 320	1.16ϕ x 12.46
Pods	4	52.5ϕ x 40	2.04ϕ x 1.55
Pontoons	2	30ϕ x 117.5	1.16ϕ x 4.56
Platform	1	220 x 220	8.54 x 8.54
1ft = 0.3 m; ϕ =diameter			

While these strength members should be avoided under water where they will experience wave loads, they could not be avoided in this case since rigidity was considered more important. More acceptable method to insure rigidity is to use internal members which are not directly exposed to waves.   Both the horizontal pontoons and braces were reinforced internally for strength and rigidity with rectangular box beam tubing. The box beam was welded to the pods and vertical columns which provided additional rigidity to prevent racking of the structure. The tubing was reinforced with "TEE" sections made from 12.7 mm (1/2 in.) steel plate. The resulting section is shown in Fig. 3.2. In addition, 9.5 mm (3/8 in.) diameter smooth rods were used as diagonal braces on all 4 vertical faces of the baseline model. These members provided the required rigidity in the platform without altering the global wave loads by a measurable amount.

FIGURE 3.1
LOAD MEASURING MODEL ON A GENERIC SHELF-MOUNTED
OTEC

Since the platform deck was above the water surface and outside the reach of waves, it was designed to provide rigidity. The deck was framed out of 2 sizes of rectangular steel tubing. The tubing was 76.2 mm x 38.1 mm (3 x 1-1/2 in.) with a 6.3 mm (1/4 in.) wall thickness. A diagonal cross brace made of 63.5 mm x 63.5 mm x 4.7

mm (2-1/2 x 2-1/2 x 3/16 in.) steel angle and 63.5mm x 4.7 mm (2-1/2 x 3/16 in.) flat bar was provided for rigidity. The deck was bolted directly to the top of the 4 columns.

The columns and pontoons were made of 356 mm (14 in.) outside diameter (12.2 m or 40 ft prototype) steel tubing with a wall thickness of 0.20 in. The circular pods were made from rolled 3.2 mm (1/8 in.) thick steel plate.

Two horizontal plate stiffeners acting as the upper and lower pod surfaces were used to secure a column to a pod. These plates were cut from 4.7 mm (3/16 in.) plate. The pods and columns were cut to allow for the passage of the reinforcing box beam. Figure 3.2 shows a typical pod/column assembly.

The OTEC model consisted of the welded steel frame over which a variety of submerged power modules (modeled as boxes) were mounted. All weld seams were ground smooth and the entire model was painted with a high visibility yellow paint. Selected reference points were marked in colored tape to provide easy identification of model components when viewing the test videotape.

FIGURE 3.2
LOAD CELL MOUNTING DETAIL

Note that a large environmental load model is typically supported on three load cells rather than four for ease in alignment. The frame was supported on three load cells, each capable of measuring forces in three orthogonal directions. One cell was mounted just outside of each pod on the west side of the north-south vertical plane of symmetry (see Fig 3.3). The remaining cell was located at the midpoint of the horizontal brace on the east side of the plane of symmetry. Each load cell was attached to the box beam frame which runs through the horizontal pontoons and braces. Figure 3.2 shows a cut-away of a typical load cell assembly.

The model was fitted on the load cells in the dry, and 19.1 mm (3/4 in.) diameter holes were cut in the model's box beam frame for passage of the 12.7 mm (1/2 in.) threaded rod used to attach the model to the load cells. The top of the vertical load cell was slightly recessed inside the horizontal cells. Three flat washers and a spherical washer were installed between the top of the vertical cell and the bottom of the box beam. The spherical washer was used to avoid any misalignment.

In order to install the model with load cell assembly on the tank floor, a baseplate 3 m x 3 m x 19.1 mm (10 ft x 10 ft x 3/4 in.) thick was installed prior to the installation of the OTEC model. This plate may be seen in Fig. 3.2 on which the load cells were mounted. The weight and size of this plate provided enough friction to keep the plate and the model fixed in the tank in waves. This is a quite common method of model mounting. The thickness of the steel plate does not significantly affect the water depth. Installation can be accomplished in the dry, and in-place calibration can be performed outside the tank. Then any problems with the load measurement or the set up may be corrected easily. However, care should be taken that the load cells are capable of withstanding the dry weight of the model. If the load cells are not capable of holding the model weight in the dry, installation must be made in the water (which may be made shallow for ease of handling). In this case the in-place calibration is carried out under water by loading the model in different directions with known weights and recording the load cell readings along the axis. The cross-axis sensitivity of the load cell (see Chapter 6) is also established at the same time. Adjustments may be necessary if the cross-talk is unacceptably high.

The load cells are bolted to the baseplate. Additional lead weights may be placed on the corners of the baseplate if needed to hold the model in place. The baseplate may also be "stitch welded" to a steel section of the wave tank or bolted to the concrete floor to prevent it from sliding in waves.

3.4   SEAKEEPING MODEL

Unlike fixed models for measuring environmental loads, the seakeeping models are allowed to respond with the environment. Therefore, in addition to the

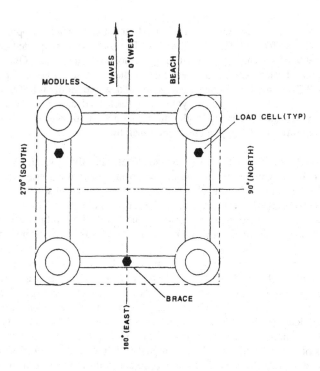

FIGURE 3.3
LOAD CELL LOCATIONS

outside geometry, the dynamic properties of the structure must be modeled using Froude's law. Some of these properties include the displacement, center of gravity, moments of inertia and natural periods of oscillation. The scaling factors for these properties have been given in Table 2.2. Typical acceptable tolerances for these parameters are as follows:

- Weight, length and center of gravity (G) within 3 percent;

- Displacement, center of buoyancy, (B) and the distance GB within 3 percent;

- Stiffness (i.e., EI factor) within 5 percent; and

- Moments of inertia within 5 percent.

There are two types of floating vessels: transport vessels and moored vessels. The transport vessels include ships and barges which move from one location of the ocean to another carrying goods and services. The other type of structures includes articulated towers, tension leg platforms and moored tankers. They are connected to the ocean bottom by some mechanical means. Therefore, modeling these structures involve modeling not only the floating structure but also the mooring system. Here, construction of three seakeeping models and one launching model is exemplified.

3.4.1 Tanker Model

A model scale is chosen first based on several factors already discussed. For the type of tanker, the model plans, namely, body, half-breadth and profile drawings, and a table of offsets, are obtained or developed. Next, an estimation of the target weight of the finished model without ballast is made. This is done to ensure that the ballasting will achieve both the static and dynamic characteristics of the tanker. Tanker models are built with a variety of materials, e.g., wax, wood, fiberglass, plastic, etc. The choice of the material is based on economy, usage, applicability, and longevity of the model as well as model construction capabilities.

It is preferable to build the model upside down on a perfectly flat worktable. This allows a close control of the model dimensions and facilitates dimensional check during construction.

3.4.1.1 Wood Construction of Model

The general method used to produce the wood model is to prepare a laminated wood block that approximates the shape of the ship by cutting each laminate along waterplane lines. The final shape of the model is achieved by removing excess material by planing, spokeshaving and scraping. If a numerically controlled (NC) machine is available and suitable for the model size, it provides the final shape. Final form and fairness are verified with metal station and waterplane templates.

From the lines drawing or table of offsets of the model, a half-breadth set of thin sheet aluminum female station and half station templates are constructed. The templates are rough cut with a band saw and machined to a scribed line. Each template is marked for centerline and deckline and for benchmark (inspection) references. The tolerance on all templates generally is 0.1 percent or less. Several full-breadth templates are also constructed for inspection of twist and flatness of the construction. All templates are made with a reference edge so that assembled alignment can be inspected and aligned by a long straight edge or machinists' optical transit.

Since the basic manufacturing method with wood utilizes a cut-out and laid up block, the direction of wood grain is an important consideration to minimize water contact and absorption through the end grain. Therefore, all wood is examined and the components are laid with the grain in one direction. Waterplane lines are then transferred to a selected plank and rough cut, 1.6 - 3.2 mm (1/16 -1/8 in.) oversized based on model size. Interior cuts of the waterplane plank are made in order to arrive at a full thickness of not less than 10% of the ship beam with an additional thickness allowed at the bow. Successive layers of waterplane line laminates are then built up by applying glue to both bonding surfaces. Joints in adjacent laminates, if required, are alternated to opposite ends so that successive layers will not have joints in close approximation. Construction of the bottom of the ship is by lamination, with a minimum of 2 layers. No metallic fasteners are used in the fabrication of the block to facilitate machine use and finishing. Bar clamps with transverse strongbacks are used to ensure uniform clamping pressure.

Reduction of the rough block is performed by hand planing, spokeshaving, scraping and sanding (Fig. 3.4). Alternatively, a numerically control machining device is used, if available. During the rough and final finishing, the hull is checked frequently with the previously prepared station templates. Final sanding is performed with successively finer grits until a smooth fair surface (such as 63 RMS) is achieved.

FIGURE 3.4
WOOD CONSTRUCTION OF U.S.S. MIDWAY MODEL

Two coats of penetrating wood sealer are applied to the inside and outside of the model. While the outside is still wet with sealer, the hull is lightly machine sanded. This technique fills slight surface imperfections and any open grains to produce a higher grade surface finish. The final outside sealer coat is wiped to ensure uniform absorption and penetration. All holes or imperfections are filled with putty and painted with one coat of epoxy primer. Two coats of finish paint are applied by spraying the model.

The completed model is inspected by positioning the station templates on the reference drawing and over the model; permissible tolerance is determined with a feeler gauge.

3.4.1.2 Fiberglass Construction of Model

The fiberglass model may be constructed from a prepared polyurethane foam block that approximates the shape of the model. Fiberglass is applied to this form using a hand lay-up method to produce a hull 3.2 - 6.3 mm (1/8 - 1/4 in.) thick, depending on model size. Final hull form is achieved by filling and sanding the surface to a smooth finish. The procedure for this construction method is described below.

Male aluminum station and half station templates cut 3.2 - 6.3 mm (1/8 - 1/4 in.) smaller than the finished dimension are prepared, aligned and stack drilled. Large diameter tubular aluminum spacers are cut to station lengths taking into consideration the template thickness. The male templates and station spacers are strung on threaded rod to form a skeleton frame of the model. Figure 3.5 shows the fabrication of a barge model. Aluminum split plates are attached to this frame which serve as full size station templates.

Medium density polyurethane foam is sprayed between the station templates and allowed to rise over the templates. When the foam has cured, excess foam is removed until the edge of the male templates appear and a smooth and fair hull form is obtained. Several coats of fiberglass epoxy resin is applied to the foam form to produce a gel coat. Fiberglass cloth layers pre-tailored to the mold are then applied. Laminating resin is applied to the cloth by brush, roller or spray and worked into the cloth until no air pockets or resin-starved areas exist. The exterior of the hull is covered by not less than three layers of the fiberglass material and saturated with the gel coat. Final form of the model is achieved by hand and machine sanding until a fair and smooth model is obtained. All holes and imperfections are filled with a commercial epoxy fiberglass compound. Final form is determined by checking with female station and waterplane templates (Fig. 3.6).

FIGURE 3.5
FIBERGLASS MODEL CONSTRUCTION OF HEAVY WORK BARGE

If the model hull form is essentially a faired, single curved surface, the model may be constructed of thin wood or polyurethane strips and applied to small buttock frames representing the hull shape along its length. Fiberglassing would then proceed as presented in the previous paragraphs to build up the final thickness.

The model is usually constructed and finished keel-up on a centerline and station reference drawing. Therefore, the final smooth and fair form of the model is ensured. Also, the application of station templates always occurs at the correct interval and is perpendicular to the line of the keel. The use of accurate metal form templates permits easy, accurate, and reproducible inspection. The hull points are checked with surface height gauges and optical transit sighting.

3.4.2 Submergence Model

The purpose of the submergence tests is to verify the stability of the structure during the submergence procedure. The model used for the submergence tests should not only be able to satisfy the dynamic properties of the structure but also be provided with ballast compartment to allow prescribed and systematic ballasting of the model. In this case, the outside dimensions of the model as well as the internal dimensions of the ballast tanks must be properly modeled. In the following, the construction procedure adopted for the submergence model of a submersible exploratory drilling structure is described.

FIGURE 3.6
FIBERGLASS MODEL OF A 300,000 DWT TANKER; DIMENSIONAL
CONTROL CHECK WITH FULL TEMPLATE

3.4.2.1 *Construction Technique*

The general layout of the model is shown in Fig. 3.7. The model consists of a large foundation mat, three large columns holding the platform and several tubular structural members.

The model's mat is constructed with watertight bulkheads to divide the mat into several separate chambers representing the prototype mat's ballast tanks, fuel tanks and drill water tanks. The mat is constructed of clear lexan sheet material except for the curved bow section which is machined from a block of clear plexiglas. The use of clear material allows the ballast level in all tanks to be observed during the testing. The geometry of the mat as well as the wall thickness are correctly scaled for similarity. The model mat is glued together at all joints except where the mat bottom attaches to the walls and bulkheads. This joint is screwed together and sealed with silicone caulking to allow for any required access to the tanks.

The model caissons are constructed to properly model the outer dimensions of the prototype caissons. Internal details are not required except that any ballast tanks inside are provided. These tanks can be filled through a hole in the top of the caisson if required for testing. The tubular braces on the model are of plastic material and are sized to model the outer diameter of the prototype braces.

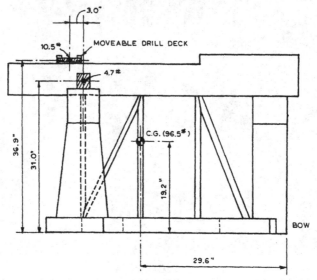

FIGURE 3.7
LAYOUT OF A SUBMERSIBLE DRILLING MODEL (SCALE 1:48)

The detailed superstructure on the model is not required for the submergence test. However, a simple platform is needed so that ballast weights may be attached to it for proper scaling of the platform properties.

In order to submerge the model according to a prescribed procedure, water ballast must be added to the mat chambers in a specified sequence. An elaborate ballasting system including plastic tubings to each ballast compartment for water transfer and air venting may be added to accommodate an external ballasting sequence. To accomplish a simpler but controlled ballasting in the current model, a valve consisting of an inner and an outer sleeve is installed in each of the mat compartments. When the inner sleeve of the valve is oriented such that the inner and outer sleeve holes are not aligned, no water may enter the chamber. To insure water tightness, the inner sleeve is coated with high vacuum silicone grease so that it seals to the inner wall of the outer sleeve. To further insure a positive seal, a grease retainer is glued to the bottom of the mat. To fill the ballast tank, the inner sleeve is rotated so that the inner and outer sleeve holes are aligned. With the valve in this position, water enters the chamber through the lower sleeve holes and the displaced air is vented through the upper holes.

3.4.2.2 Static and Dynamic Properties

The model in this case may be constructed with the mat, caissons and superstructure all being reasonably represented. Therefore, the weight distribution of the model approaches the weight distribution of the prototype structure, and the model tends to approximate the dynamic characteristics of the prototype. Any variation may be adjusted with additional weights.

After the model is constructed, its weight and C.G. location are determined. The longitudinal C.G. position may be found by balancing the model on a long 12.7 mm (1/2 in.) diameter rod. The vertical C.G. location may be determined by hanging the model by the bow caisson so that it balances with the mat's base in a vertical plane. Lead weights are added to the model as necessary to model the prototype weight of the lightship (unloaded) plus the upper deck variables (pipe, machinery, etc.). The weights are positioned so that the C.G. of the lightship and upper variables is correctly modeled (Fig. 3.7).

The verification of the dynamic properties of such models will be described in conjunction with the description of the following structure model.

3.4.3 Tension Leg Platform Model

The complete Tension Leg Platform (TLP) consists of a floating steel (or concrete) platform attached to a bottom founded template. Each of the columns of the TLP platform is attached to the ocean floor by means of a tendon. For the purpose of seakeeping tests, the hull and the tendons are modeled without the bottom template. The tendon models are attached to the floor of the wave tank. Similarly, the deck structure need not be modeled, but structural support is provided for the integrity and rigidity of the model. The following describes the construction of a TLP model.

3.4.3.1 TLP Hull

The structural components of the TLP platform can be divided into four basic categories: (1) the deck support, (2) the columns, (3) the pontoons and (4) the bracing. The chosen scale and various dimensions of a TLP in prototype and model scale are given in Table 3.2. The structural components of the platform model are fabricated from carbon steel.   Figure 3.8 shows a photograph of the finished model prior to placement in the wave tank.

The deck support consists entirely of 9.5 mm (3/8 in.) thick steel rectangular tubes. Four large rectangular tubes run from column to column forming the square perimeter of deck support.   Smaller rectangular tubes bisect the deck span in both the longitudinal and transverse directions.

TABLE 3.2
PROTOTYPE AND MODEL TLP
SCALE = 1:16

COMPONENT	PROTOTYPE	UNIT	MODEL	UNIT
Column	40	ft	30	in.
Pontoon	28	ft	21	in.
Vertical Braces	8	ft	6	in.
Inclined Braces	6	ft	4.5	in.
Displacement	61,440	kips	15,000	lbs
Heave Period	8	sec	2	sec

The columns were 1 m (39 in.) in diameter and 3.7 m (12 ft 1-1/2 in.) long made out of 4.7 mm (3/16 in.) thick steel plate.   The bottom cap plate was also 4.7 mm (3/16 in.) thick. The columns were stiffened internally with four equally spaced ring stiffeners and longitudinal stiffening between the upper two ring stiffeners. In order to accommodate the tendons a 25.4 mm (1 in.) diameter schedule 80 pipe was run vertically through the center of each column extending 25.4 mm (1 in.) below the lower face of the cap plate.   At the top of each column formed a "chair" consisting of two parallel beams spanning the columns to hold the tendons.

The pontoons were 533.4 mm (21 in.) diameter circular tubes formed from 4.7 mm (3/16 in.) thick steel plate. The braces were constructed entirely out of schedule 40 pipe. The outside diameter of the bracing ranged from 133.4 mm (5-1/4 in.) diameter to 190.5 mm (7-1/2 in.) diameter.

All joints were welded completely around and ground smooth.   To insure that the flotation pontoons were air tight, each pontoon was individually pressure tested. Subsequent to welding and pressure testing, the model was given one coat of primer and two coats of yellow paint.

3.4.3.2 Tendons and Tendon Attachment Joints

At a scale of 1:16, the typical water depth for a TLP is on the order of 30.5m (100 ft) in the model scale. Without modeling this depth it is not possible to model both axial and bending stiffness of the tendon. For the TLP dynamics, the axial stiffness is considered more important than bending stiffness. The tendon's length must be distorted in the model due to limited tank depth available. The axial stiffness was modeled by a suitable plastic rod. The bending stiffness was higher in the model compared to scaled prototype value. The tendons consisted of a 15.9 mm (5/8 in.)

FIGURE 3.8
A TENSION LEG PLATFORM MODEL FLOATING IN WAVE TANK

diameter steel rod end-connected to a 19.1 mm (3/4 in.) diameter acetal plastic rod. The end to end connection of the steel to plastic rod was provided by a 8900N (2000 lb) load cell for the measurement of axial load. The steel rod was threaded at the ends and was 4.07 m (13 ft 4-1/8 in.) long. It extended approximately 76.2 mm (3 in.) above the top of the tendon chair. The rod passed through the column (through the 25.4 mm or 1 in. diameter schedule 80 pipe) and terminated at the submerged load cell approximately 158.8 mm (6-1/4 in.) below the bottom of the column. The lower end of the steel rod threaded directly into the top of the load cell. The upper end of the acetal plastic rod was capped with a 19.1 mm (3/4 in.) tube to 19.1 mm (3/4 in.) Swagelok end connector. The end connector threads into a 19.1 mm (3/4 in.) pipe cap that has a 12.7 mm (1/2 in.) fine thread stud welded to its outside face. The 12.7mm (1/2 in.) stud threads into the bottom of the load cell. The lower end of the plastic rod also has a 19.1 mm (3/4 in.) tube Swagelok end connector. This Swagelok threaded directly into the bottom plate located in the steel floor slab. Three nuts were located at the top end of the steel rod. Once the TLP was in the operational mode, the three nuts were used to vary the pretension acting on the tendons. By turning the lower nut against the tendon chair (while preventing the upper two nuts from rotating), the column could be pulled down against the tendon or eased up from the tendon, thus changing the static pretensile load on the tendon.

3.4.3.3 TLP Model Deployment

Once the platform had been constructed, the wave tank was filled with water and the platform was placed in the tank. The platform was then floated to a convenient position, the tank drained and the model was allowed to settle on the wave tank floor. The dry model calibrations (described later) were then performed. Subsequent to the dry calibrations, the tank was partially filled and the floating natural period tests and wet inclination tests were done. After these tests were complete, the steel and plastic tendons were cut to length and the wave tank was filled further. At an appropriate water level, the filling of the wave tank stopped, the steel rods were dropped through the columns from above and then divers installed the plastic tendons from below. With the tendons in place, the water level was raised further until the TLP reached the desired testing draft of 2.1 m (6 ft 10-1/2 in.). To remove the model, the tendon installation procedure was reversed.

3.4.4 Jacket Launching Models

Jacket structures are generally transported to the installation site by barges. Model tests are required to simulate transportation, launch from the barge and up-ending of a barge mounted jacket. The transportation phase of the testing program is conducted to identify the dynamic characteristics of the jacket/barge combination as well as to determine the tie down forces and towing loads. The results are generally compared with numerical predictions of the jacket/barge behavior.

3.4.4.1 Jacket Model

A Froude scale model of the jacket is constructed to the external dimensions of the prototype. All structural members, including skirt pile guides and conductor guide framing are modeled to the specified tolerances. Mud (foundation) mats, barge bumpers, etc., are usually modeled in simplified form.

The model jacket structure is often of composite construction in that the main structural members (legs, horizontals and X-bracing) may be constructed of thin walled aluminum tubing. All of the other structural members may be fabricated of plastic tubing. Because of the complexity of the model, a welding torch can not reach some of the smaller, more complicated tube intersections. These intersections may be held in place with epoxy glue. All ends of the plastic tubular members are plugged internally and tested for leaks before attachment to the main structure. The model is tested for leaks by coating all joints with a soap film and then applying air pressure to the aluminum tubing. This form of construction provides a rigid and strong model with the specified mass, center of gravity and buoyancy. Lifting eyes are provided for handling the model. The model is painted with high visibility paint and reference points marked with colored tape.

The model is constructed to permit individual flooding of the volumes within the jacket legs during upending. Provisions are also made for the addition of lead shot, etc., internally in the event scaled weights of ballast water can not be provided. Alternatively, external lead weights are provided for attachment to the structure for simulation of added ballast during up-ending tests. All of the model properties are determined, verified and documented.

3.4.4.2 Barge Model

A barge model for jacket launching may be fabricated principally of marine plywood and sheet aluminum. Double curved surfaces may be formed from polyurethane foam. The entire outer hull is then covered with several layers of fiberglass cloth and resin. Internal hull construction is provided with suitable framing to provide adequate strength and rigidity for withstanding handling and testing. Water tight internal compartments are also provided to allow for adjustments in draft and trim. Sealed access openings are provided in the deck to allow for adjustment of draft, trim and mass properties, and for simulation of ballast conditions as specified.

The barge model is provided with the launch ways, tilt beam assemblies and mooring points scaled from the prototype. A variable speed electric winch of suitable size is installed in the barge model. Friction brakes consisting of pivoting roller assemblies are provided as an adjustable friction device between the jacket skids and the barge tilt beams. Instrumentation is provided for the measurement of line pull, rocker arm loads and motions of the barge. The barge model is painted with a color contrasting with that of the jacket. Different draft lines are marked as required. The launch barge details have been further discussed in Chapter 8.

3.5   CONSTRUCTION OF MOORING SYSTEM

Several types of mooring systems are used with floating structures. The most common of these are mooring chains and mooring hawsers. Examples are provided here for the construction of modeled mooring chains and hawsers.

3.5.1 Mooring Chains

The mooring chains are modeled to prototype characteristics by dividing by the proper scale factor (Table 3.3). The chains used in the tests are selected on the basis of their dry weight as this is the characteristic which is considered most important and can be easily modeled.

TABLE 3.3
MOORING CHAIN CHARACTERISTICS (SCALE 1:60)

DESCRIPTION	PROTOTYPE	MODEL	
		Scaled Prototype	Actual
Number	12	12	12
Pairs	6	6	6
Angle Between Pairs	60 deg	60 deg	60 deg
Angle Between Chains in Pairs	2 deg	2 deg	2 deg
Length (Fore & Aft Chains)	2,600 ft	43.33 ft	43.33 ft
Length(Side Chains)	2,600 ft	43.33 ft	19.75 ft
Pretension Angle to Horizontal (with 100% Tanker)	60 deg	60 deg	60 deg
Linear Weight (in Water)	305 lbs/ft	0.085 lbs/ft	0.084 lbs/ft
Elasticity (for Full Length)	166 kips/ft	3.84 lbs/in.	3.85 lbs/in.
Soil Friction Coefficient	1.0	1.0	0.42

The stiffness of the chain is determined by placing a length of the chain in a tension test machine and measuring its elongation at various loadings. Table 3.3 lists the prototype chain characteristics, theoretical model chain characteristics and the actual chain characteristics for a given mooring system. Due to limited tank width, the model chains in the transverse direction are shorter than the scaled prototype value. If the stiffness of the actual chain is greater than what is desired, springs are installed in the system to decrease the spring constant. The required spring stiffness is determined by the following equation:

$$\frac{1}{K} = \frac{1}{K_1} + \frac{1}{K_2}$$

(3.1)

where K = desired spring constant of the system; K1 = spring constant of the chain; and K2 = required spring constant in the added spring system. For example, if K = 980.1 N/m (67.15 lbs/ft) and K1 = 10,319 N/m (707 lbs/ft), this results in a K2 of 1080 N/m (74 lbs/ft). The length of these springs should be chosen so that the spring range is not exceeded during the entire loading sequence. For the above example, if each spring chosen has a constant of 388.3 N/m (26.6 lbs/ft) and six parallel sets of two springs in series are used, it results in a composite spring constant of 1164.8 N/m (79.8 lbs/ft) and provides the needed extension in the springs during testing.

A static in-place calibration of the installed mooring system is performed by hanging a series of known weights on a string from the bow and stern of the ballasted tanker over a pulley and measuring the displacement of the tanker. An example data set is plotted in Fig. 3.9.

3.5.2 Mooring Hawsers

Let us consider mooring hawsers based on a prototype 610 mm (24 in.) circumference nylon hawser with a breaking strength of 646.5 metric tons (1425 kips). Load-strain characteristics of the prototype hawser is defined by the equation:

$$F = 0.125[(100)(\varepsilon)]^2 \qquad\qquad\qquad (3.2)$$

where F = load as a percent of breaking strength and ε = percent strain. The actual load at any given strain, then, is the breaking strength times F. Table 3.4 contains the prototype and desired model characteristics (at a scale of 1:48) of the hawser for the example system.

TABLE 3.4
NYLON HAWSER CHARACTERISTICS

ITEM	PROTOTYPE	MODEL
Circumference	24 in.	1/4 in. Rubber Bands
Breaking Strength	1425 kips	12.9 lbs
Length	344.5 ft	7.2 ft
Actual length of rubber band strand is about 15 in.		

Since the prototype hawser is nonlinear, the model hawser is also made nonlinear. One method of manufacturing the model hawser is to use rubber bands of unequal lengths banded together at the ends. For convenience in modeling, the rubber bands are usually chosen to have equal width. The load elongation curve of the prototype hawser is segmented into several parts, each of which may be represented by a mean

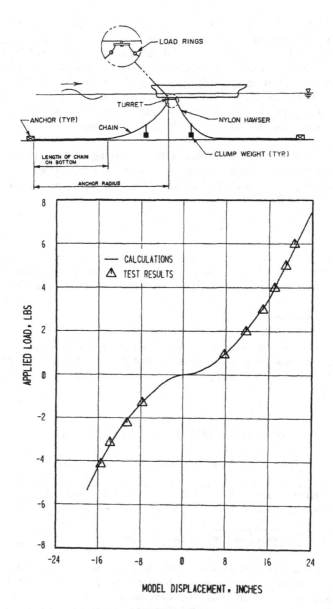

FIGURE 3.9

LOAD DISPLACEMENT CURVE MEASURED ON MODEL IN PLACE

slope. The initial slope (stiffness) at the lowest strain is matched by a rubber band of a given width and length such that this length of a rubber band can stretch to the full elongation required of the model hawser. Then the width and length of the band represents the first segment of the load-elongation curve. For the same width, the second length of the rubber band is chosen in such a way that it provides no load (i.e., remains slack) over the first segment, but begins to stretch to pick up load over the second segment. Thus, there is a corresponding rubber band length for each segment of the stress-strain (i.e. load-elongation) curve. Additional lengths of band are added, which correspond to the strain for each loading point and the combined strain of the rubber bands match the slope of the particular segment. Sometimes, multiple bands or bands of unequal width are needed to match the segment load versus elongation characteristics of a particular mooring hawser. As the hawser is stretched, a different number of bands are loaded. Since the number of bands loaded at one time varies with the applied load and resultant strength, the response of the model hawser is nonlinear.

The load-elongation characteristics of a prototype hawser is given in Table 3.5. The characteristics of the model hawser (scale 1:42) made of rubber bands are checked by applying a weight to one end and measuring the displacement. Figure 3.10 shows a comparison of theoretical and actual hawsers for the example problem.

TABLE 3.5
LOAD-ELONGATION CHARACTERISTICS OF A HAWSER
(SCALE = 1:42)

STRAIN	PROTOTYPE LOAD (kips)	MODEL LOAD (lbs)
.02	7.13	.096
.04	28.50	.385
.06	64.13	.866
.08	114.00	1.539
.10	178.13	2.404
.125	278.32	3.757
.150	400.78	5.410
.175	545.51	7.363
.20	712.50	9.619

3.6   MODEL CALIBRATION METHODS

Special calibration procedures are needed to determine the model properties of a floating platform and the corresponding mooring system. The platform calibrations are done to determine the following properties:

- weight,
- center of gravity,
- mass moments of inertia, and
- metacentric height.

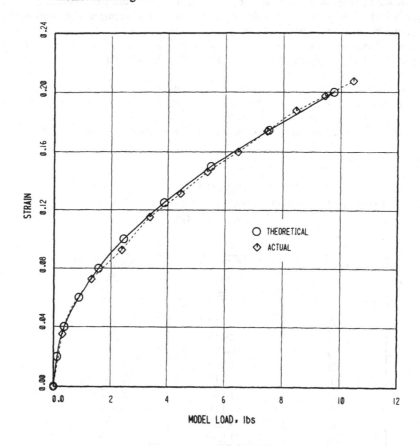

FIGURE 3.10
MODEL LOAD VS. STRAIN FOR A MOORING HAWSER ARRANGEMENT

The platform model is usually designed so that it is quite light compared to the required displacement. In order to model the specified weight, (lead and/or water) ballast is placed on the model at various positions. The positioning of the weight placement is dictated by the results of the center of gravity and moment of inertia tests. The term "unballasted model" refers to the platform without ballast, mooring components and instrumentation, and the term "ballasted model" refers to the model with ballast, mooring system and instrumentation. After the model construction is complete, the C.G. test and the mass moment of inertia test are performed on the unballasted model.

Calculations are then made to determine the lead weight placement necessary to bring the unballasted model C.G. and moments of inertia to the required values. The calibration method for the properties of a tanker has been described through worked out examples by Munro-Smith (1965). Bhattacharyya (1978) also presented several cases of model calibration.

3.6.1 Platform Calibrations

The dry calibration of a dynamic model platform is performed after the model construction is complete, but before the ballast weights are placed. The amount and location of the ballast weights is determined based on the initial calibration. A final check calibration is usually performed after the placement of the ballast weights. It is always wise to perform a wet calibration of the platform before the starting of the test runs.

3.6.1.1 Weight Estimate

Several evaluations may be made during various stages of model construction to estimate the weight of the unballasted platform model. These evaluations can be summarized as follows:

a. Hand calculations of each structural component weight during design stage.
b. Measurement of the component weights once they are available.
c. Measurement of free floating draft of the completed model.
d. Hanging model with a scale or a load cell inline with the cable and recording the model weight (based on the load cell output). The setup for this method is shown in Fig. 3.11.

Each weight estimate resulting from the above four methods is adjusted for components not included in the original calculation. An example of the results of the four different weight evaluations for a TLP model is given in Table 3.6.

3.6.1.2 Center of Gravity Estimate

Small models may be balanced on a knife-edge along a particular model axis for the determination of the center of gravity along that axis. A round rod may also be used for this purpose instead of a knife-edge. This method works quite well for long models that are easy to handle. The model is placed transverse on its sides on the rod and moved until the two sides tend to balance and a small displacement on either direction of the rod provides a bias in that direction. Then the distance from the edge of the model to the center of the rod gives the location of the C.G.

The setup for the center of gravity estimation for large models is shown in Fig. 3.11. This setup shows the model hanging from a universal joint such that it is free to swing in the roll and pitch directions. By lifting the bow of the model and simultaneously recording the lifting load (using a load cell) and the angle of inclination (using an inclinometer), the C.G. is calculated. The formula used to calculate the C.G. arises from the static equilibrium of moments as follows:

$$\sin\theta = \frac{Fd_1}{Wd_{cg}}$$

(3.3)

TABLE 3.6
MODEL TLP PLATFORM WEIGHT ESTIMATES

METHOD	PREDICTED WEIGHT (lbs)	% DIFFERENCE FROM METHOD d
a	14845	-2.1
b	15062	-0.7
c	15209	0.3
d	15162	0.0

where θ = model's angle of inclination, F = lifting force, d_1 = horizontal moment arm from lifting point to rotational point, W = model weight and d_{cg} = distance from C.G. to rotational point (e.g., universal joint axis). Since all other quantities are known or measured, d_{cg} may be computed from Eq. 3.3. Generally, several different inclination angles are applied to the model, and a line of best fit to these lifting load/angle points is obtained.

The rotational point in this example of a TLP is 3537 mm (139.25 in.) above keel. Therefore, the best fit line to the dry inclination curve (for small angles) is expressed as

$$\theta \approx \sin \theta = \left\{ \frac{d_1}{W(139.25 - KG)} \right\} F \qquad (3.4)$$

where θ is in radians and the term in brackets is the slope of the best fit line. Knowing d_1 (in inches), W and the slope, one can solve for KG (in inches), the distance from the structure keel to the C.G.

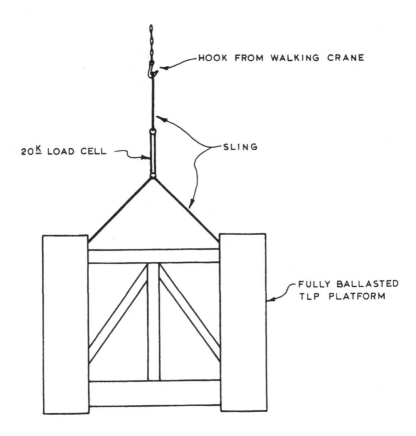

FIGURE 3.11
PLATFORM CALIBRATION TEST SET-UP –
C.G. AND MOMENTS OF INERTIA TESTS

3.6.1.3 Estimate of Moments of Inertia

The analysis of the motions of a floating body requires that the model be scaled properly for the rotational properties of the prototype. The moments of inertia (or the radii of gyration) of a floating model for the roll, pitch or yaw motion may be determined from a swing frame built from structural members (e.g., I-beams). The weight, center of gravity and moment of inertia of the frame alone are known (i.e., measured) a priori. The model is mounted on the frame, which is placed on two knife edges.

The vertical C.G. of the model is estimated by measuring the angle of tilt of the frame with an angle sensor by applying a known moment statically. By balancing the moments on the knife edge, the vertical C.G. is estimated.

The model may be swung freely in roll, pitch or yaw. The periods of motion of the model are determined with a stopwatch or the recording of an angle sensor. The moments of inertia of the system for a particular degree of freedom may be estimated from the natural period.

An alternative method uses the center of gravity test setup described in the previous section (Fig. 3.11). A sonic wave probe is mounted horizontally on a stand adjacent to the model. The model is given an initial rotational displacement and then allowed to swing freely about the point of rotation (e.g., the universal joint axis). By measuring the relative displacement of the swinging model with the probe, the natural period of displacement time history is determined from the average time period between each successive pair of crests. The mass moment of inertia of the model is then calculated based on the oscillation "swing" period of the model.

The calculation of the moment of inertia is given by the following formula:

$$I_g = \left(\frac{T_N}{2\pi}\right)^2 W d_{cg} - \frac{W}{g} d_{cg}^2 \tag{3.5}$$

where I_g = mass moment of inertia about pitch axis through the C.G. (kg - m^2), T_N = measured natural period of swing (s), and W = model weight (N).

3.6.1.4 Righting Moment Calibration

A wet inclination test is performed to measure the metacentric height (GM) of the platform in the pitch direction. The platform is floated freely and known weights are placed on the center of the bow deck support. The trim angle of the model as measured by an inclinometer is recorded. The applied heeling moment equals the amount of added weight times its moment arm; thus the equilibrium of static moments requires the following:

$$Fd_1 = W(GM) \sin\theta \qquad (3.6)$$

where the right hand side is the righting moment term. For small angles,

$$\theta \approx \sin\theta = \left[\frac{d_1}{W(GM)}\right] F \qquad (3.7)$$

where θ = trim angle in radians, d_1 = horizontal moment arm for applied weight, W = model weight, GM = metracentric height and F = applied weight. Several angles of trim are applied to the model. Then, the wet inclination curve is a line of best fit through these points. Similar to the dry inclination curve, the term in the bracket of the above equation represents the slope of the best fit line to the wet inclination curve. Knowing the slope, "d_1" and "W", the GM may be calculated.

An example of a submersible drilling rig is included here. The intent of the test run was to confirm the righting moment characteristic curves for the drilling rig. The test was performed on a 1:48 scaled model. Two horizontal cables, one connecting the right side of the mat's bottom to an anchor located at one end of the tank and the other connecting the left side of the deck to a weight over a pulley on the opposite end of the tank were used to induce the righting moment on to the rig. The two points of load application are on the plane of the longitudinal center of gravity, located 743 mm (29.25 in.) from the bow of the model. Depending on the tilt angle (θ), the moment arm (D) in inches varied:

$$D = [29.25 - (L_1 + L_2)\tan\theta]\cos\theta \qquad (3.8)$$

where L_1 = distance (in inches) between point of load application at the mat and the centerline of the model and L_2 = distance (in inches) between point of load application at the deck and centerline of the model. The tilt angle was measured by an inclinometer located at the top of the deck.

The data are tabulated on Table 3.7. The relationship of tilt angle to the horizontal distance GZ is plotted on Fig. 3.12. The model properties are adjusted to satisfy the theoretically computed curves for the prototype.

3.6.2 Tendon Calibrations

Unlike fixed or floating models, the anchoring system requires different modeling and calibration procedures. Mooring chain and hawser models have already been described. Another example is the TLP tendon. Risers fall in the same category. Tendons are generally modeled using plastics. Plastic rods experience creep which should be stabilized before testing with tendon rods should commence. The tendon calibrations are done in wet and dry environments to determine:

- The dry creep characteristics;
- The dry static stiffness;
- The dry dynamic stiffness and damping;
- The effect of hysteresis under dry dynamic loading; and
- The wet in-place static stiffness.

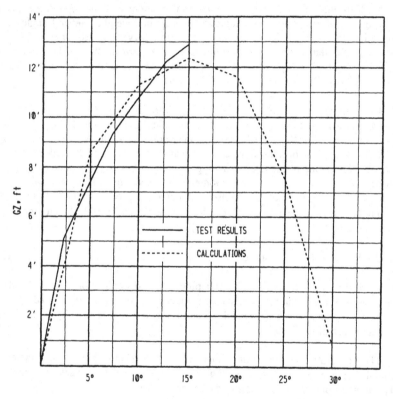

FIGURE 3.12
RIGHTING MOMENT CURVE FOR UNDAMAGED SUBMERSIBLE

During the dry tendon tests, data is collected to determine the creep, stiffness and damping properties of individual tendons which are loaded in the axial direction. A dry tendon test setup is presented in Fig. 3.13, showing a tendon hanging vertically from a roof beam with a load cart attached to the lower end of the tendon. The axial load in the tendon is varied by changing the amount of weight contained in the cart.

TABLE 3.7

**TILTING CHARACTERISTICS OF
A SUBMERSIBLE MODEL (SCALE 1:48)**

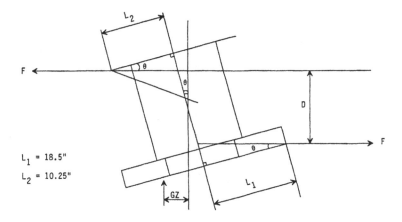

$L_1 = 18.5"$

$L_2 = 10.25"$

$D = [29.25 - (L_1 + L_2) \tan \theta]\cos \theta$					
Model					Prototype
F (lbs)	θ (deg)	D (in.)	M (in. lbs)	GZ (in.)	GZ (ft)
5	2.15	28.15	140.76	1.29	5.15
10	7.21	25.40	254.08	2.33	9.30
12	9.57	24.06	288.79	2.64	10.57
15	12.67	22.23	333.44	3.05	12.20
17	15.08	20.76	352.98	3.23	12.92

3.6.2.1 Dry Creep Characteristics

Since the tendons are under tension for an extended period of time while the TLP is in the water, it is necessary to determine the rate of creep of the tendon material. Vendor supplied information indicates that stiffness of plastic rod is influenced by the extension of the material due to creep; hence, it is necessary to perform stiffness tests with the effects of creep eliminated. Because of this, many of the dry tendon tests are done after the tendons have been loaded for several days. In a test of a tendon under load, approximately 7 days were required to achieve 95 percent of the material creep, although this may vary considerably depending on the material being used. For this particular test, DELRIN was used because of its desirable overall characteristics.

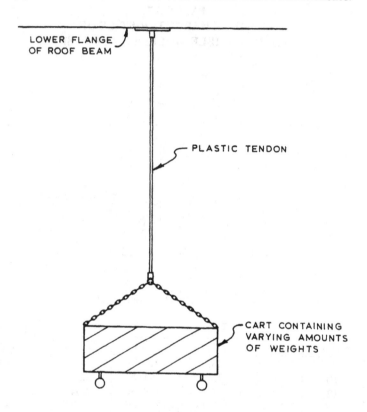

FIGURE 3.13
TENDON CALIBRATION TEST SET-UP

3.6.2.2 Dry Static Stiffness

Stiffness tests under static conditions are accomplished by initially loading the tendon with a given pretensile load and then incrementally increasing the load to a prescribed maximum. Upon reaching the maximum, the load is incrementally decreased to a prescribed minimum (less than the pretensile load), and then the load is increased back up to the pretensile load. These load changes are done in equal increments. By measuring the tendon length at each increment (using a measuring tape), the tendon stretch as a function of tendon axial load is determined. This procedure also identifies any hysteresis present in the material. The load range used for the static stiffness tests is chosen based on the tendon loads expected during operation.

3.6.2.3 Dry Dynamic Stiffness and Damping

The dry dynamic stiffness and material damping tests may use the same setup as the dry static stiffness tests. Instead of incrementally loading the rod (as was done for the static stiffness test), a weight may be simply dropped on the (pre-loaded) weight cart which is attached to the bottom of the hanging tendon (Fig. 3.13).

The resulting free oscillation of the axial load (e.g., strain) is recorded from a load cell or a strain gage. To conduct damping tests using load cells, the load cell is connected to the top of the hanging tendon so that the measurement axis of the cell is in line with the tendon axis. For dry material damping tests, a strain gauge may be mounted directly on the plastic rod. Both methods give a time history of the oscillating tendon load from which the stiffness and damping values are calculated. Strain gauges are preferred over load cells to avoid the inclusion of the stiffness of the load cell in the measurement.

3.6.2.4 Hysteresis Effect Under Dry Dynamic Loading

The dry static stiffness tests show that the plastic material (DELRIN) does possess some hysteretic properties. Therefore, it is necessary to determine the significance of the hysteresis under dynamic loading conditions. Stiffness and damping tests which use both a strain gauge and a load cell together may be conducted to make this determination. The effect of hysteresis is such that as the tendon axial load decreases to zero, the corresponding strain decreases to some value greater than zero. By simultaneously recording the rod stress and strain (using the load cell and the strain gauge) during testing, the stress and strain during dynamic loading and unloading may be obtained.

3.6.2.5 Wet In-Place Static Stiffness

The wet tendon tests are conducted on all four tendons simultaneously with the TLP in the wave tank, and they determine the total effective stiffness of the tethered TLP in the surge and heave directions. The setups required for the wet tests are shown in Figs. 3.14 and 3.15, respectively.

If the plastic (DELRIN) tendons are used, the static stiffness of the four tendons should be estimated about seven days after the TLP had been deployed in the wave tank in order to account for material creep. The static stiffness of the four tendons acting together in the surge direction is determined by horizontally loading the structure using a pulley system (Fig. 3.14). Known horizontal loads are applied to the structure by placing known weights on the platen. The surge displacement of the structure may be measured by an optical system.

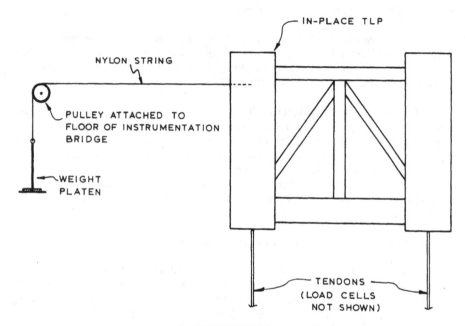

FIGURE 3.14
TENDON CALIBRATION TEST SETUP
IN THE SURGE DIRECTION

With the TLP in the operating mode, the axial force in the tendons is varied by changing the water level in the wave tank (e.g., changing buoyant force). The loads in the tendons are measured by the load cells. By taping rulers (0.25 mm or 0.01 in. resolution) to each column and using a surveyor's level (Fig. 3.15), the change in elevation of each column may be measured. This change in elevation is a direct reflection of the tendon elongation. By incrementally varying the water level, the desired load range in the tendons can be achieved.

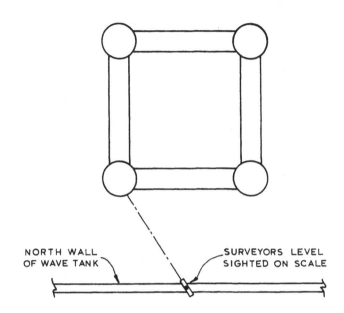

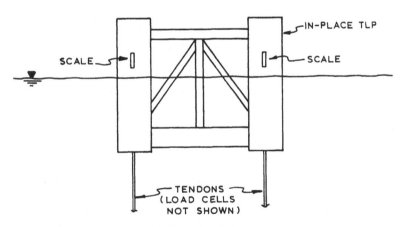

FIGURE 3.15
WET TENDON CALIBRATION TEST SETUP IN THE HEAVE DIRECTION

3.7   REFERENCES

1.   Bhattacharyya, R., <u>Dynamics of Marine Vehicles</u>, John Wiley and Sons, New York, New York, 1978.

2.   Munro-Smith, R., <u>Notes and Examples in Naval Architecture</u>, Edward Arnold (Publishers) Ltd., London, England, 1965.

CHAPTER 4

MODEL TESTING FACILITY

4.1   TYPE OF FACILITY

The facility for model testing of an offshore structure generally consists of a wave generating basin, a towing tank and a current   generating   facility. It   is advantageous to simulate the wave and current generation in a single basin so that their combined interaction with the structure model may be investigated. Sometimes, wind generating capability is also required. The wind generation in a wave basin is often accomplished using a series of blowers located just above the water surface near the model.

Most   of the wave tanks built pre-eighties are two-dimensional, i.e., they are capable of generating waves that travel in one direction only. The long period waves in the ocean may exhibit unidirectional behavior. However, wind-generated ocean waves are generally multi-directional. In order to generate multi-directional waves, the wave basins must have widths comparable to their lengths. Many modern facilities are capable of producing multi-directional waves.

This   chapter will   describe different   types of wave generators,   their design methods and the generation capabilities of many commercial wave testing facilities.   The wind and current generating techniques will also be discussed.

4.2   WAVE GENERATORS

In the natural sea state, the period range of the power spectrum of waves having appreciable energy contents varies from about 5 seconds to 25 seconds. If a largest recommended scale ratio for model testing in a wave tank is 1:100, then the shortest wave required in the tank is 0.5 second (2 Hz). Therefore, the wavemaker [Biesel (1954)] in the tank should be capable of producing waves of frequencies up to 2 Hz.

There are many types of wave generating devices employed in model basins. They may be classified in two general categories: active and passive. The active generators consist of mechanical devices of various sorts displacing the water in direct contact with the generator. By controlling the movement of the device, the wave form is created. On the other hand, the passive wavemakers have no moving parts in contact with the water. They use air pressure to generate oscillations of the water. Various types of active and passive generators are summarized in Fig. 4.1. All wavemakers in the figure are of the active type except for the types R and S.

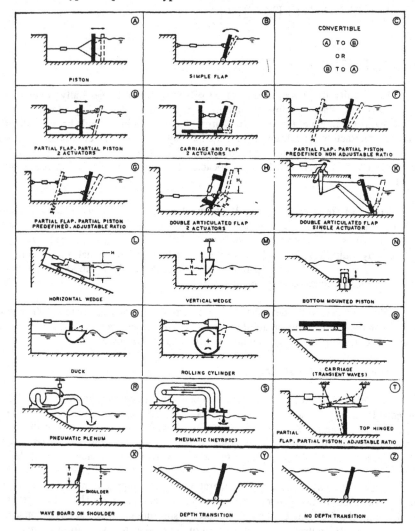

FIGURE 4.1
VARIETY OF WAVE GENERATOR SCHEMATICS
[Ploeg And Funke (1980)]

The wedge shaped wave generator is best suited for high frequency waves but has limited success in generating low frequency waves. On the other hand, the pneumatic wavemaker can generate low frequency waves quite well but is limited to high frequencies of about 1 Hz due to the quick response time required to form these waves.

A wide frequency range (say, 0.5 – 10 seconds) can be best accomplished by a double articulated mechanical flapper. The top flapper generates waves between the frequencies of 1.0 Hz to 2.0 Hz, while the bottom flapper produces the lower frequency waves. Note that a random wave of broad-banded spectrum requires a large frequency range. In this case, both flappers may be needed to produce this spectrum. A schematic of the double flapper arrangement is shown in Fig. 4.2. Typical dimensions for a water depth of 5.5 m (18 ft) are shown on the figure. Modes 1 and 2 are for high and low frequency waves respectively, while the combination of the two (Mode 3) is needed for a wide banded wave spectrum.

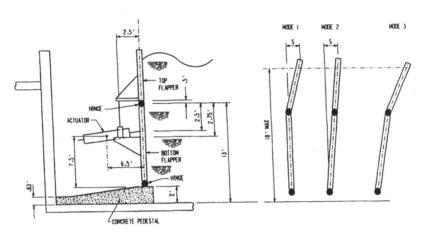

FIGURE 4.2
DOUBLE-HINGED WAVE GENERATOR

4.3 MECHANICAL WAVEMAKER

There are two main classes of mechanical type wavemakers that are installed in facilities to generate waves [Patel and Ionnau (1980)]. One of them moves horizontally in the direction of wave propagation in a simple harmonic motion (to generate sinusoidal waves) and has the shape of a flat plate driven as a paddle or a piston. The other type moves vertically at the water surface in a simple harmonic motion and has the shape of a wedge [Bullock and Murton (1989)]. In the design of these wavemakers [Gilbert, et al. (1971)], a boundary value formula is developed based on the type of wave generator and its motion.

4.3.1 Hinged Flapper Wave Theory

Consider a flapper hinged at the bottom spanning the width of the wave tank [Galvin (1966), Hudspeth, et al. (1981)]. For the purposes of generating the theory, the flapper is assumed to undergo small simple harmonic motion. The problem is treated as two-dimensional. The water is assumed incompressible and inviscid and flow irrotational. Thus, a velocity potential, ϕ may be assumed to describe the flow. This wave generator theory mainly follows the work done by Biesel and Suquet (1953). Also see Hyun (1976).

The velocity potential $\phi(x,y,t)$ for small amplitude waves generated by a simple-harmonic flap motion about $x = 0$ is governed by the Laplace equation, where (x,y) is the coordinate of a point in a two-dimensional space and t is time:

$$\frac{\partial^2 \phi}{\partial x^2} + \frac{\partial^2 \phi}{\partial y^2} = 0, \ 0 \le x < \infty, 0 \le y \le d \tag{4.1}$$

where d = still water depth at the wavemaker.

Knowing the expression for ϕ, the wave form for a given ω may be evaluated. Along with the differential equation, the following boundary conditions must be satisfied. The boundaries include the free surface, the plunger surface, and the bottom.

$$\frac{\partial \phi}{\partial y} = 0 \qquad \text{for } y = 0, \tag{4.2}$$

$$\frac{\partial^2 \phi}{\partial t^2} + g \frac{\partial \phi}{\partial y} = 0 \qquad \text{for the surface condition } (y = d) \tag{4.3}$$

which, in first-order theory reduces to

$$\omega^2 \phi + g \frac{\partial \phi}{\partial y} = 0 \tag{4.4}$$

and

$$\frac{\partial \phi}{\partial x} = \omega \xi(y) \cos \omega t \tag{4.5}$$

The last equation represents the boundary condition at the moving wall (x = 0), where ξ (y) denotes the prescribed maximum displacement of the flap surface such that x = ξ(y) sin ωt.

The solution for the velocity potential, ϕ, can be shown to be

$$\phi = -\frac{\omega}{k} A \cosh ky \sin(\omega t - kx) - \sum_{n=1}^{\infty} \frac{\omega}{k_n} A_n \cos k_n y \exp[-k_n x] \cos \omega t \tag{4.6}$$

where k = 2π/L is the wave number that satisfies the dispersion relation

$$\omega^2 = gk \tanh kd \tag{4.7}$$

and k_n are the positive real roots of

$$\omega^2 = -g \, k_n \tan k_n d \tag{4.8}$$

The coefficients, A, and A_n can be determined by the boundary condition given by Eq. 4.5, such that

$$A \cosh ky + \sum_{n=1}^{\infty} A_n \cos k_n y = \xi(y) \tag{4.9}$$

Hence, we have

$$A = \frac{2k \int_0^d \xi(y) \cosh ky \, dy}{\sinh kd \cosh kd + kd} \tag{4.10}$$

$$A_n = \frac{2k_n \int_0^d \xi(y) \cos k_n y \, dy}{\sin k_n d \cos k_n d + k_n d} \tag{4.11}$$

By means of the velocity potential ϕ(x,y,t), the displacements of the water particle in the x and y directions are obtained.

$$\Delta_x = A \cosh ky \sin(\omega t - kx) + \sum_{n=1}^{\infty} A_n \cos k_n y \exp[-k_n x] \sin \omega t \qquad (4.12)$$

$$\Delta_y = A \sinh ky \cos(\omega t - kx) + \sum_{n=1}^{\infty} A_n \sin k_n y \exp[-k_n x] \sin \omega t \qquad (4.13)$$

By inspecting the above two equations, we can conclude that the fluid motion is composed of two parts:

- A propagating part in the positive x direction; and
- An evanescent part as expressed by $e^{-k_n x}$.

The exponentially decaying part approaches zero when $x > 3d$. Thus, at $x > 3d$,

$$\Delta y = A \sinh ky \ \cos(\omega t - kx) \qquad (4.14)$$

which we can rewrite for the still water surface as

$$\Delta y = a \cos(\omega t - kx) \qquad (4.15)$$

where a is the wave amplitude $= A \sinh kd = H/2$.

With the aid of Bernoulli's equation, the pressure up to first order is given by

$$p = -\rho \frac{\partial \phi}{\partial t} + \rho g(d - y)$$

$$\qquad (4.16)$$

$$= p_w + p_i + \rho g(d - y)$$

where

$$p_w = \rho g a \frac{\cosh ky}{\cosh kd} \cos(\omega t - kx) \qquad (4.17)$$

and

$$p_i = \sum_{n=1}^{\infty} \rho g A_n \tan k_n d \cos k_n y \exp[-k_n x] \sin \omega t \qquad (4.18)$$

At $x = 0$, Eq. 4.17 becomes

$$p_w = \rho g a \frac{\cosh ky}{\cosh kd} \cos \omega t = C_1 \cos \omega t \qquad (4.19)$$

and Eq. 4.18 becomes

$$p_i = \sum_{n=1}^{\infty} \rho g A_n \tan k_n d \cos k_n y \sin \omega t = C_2 \sin \omega t \qquad (4.20)$$

Note that p_w and p_i are 90° out-of-phase with respect to each other.

Thus, the maximum value of $C_1 \cos \omega t + C_2 \sin \omega t$ is $\sqrt{C_1^2 + C_2^2}$ whereas its phase is

$\tan^{-1} \dfrac{C_1}{C_2}$.

For a flap-type wave generator, the displacement of the wavemaker is written in terms of its stroke at the still water level, S, as

$$\xi(y) = \frac{S}{D}(y - \delta d), \text{ for } \delta d \le y \le d \qquad (4.21)$$

$$\xi(y) = 0, \text{ for } y < \delta d \qquad (4.22)$$

where D is the flap draft and δd is the elevation of the hinge point ($\delta d = d - D$). Substituting the above equation into Eqs. 4.10 and 4.11

$$A = 2S \frac{kD \sinh kd - \cosh kd + \cosh k\delta d}{kD(\sinh kd \cosh kd + kd)} \qquad (4.23)$$

$$A_n = 2S \frac{k_n D \sin k_n d + \cos k_n d - \cos k_n \delta d}{k_n D(\sin k_n d \cos k_n d + k_n d)} \qquad (4.24)$$

The relationship between wave height (H) and stroke (S) can be established by combining Eq. 4.23 and the expression for 'a' in Eq. 4.15.

$$\frac{S}{H} = \frac{kD[\sinh kd \cosh kd + kd]}{4 \sinh kd [kD \sinh kd - \cosh kd + \cosh k\delta d]} \qquad (4.25)$$

Substituting this relationship into Eqs. 4.24 and 4.20, we can write an expression for p_i as a function of H and T.

For the case of random waves [Hudspeth, et al. (1978)], the wave profile can be represented by the sum of an infinite number of sine waves as

$$\eta(t) = \sum_n \frac{1}{2} H_n \sin(\omega_n t + \varepsilon_n) \tag{4.26}$$

The wave height at a particular period corresponding to ω_n over a frequency band $\Delta\omega$ can be determined from

$$H_n = 2\sqrt{2S(\omega)\Delta\omega} \tag{4.27}$$

where the energy density spectrum may be represented, for example, by the Pierson-Moskowitz spectrum model,

$$S(\omega) = 0.0081 g^2 \omega^{-5} \exp\left[-1.25(\omega/\omega_0)^{-4}\right] \tag{4.28}$$

where $\omega_0 = \sqrt{0.161g/H_s}$ and H_S = significant wave height. The solution for the flapper motion is obtained by the linear superposition of the individual H-T pair in Eq. 4.26 and relating it to Eq. 4.25.

In addition to the forces needed to create the wave propagation, additional forces are required to counteract the static water pressure and to overcome the inertia of the flapper while in motion.

4.3.2 Wedge Theory

Consider a plunger of arbitrary triangular section and of width, b, and depth, D, at the still water level (Fig. 4.3). In developing the theory, the plunger is assumed to move harmonically at a frequency, ω, in small amplitudes in the vertical direction. Since the wedge spans the width of the tank, the problem is treated as two-dimensional. The wedge is assumed to travel a short distance so that linear theory may be applied. This allows the formulation of the problem [Wang (1974)] in terms of a velocity potential ϕ that satisfies Laplace's equation (Eq. 4.1).

The free-surface linearized boundary condition in this case is written as

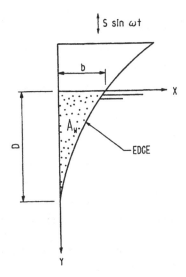

FIGURE 4.3
COORDINATE SYSTEM FOR A PLUNGER

$$\frac{\partial \phi}{\partial y} + \frac{\omega^2}{g}\phi = 0 \text{ at } y = 0 \tag{4.29}$$

The normal fluid velocity at the plunger surface is equal to the velocity component, v_n, of the forced oscillation of the plunger in the same direction,

$$\frac{\partial \phi}{\partial n} = v_n \tag{4.30}$$

In addition, the radiation condition dictates that at a large distance from the plunger, the waves are outgoing progressive waves.

The solution is obtained through a conformal transformation from the (plunger) z-plane and its image in the free surface to a reference ζ-plane of a unit circle (Fig. 4.4) in which $z = (x,y)$ and $\zeta = (r,\theta)$ and y is measured positive upwards from the still water level.

$$\frac{z}{a_0} = \zeta + \sum_{n=1}^{N} a_{2n-1}\zeta^{-(2n-1)} \tag{4.31}$$

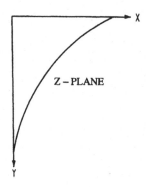 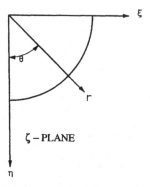

FIGURE 4.4
CONFORMAL TRANSFORMATION

where z and ζ are complex variables and the coefficients a_n are real numbers. In terms of coordinates (x, y) and (r, θ), we have

$$\frac{x}{a_o} = r \sin\theta + \sum_{n=1}^{N} (-1)^{n-1} \frac{a_{2n-1}}{r^{2n-1}} \sin(2n-1)\theta \tag{4.32}$$

$$\frac{y}{a_o} = r \cos\theta + \sum_{n=1}^{N} (-1)^{n-1} \frac{a_{2n-1}}{r^{2n-1}} \cos(2n-1)\theta \tag{4.33}$$

Note that the first term (only a_0) in Eqs. 4.32 and 4.33 provides the fundamental shape of the plunger as a circular cylinder. If additionally a_1 is considered nonzero (only a_0 and a_1), it yields a family of elliptic cylinders. The solution for ϕ is constructed as a superposition of the source and multipole potentials at the origin in terms of these coefficients, having the form

$$\phi = \omega S a_0 \left[\left(A\phi_s - B\phi_c + \sum_{m=1}^{\infty} P_{2m}\phi_{2m} \right) \sin\omega t + \left(B\phi_s + A\phi_c + \sum_{m=1}^{\infty} Q_{2m}\phi_{2m} \right) \cos\omega t \right] \tag{4.34}$$

in which the source potential

$$\phi_s = \pi e^{-ky} \sin kx - \int_0^\infty \frac{e^{-\kappa x}(\kappa \cos\kappa y - k \sin\kappa y)}{\kappa^2 + k^2} d\kappa \tag{4.35}$$

$$\phi_c = \pi e^{-ky} \cos kx \tag{4.36}$$

and the multipole potentials are expressed as

$$\phi_{2m} = \frac{\cos 2m\theta}{r^{2m}} + ka_o\left[\frac{\cos(2m-1)\theta}{(2m-1)r^{2m-1}} + \sum_{n=1}^{N}(-1)^{n-1}\frac{(2n-1)a_{2n-1}\cos(2m+2n-1)\theta}{(2m+2n-1)r^{2m+2n-1}}\right] \qquad (4.37)$$

The coefficients A, B, P$_{2m}$ and Q$_{2m}$ are determined from the normal velocity boundary condition at the body. All other boundary conditions including the radiation condition are automatically satisfied by ϕ. Note that the solution is linear with respect to the plunger amplitude, S. The function ϕ_S is a two-dimensional source, while ϕ_{2m} are multipole functions which represent only local disturbances in the vicinity of the (plunger) body and contribute no waves at infinity. ϕ_S and ϕ_C behave like outgoing progressive waves on the free surface and approach regular sinusoidal waves asymptotically as x increases. The above derivation is essentially taken from Wang (1974). After the velocity potential ϕ is determined, the form of the free surface is obtained from the linearized condition

$$\eta(x,t) = -\frac{1}{g}\frac{\partial\phi}{\partial t} \text{ at } y = 0 \qquad (4.38)$$

At the point of interest away from the plunger, the waves will have the form

$$\eta(x,t) = ka_o\pi S\sqrt{A^2 + B^2}\,\sin(\omega t - kx + \varepsilon) \qquad (4.39)$$

where

$$\varepsilon = \tan^{-1}\frac{B}{A} \qquad (4.40)$$

The nondimensional amplitude of the wave, $\overline{A}$ is given as

$$\overline{A} = \frac{a}{S} = kb\pi\sqrt{A^2 + B^2}\,/R \qquad (4.41)$$

where

$$R = \frac{b}{a_0} = 1 + \sum_{n=1}^{N}a_{2n-1} \qquad (4.42)$$

Thus, $\overline{A}$ is dependent on the frequency parameter kb. Since potential ϕ on the body is known, the hydrodynamic pressure from Bernoulli's equation and corresponding forces on the plunger may be determined.

Wang (1974) presented many results on the amplitude ratio $\bar{A}$ for various plunger geometries. Typical plots of wave amplitude as a function of kb are shown in Fig. 4.5. As the ratio of depth to width decreases, the wave amplitude increases for the same stroke. Also, there is a general tendency of increase of amplitude with kb even though there is an optimum value at higher D/b values. The plot represents a plunger with σ (= A_W/bD) having a value of 0.5, where A_W is the cross-sectional area of the submerged wedge (Fig. 4.3).

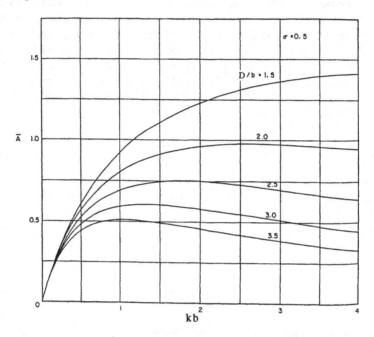

FIGURE 4.5
DESIGN CURVES FOR A WEDGE-SHAPED WAVEMAKER [Wang (1974)]

4.4   PNEUMATIC WAVE GENERATOR

Unlike the flapper or the plunger, a pneumatic wave generator has no moving parts in direct contact with the water. The heart of a pneumatic wave generator is a blower with a low pressure head. The blower is connected to a partially immersed plenum chamber which is open at the bottom and situated at one end of the tank (Fig. 4.6). A flapper valve is placed at the bifurcation between the outside vent duct and the plenum chamber duct and connects alternately the discharge and the intake of the blower to and from the plenum chamber. This introduces a pressure differential in the plenum chamber which alternately draws water up into the plenum chamber and then pushes it down. The plenum contains baffles to minimize transverse waves as well as to help distribute the air uniformly in the chamber. Usually, there are several openings to the plenum from the flapper valve for an even distribution of air. The cyclic motion of the water in the plenum chamber generates the waves in the tank.

The position of the flapper valve is controlled by an electric or hydraulic servo system. The system accepts both a flapper position feedback signal from a transducer at the flapper as well as a reference signal, and operates an actuator to cause the flapper position to match the reference. The amplitude and frequency of the generated waves are directly related to the amplitude and frequency of the reference signal.

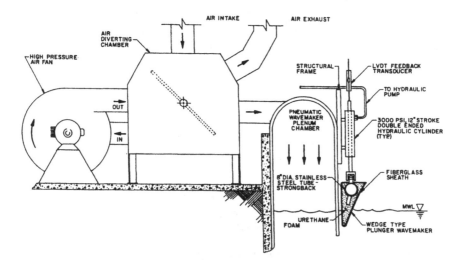

FIGURE 4.6
PNEUMATIC AND PLUNGER WAVEMAKERS

Because air is used to produce the wave form, the control mechanism for the flapper at the low air pressure as well as the air compressibility make the transfer function quite soft. A plot of the magnitude of the reference signal (or equivalently, the amplitude of the flapper motion) versus the generated wave in the tank is shown in Fig. 4.7. The direct drive wavemaker uses a similar signal to produce wave form (Fig. 4.8), but the relationship of input-output signal is different. In fact, the waves formed pneumatically grow slowly at the lower amplitudes and increase in height rapidly only as the signals approach maximum levels. Moreover, this phenomenon also limits the height of the high frequency waves. The pneumatic wave generators are not particularly suitable for high frequency waves (≥ 1 Hz), since the flapper valve is usually incapable in handling the passage of high air volumes at these high frequencies.

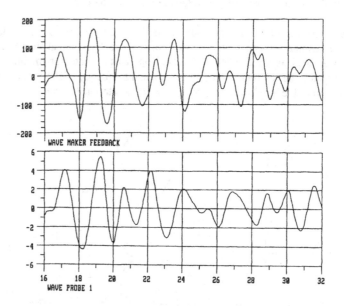

FIGURE 4.7
COMPARISON OF PNEUMATIC WAVEMAKER SIGNAL
AND WAVE PROFILE AT MID-TANK

4.5   DESIGN OF A DOUBLE FLAPPER WAVEMAKER

A double flapper wavemaker consists of two pivoted flappers, an actuation system, driven hydraulically, and a control system (Fig. 4.2). In accordance with the wave generation theory, the force and velocity requirements of the actuating devices of both flappers can be determined. The hydraulic demand depends on the work done by the actuator, which is obtained as the product of the acting force and its displacement or stroke. By superimposing the force requirements of the upper and lower flappers, the maximum output demands of the hydraulic units are computed. Usually, the lower flapper has a much higher input requirement due to higher imposed force as well as a greater stroke length.

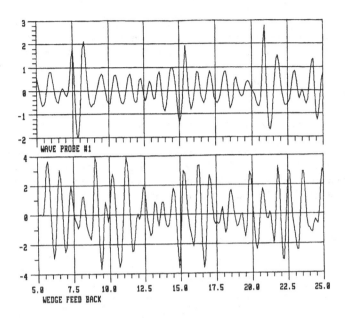

FIGURE 4.8
COMPARISON OF DIRECT DRIVE WAVEMAKER SIGNAL
AND WAVE PROFILE AT MID-TANK

Although it is feasible to have water on both sides of the flapper, the side of a double flapper wavemaker away from the waves is often maintained dry. Both types of system exist among existing facilities. For example, the facility at the Stevens Institute of Technology at Hoboken, New Jersey, is wet on both sides. On the other hand, the MARINTEK facility at Trondheim, Norway, has a dry back side. Both systems have advantages and disadvantages. The wet-wet system does not require the power to balance the static water pressure on the flapper. The possibility of leakage at the ends of the flapper is not a concern. The drawback is in the area of regular maintenance of the hydraulic units which are submerged for this system. Having water on both sides, it requires an additional force for wave generation since it must push water in both directions. Also, it demands a second wave absorber to dissipate the waves on the back side of the flapper.

For the wave generator with water on one side, most of the force requirement is in counteracting the hydrostatic pressure. This force is a function of the water depth and is invariant of the type and size of waves generated. A separate system (from the hydraulic driving system), such as a pneumatic or nitrogen gas system, can be used as a medium to generate the force required to counteract the static pressure. Then, the flow and power requirements of the hydraulic system can be greatly reduced.

4.5.1   Wetback and Dryback Design

A dryback design is generally considered better in terms of the quality of wave generation and active absorption of the wave incident on the flapper. For a dryback design, a seal keeps the water from entering the back side of the flapper or wave board. A "rolling" seal is suitable to withstand many cycles of flexing and deformation from the motion of the wave board. It is more economical to operate than the wetback design even though additional mechanical or hydraulic equipment are needed to withstand the additional static load. It is, however, not very suitable for translational mode or multimode operation.

4.5.2   Hydraulic and Pneumatic Units

An open or a closed loop hydraulic system may be chosen to drive the mechanical flappers. A closed loop system, however, is more accurate. A variable displacement pump is used to move the pumping element to the opposite position, reversing the direction of oil flow with the shaft being driven in the same direction. The shifting is regulated by a servo valve mounted on the pump (or on the actuator) with the hydraulic fluid supplied by a pressure compensated pump. Several pumps may be required to ensure a good response for the servo valve. The servo valves are controlled by a simple control signal. To assure equal displacement of the actuators, a set of flow control valves are provided which are controlled from the feedback of the actual movement of the actuator. The pressure of the hydraulic system is regulated by relief valves located inside the main housing.

An appropriately sized pressure vessel may be used as an accumulator for the control of the pressure of a pneumatic or nitrogen system to overcome static head. With a higher pressure source (cylinders for feeding) on one end and a lower pressure source (cylinders for drawing) on the other, the pressure of the pneumatic system can be adjusted to accommodate various water depths.

4.5.3   Control System for a Two-Board Flapper

The flapper is driven by an electrically controlled hydraulic system. The control system consists of a command signal generator, a cross-over network and dual servo-amplifiers to supply current to the electrically controlled valves in the hydraulic flapper drive system.

The command signal generator outputs an analog signal representing the desired flapper position for each point in time during operation. The generator has the capability to provide both regular and random command signals. Signal generators are available commercially that are specifically designed for this purpose. A programmed micro-computer may also be used to generate the command signal. After the signal is output, it is fed to a cross-over network where it is split into high and low frequency signals to drive the upper and lower flappers respectively. This task may be accomplished by a computer. The servo-amplifiers in the system accept the outputs of the cross-over as command signals as well as a position feedback signal from the flapper to produce a drive signal to the servo-valves that is proportional to the difference between the command and feedback signals, thus reducing any error in the generating wave.

4.5.4 Waveboard Sealing and Structural Support System

For a dryback system, a seal is required between the ends of the flappers and the side walls of the tank. Usually, a stainless steel wear plate is mounted on the concrete side wall. A plastic channel which is held in place by an inflatable air seal runs against the plate (Fig. 4.9). The air bag is maintained at a given pressure provided by an air cylinder and a compressor system. For the joint between flappers, a urethane/butyl fabric can be applied to prevent leakage. O-rings are used to seal both ends of the fabric. Several facilities, whether single or double flapper, use this type of system, including the U.S. Naval Academy at Annapolis, Maryland, and Oregon State University at Corvallis, Oregon. In a simple design, a rubber fabric folded between the flappers and the wall has been used with some success.

To relieve the nitrogen counter-balance system, a structural system is used to support the actuators. It also provides structural constraint to the flappers in the case of the hydraulic system. The load on the flappers is transmitted through the structure onto the back wall and the floor of the wave tank (Fig. 4.2).

4.6   A TYPICAL WAVE TANK

The wave tank at CBI's Marine Research Facility is 76.2m (250 ft) long, 10m (33 ft) wide and 5.5m (18 ft) deep. A 2.4m (8 ft) diameter pit is located in the center of the tank and extends an additional 4.1m (13.5 ft) below the tank floor. The layout of the wave tank showing the various components of the facility is shown in Fig. 4.10.

The rail mounted traveling bridge is the test control center. The bridge supports the signal conditioning and data acquisition equipment and the controls for wave generator operation. A built-in hydraulic system provides power for transporting the bridge along the length of the tank. An electro-mechanical servo valve controls the hydraulic system and maintains the bridge at a constant speed. The maximum velocity of the bridge is 1.4m/s (4.5 ft/sec), which is suitable for low-speed towing tests. A second bridge on the same rails generates a speed of up to 3.65 m/s (12 ft/sec) and may be used for higher speed towing of models.

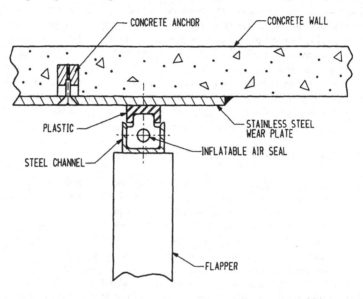

FIGURE 4.9
SEALING ARRANGEMENT OF A DRY BACK FLAPPER SYSTEM -
PLAN VIEW

Because the water depth in relation to the wave length has a significant influence on the wave kinematics, it is often necessary to vary the water depth in the wave basin based on the scale chosen and the prototype simulation depth desired.   There are two basic approaches to achieve this. One is to have an adjustable basin floor (e.g., CBI, and MARINTEK). The other is to build a wave generator that can be adjusted vertically along the basin wall and then change the water level in the basin [e.g., National Research Council (NRC), Canada].

The customary method of changing the water depth in a tank is by draining the water from the tank. However, the pneumatic type of wave generator installed at a fixed elevation in the CBI tank, does not permit the water depth to be varied by lowering the water level. The elevation of moveable false floor sections in the tank (Fig. 4.10) can be adjusted to simulate ocean bottom conditions. The level of water intake at the pneumatic wavemaker and the high frequency wave generator is maintained for all floor

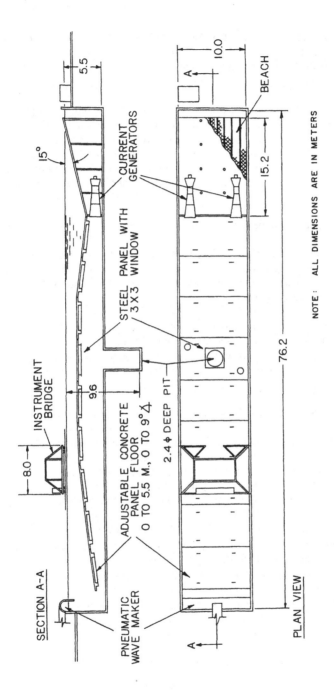

FIGURE 4.10
LAYOUT OF CBI WAVE TANK

configurations. Therefore, the height of the waves generated at these wavemakers is expected to be about the same for all water depths. The waves are, of course, modified by the bottom conditions (arrangement of the floor sections) as the waves travel down the length of the tank.

The wave absorbing beach is graded stone in a rack covered with a plastic mesh. The rack is inclined at an angle of 15° to the still water surface and extends 3m (10 ft) below the surface. The wave energy absorbing efficiency of this beach is quite good having a reflection coefficient of 5 to 10 percent depending on the wave period. This beach in conjunction with the moveable floor slabs can be used to simulate near shore conditions. This is particularly useful for performing seakeeping tests of vessels in the surf zone.

4.6.1 Low Frequency Wavemaker

This wave generator at this facility is a pneumatic type, consisting of an open bottom plenum chamber partially immersed in water and a blower with a low pressure head. Both the suction side and pressure side of the blower are connected to the plenum chamber by a flapper valve (Fig. 4.6).

The system has maximum response capabilities between periods of 2.5 and 3.5 seconds, where maximum wave heights of 508 mm (20 in.) to 559 mm (22 in.) can be produced. The response capability of the wave generator declines at lower periods. At a period of 1 second, the maximum wave height that can be generated is approximately 51 mm (2 in.).

Types of waves generated in commercial two-dimensional tanks are similar, differing only in their ranges. Regular waves, wave groups and random waves can be generated in the tank at this facility. The signal generation for driving the wavemaker is also typical. The wave form reference signal is generated by a mini-computer. For regular waves, the software generates a sinusoidal signal of constant amplitude and period. The maximum response of the pneumatic wavemaker at various wave periods between 1 and 4 seconds is shown in Fig. 4.11. For random waves, the software decomposes the desired spectrum into a large number of frequency components. The wave reference signal is initially calculated and stored as a digital time series. To generate the analog reference signal required by the servo controller, the stored time series is output through a digital to analog converter under the control of a dedicated microcomputer. The generated wave is then analyzed. The resulting energy spectrum is plotted by the data acquisition system and compared with the desired theoretical spectrum. The reference signal can then be modified as required to insure a satisfactory correlation between the generated and desired spectra. The modified reference signal is stored as a time series and can be repeated on demand, thus insuring consistency between test conditions.

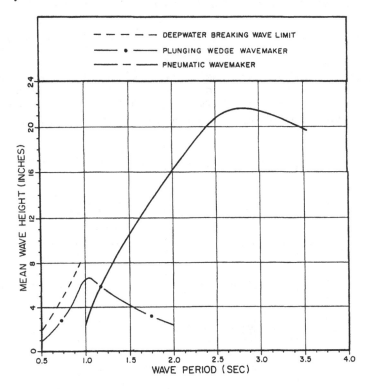

FIGURE 4.11
HEIGHT OF REGULAR WAVES BY THE CBI PNEUMATIC WAVEMAKER

Figure 4.12 is a comparison of a measured model tank spectrum and a theoretical wave spectrum; in this case, a JONSWAP type wave spectrum was used. As described above, a digital time series was executed to model the theoretical frequency distribution, which is plotted with the heavier line on Fig. 4.12. The recorded signal from a wave probe in the tank was analyzed in the frequency domain to produce the curve plotted with a lighter line on the same figure.

The probabilistic estimates of random waves are made by assuming that the wave profile follows a stationary Gaussian random process and that the corresponding wave heights follow a Rayleigh distribution. Figure 4.13 shows the wave height distribution for the wave analyzed in Fig. 4.12. The solid line represents the cumulative probability of the normalized wave height (H/H$_{rms}$, H$_{rms}$ = root mean square wave height) with the Rayleigh distribution. The wave tank data shown on the plot is grouped in small bands of wave height. The probability of occurrence is computed for each band.

4.6.2 High Frequency Wavemaker

For generating the higher frequency waves than can be produced with the pneumatic generator, a plunging wedge-type wave generator is used (Fig. 4.14). This system consists of a wedge shaped plunger that spans the width of the tank and is vertically driven by two parallel hydraulic cylinders. This wave generator is controlled by the same dedicated microcomputer that controls the pneumatic generator.

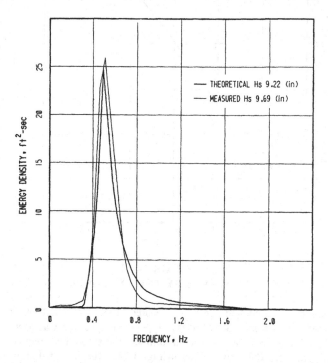

FIGURE 4.12
**COMPARISON BETWEEN A JONSWAP TARGET SPECTRUM WITH A
MEASURED WAVE SPECTRUM GENERATED BY PNEUMATIC
WAVEMAKER**

An example of a comparison between the energy density spectra of a wave produced and the Bretschneider model is shown in Fig. 4.15. The model has a significant wave height of 66.5 mm (2.62 in.) and spans the high frequency segment of the wave generation capability.

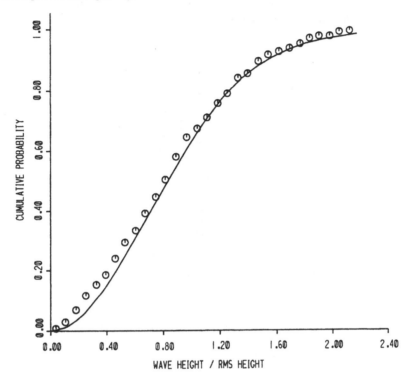

FIGURE 4.13
RAYLEIGH DISTRIBUTION OF A SINGLE WAVE RECORD GENERATED BY THE PNEUMATIC WAVEMAKER

4.7   DESIGN OF MULTIDIRECTIONAL WAVE GENERATOR

A "serpentine" wavemaker is a machine in which the wave board moves in a snake-like movement and produces progressive waves that propagate at oblique angles to the front of the wave board. The initial designs, over 25 years ago, produced monochromatic, but oblique long-crested waves. They had a common mechanical drive for all segments of the serpentine, but an individual phase adjustment for each segment [Miles, et al. (1986)]. With the advent of faster and cheaper controls and digital hardware, it has been possible to provide individual articulation of each segment of the wave board, thus allowing the simulation of multi-directional ocean waves. The most common method of wave generation is with the articulated wave boards. This concept of individually controlled generator segments was first implemented in 1978 [Salter (1981]. Today, there are numerous laboratories in the world that have facilities capable of reproducing multi-directional waves.

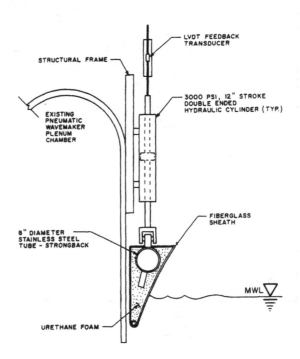

FIGURE 4.14
WEDGE TYPE PLUNGER WAVEMAKER

Table 4.1 summarizes many of the installations of multi-directional wave facilities. As shown, the majority use the flapper (i.e., rotational) mode. Others use the plunger or the piston. Some of the machines are of the "dry-back design", while the remainder are of the "wet-back design". As can be seen in the table, a few are particularly suited for offshore engineering application (as opposed to coastal zone simulations) because of their water depth and wave generation capabilities.

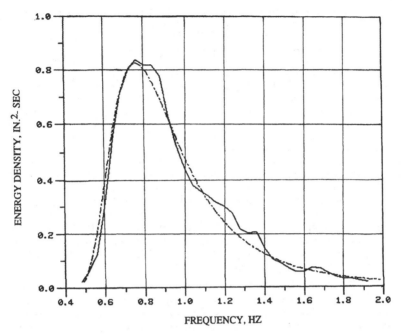

FIGURE 4.15
COMPARISON BETWEEN BRETSCHNEIDER SPECTRUM WITH MEASURED
WAVE SPECTRUM BY THE COMBINATION GENERATOR

In a large rectangular basin, wave generators may be located on two sides with absorbers on opposite faces. Generally, one set of generators are designed in segments to produce short-crested (directional random) waves, while the other is used to produce long-crested waves. This permits the generation of the interaction effect of short-crested locally generated storm with the long-crested swells. However, more recent laboratories (e.g., NRC-IMD at St. John's, Newfoundland) use segmented machines on two sides.

The design of multi-directional wave generators requires the choice of the width of the wave board segment. The smaller the segment width, the larger is the angle of obliqueness and the better is the quality of oblique waves. However, a unit having more, smaller segments has a bigger cost associated with it. On the other hand, the smaller width segments require less driving power compared to larger segments requiring more actuator power and complicated support system. Therefore, a compromise is required between the angular spreading, the power limits of commercially available actuators and the overall cost of the system.

TABLE 4.1 BASINS WITH SEGMENTED WAVE MACHINES
[Miles, et al.(1986)]

Name of Basin	Basin		No.	Mode	Segment		Stroke /2	Action	Dry/	Regular Wave
	Size (m)	Depth (m)	of Segments		Width (m)	Height (m)	(±)		Wet	Ht. (m)
U Edinburg	27x11	1.2	80	Flap	0.3	0.7	15°	Elec	Dry	.22
HRS	30x48	2.0	80	Flap	0.31			Elec	Wet	.50
NEL				Flap				Elec	Dry	
MARINTEK	50x72	0-10.	144	Flap	0.5	1.3	16.5°	Elec	Dry	.40
DHI	30x20	3.0	60	Flap	0.5	1.5	16.7°	Hydr	Dry	.50
CERC	59x111	0.76	60	Piston	0.46	0.76	0.15m	Hydr	Wet	.30
DHL	variable	1.3	80	Variable	0.33	1.28	0.40m	Elec	Wet	
NRCC	30x50	0.3-2.9	64	Variable	0.5	2.0	0.40m	Hydr	Wet	.70

In the design of wave boards, the water depth and the range of required frequencies dictate the choice of generating mode. The simplicity of design and proven performance of the hinged board generally make it the preferred choice over other designs such as the plungers [Miles, et al. (1986)]. The translational and rotational modes of a flat board offer simple and reliable calculations for wave height generation. The height of the wave board and its stroke height are determined from the maximum design wave height for the facility, assuming the translation mode of the wave board for wave generation. However, this translational mode is not appropriate for the shorter period wave generation, first because of the difficulty and inaccuracy in controlling the short stroke required for these waves. Secondly, there is a tendency of waves to develop in cross mode from the unused trapped energy at the wave board. If these modes have a vertical position control to vary the immersion depth, then they work well in all depths. KRISO in Korea has opted for this solution in their wave tank in which the water depth varies from 0.5 to 4m. However, these short period waves can be generated more efficiently in a rotational mode. The intermediate waves can then be generated by a combination of translational and rotational motion (Fig. 4.1E). A dual actuator system is necessary for these operations. A more versatile wave machine covering shallow to deep water waves can be achieved, but at an additional expense. Sometimes, a mechanical linkage is used in conjunction with a single actuator to limit cost. In this case, the generating mode has to be mechanically changed from one linkage to the other. One wave board system installed at NRC, Ottawa, Canada, is shown in Fig. 4.16.

4.7.1 Actuator and Control

The actuator for the wave board can be either an electrical or a hydraulic device. The hydraulic actuator is considered to be a better choice because it is less expensive, more reliable and has a better frequency response. The electric actuator can be economical when the force requirement per unit board length is small. The Moog actuators are most commonly used in the laboratory. In designing an actuator, the natural frequency of the board linkage and actuator system should be investigated (which should be about 5 Hz or more). The servo control for the actuator can be analog, digital or a hybrid. Most two-dimensional machines use analog control for the actuators. For a three-dimensional machine, however, it becomes more complicated and expensive. Hybrid controllers consist of an analog control loop which is tuned digitally. They are more expensive, but, at the same time, considerably flexible. Digital servo controllers, on the other hand, use software to control the system and, hence, have the greatest flexibility. Standard microcomputer components may be used for this system. Software may also be implemented in the control loop to help in the active wave absorption.

FIGURE 4.16
DEEPWATER MULTIDIRECTIONAL WAVE BASIN AT NRC, OTTAWA,
CANADA [Miles, et al. (1986)]

Details of a multi-mode segmented wave generator design may be found in Miles, et al. (1986).

4.8   A MULTI-DIRECTIONAL TANK

The multi-directional wave basin at the Hydraulic Laboratory, National Research Council of Canada, Ottawa, is equipped with a segmented wave generator capable of producing multi-directional sea states in the test basin. The basin is 50 m wide, 30 m long and 3 m deep (Fig. 4.17). The basin has a central pit 5 m in diameter and 13 m deep. The segmented wave generator occupies one side of the basin while the wave energy absorbers made of perforated layers of metal sheeting (see Section 4.12) are placed along the other side of the basin.

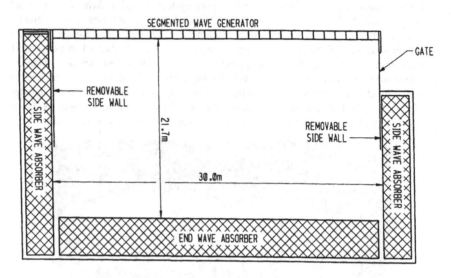

FIGURE 4.17
PLAN VIEW OF WAVE BASIN

The wave generator consists of 60 segments or wave boards driven individually by a servo controlled hydraulic system. The individual wave boards are 2.0 m high and 0.5 m wide, driven by Moog hydraulic actuators with a maximum stroke of ±0.1 m. The displacement of the actuator is mechanically amplified by a factor of 4 through a lever arm. The boards operate in the piston or translational mode (for shallow water waves), or flapper or rotational mode, or a combination of the two (for deep water waves). The machine is also vertically relocable to accommodate different water depths.

The control system for the directional wave generator is a micro-processor based digital system with a four unit modular design. Each unit is responsible for the control of 16 segments and performs several real time tasks. The drive signals are initially synthesized on a VAX workstation. The multi-directional wave generation signals are usually based on linear theory and produce a sinusoidal (serpentine or snake-like) motion along the length of the wave generator. However, other methods such as side wall reflection techniques and nonlinear wave theory are also used.

4.9 CURRENT GENERATION

The generation of a uniform (as opposed to local) current in the wave tank requires a closed loop for the circulation of water. The closed loop can be provided if the tank has a false floor. For any given system, the speed of current will depend on the water depth. It is also possible to connect the two ends of the tank with piping and a pump to draw water from one end (e.g., the beach end) and pump it back to the tank at the other end (e.g., the wavemaker end). This system is most practical for shallow water, since current is inversely proportional to water depth. Both systems can be designed to reverse the current flow, permitting the generation of current with or against the wave direction.

Several wave tanks are equipped with current generation capabilities. Some of these facilities are CBI, and MARINTEK (using water jets under the false floor). The technique of generation is similar, but the mechanical systems are unique. An example of current generation and the type of current profile generated is illustrated in the following section.

4.9.1 A Typical Current Generator

In the CBI wave tank, inline currents are generated by a 67 KW (90 HP) Lister diesel engine which drives two 762 mm (30 in.) propellers beneath the beach. Two sets of hydraulic lines are routed from the diesel to two underwater 44 KW (60 HP) reversible hydraulic motors. Each motor is mounted at one end of a converging-diverging nozzle located underneath the beach as shown in Fig. 4.10. Each nozzle can be independently controlled.

Inside the nozzles are 762 mm (30 in.) diameter, three bladed propellers mounted on roller bearings. These propellers are of the high thrust type. The current speed can be controlled by either the engine speed or the pressure controls. Current can be directed to flow either with or against the direction of wave travel by reversing the rotation of the hydraulic motors.

A series of flow straighteners are used to restrict the deviation in the mean current velocity to less than 10%. A single unit of flow straightener consists of corrugated plastic sheets bolted together and is 1.2 m (4 ft) wide, 1.2 m (4 ft) long and 1 m (3-1/4 ft) high. The units are interlocked to span the width of the tank and are located upstream of the model.

Figures 4.18a and 4.18b present the current profiles for flow in both directions. The mean value and the variation (± one standard deviation) have been plotted. Figure

4.18a presents the vertical profile for currents inline with the direction of wave propagation for maximum and half-maximum settings.

Figure 4.18b presents a similar profile for current flowing in the opposite direction. Current in this direction is slightly more uniform than the in line current because the current passes through the beach prior to impinging on the model, functioning as an additional flow straightener.

Figures 4.18c and 4.18d present the horizontal velocity distribution measured at 356 mm (14 in.) from the bottom in a 2.1m (7 ft) water depth for current in both

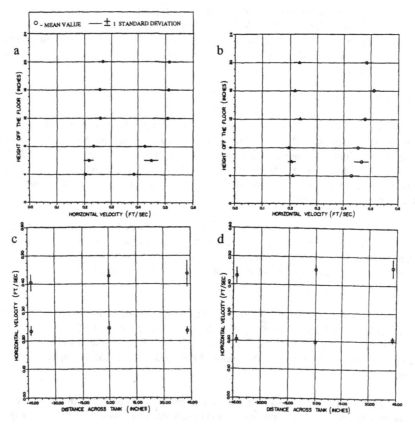

FIGURE 4.18
VERTICAL AND HORIZONTAL CURRENT PROFILES

directions. Both profiles are nearly uniform across the measured range. The distance across the tank is plotted as the abscissa and is referenced to the tank centerline.

4.9.2 *Local Current Generation*

Local currents can be generated through the use of portable electrically driven propellers. These motors can be rigidly mounted to the wave tank rail or instrumentation bridge, and the motor speed, spacing between motors and their depth can be varied to obtain the desired current velocity and profile. If the generators are mounted alongside the tank, cross current may be generated transverse to the waves (Fig. 4.19). This is often an important design consideration for a moored tanker at a site where current exists transverse to the predominant wave direction. The current generation units are extendable from the surface with long shafts in order to produce vertical velocity profiles and can actually be used to simulate shear current. The spacing between motors and distance from the test section are adjusted during calibration prior to testing. The speed of the motors is varied to allow adjustment of the current velocity once the motors are in place. Any fluctuation in the generated current may be minimized by using flow straighteners.

FIGURE 4.19
CROSS CURRENT AND WIND GENERATION SETUP IN TANK

Local and cross currents are sometimes produced by arranging pipes close to the model with flows from an external source. This method is generally inadequate for developing a uniform current flow field around the model.

4.9.3 Shear Current Generation

The shear current is a common occurrence in the ocean. Generally, the surface current velocity is higher than the bottom current. The profile with depth may be approximately represented by a linear or bilinear shear current. Near the bottom, the boundary layer effect provides a parabolic profile reducing current at the bottom to zero. In deep water in particular, the shear current may have a significant influence on a component of an offshore structure. One example of such a component is a riser, or a group of risers.

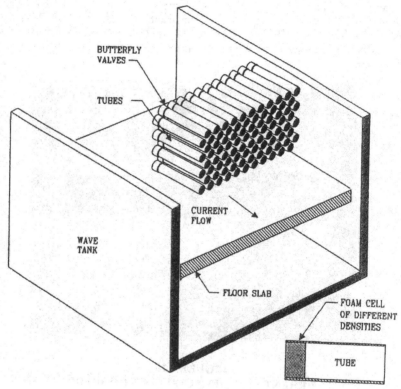

FIGURE 4.20
SHEAR CURRENT GENERATION IN TANK

In order to generate shear current of a given profile, the flow at various elevations must be controlled to provide the desired profile in the tank. One method of generation may use a flow control mechanism at the forward end of the flow straighteners. This flow control can be achieved by installing butterfly valves at the inlet to each flow straightener (Fig. 4.20). Alternately, materials of varying porosity, e.g., foam or sponge, may be inserted in the tubes (Fig. 4.20, inset) to reduce the flow by a desired amount. Thus, by restricting the flow through the rows of tubes in the flow straightener by various degrees with depth, a positive or negative (linear) shear current may be generated. A bilinear shear current or simultaneous vertical and horizontal shear current profiles can also be created using this technique.

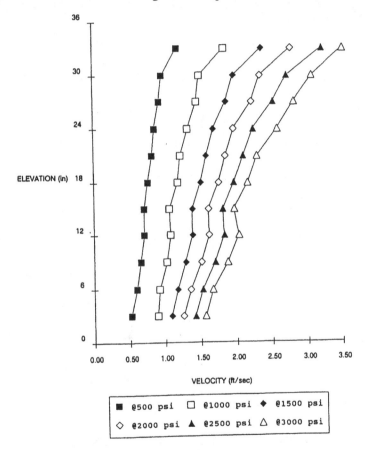

FIGURE 4.21
POSITIVE SHEAR CURRENT PROFILES INTANK

In a test at the CBI wave tank, flow straighteners consisting of smooth 102 mm (6 in.) diameter, 1.2 m (4 ft) long cylinders were placed spanning a 1.2 m (4 ft) deep, 3 m (10 ft) wide channel. Uniform fluid flow was created downstream of the flow straighteners.

Shear currents were generated by providing increasing levels of resistance to flow at prescribed elevations. This resistance was achieved by placing a mesh screen in front of the flow straighteners covering the width and depth of the tank. This mesh consisted of polyethylene cloths with a diamond shaped mesh pattern. The cloths were commercially available in 914 mm (36 in.) wide rolls in various sizes with nominal openings of 3 mm (1/8 in.), 5 mm (3/16 in.), 6 mm (1/4 in.), 13 mm (1/2 in.), 19 mm (3/4 in.) and 25 mm (1 in.).

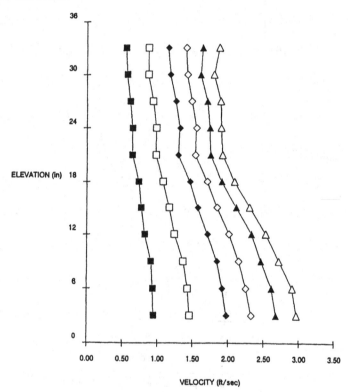

FIGURE 4.22
NEGATIVE SHEAR CURRENT PROFILE
(see Fig. 4.21 for symbols)

The cloths were cut into 152 mm (6 in.) widths, and various combinations of mesh openings were tried at each elevation. The added resistance provided by this cloth and the subsequent shear pattern developed proved to be highly nonlinear, and a trial and error approach was applied. The final mesh pattern with overlaying cloth layers is shown in Table 4.2. This mesh screen was then turned upside down with most resistance on top of the 914 mm (36 in.) depth to produce a negative shear. These shear currents are shown in Figs. 4.21 and 4.22 for various hydraulic settings of the current generator (Section 4.8.1).

TABLE 4.2
MESH CONFIGURATION FOR GENERATION OF
SHEAR CURRENT

ELEVATION (IN.)	MESH OPENINGS (IN.)			
30-36	1			
24-30	3/4			
18-24	1/2	3/4		
12-18	1/4			
6-12	3/16	1/4		
0-6	1/8	1/8	1/8	1/8
BACK-TO-BACK LAYER NO.	1	2	3	4

4.10   WIND SIMULATION

The wind load may be simulated with the help of blowers strategically placed in front of the model. They are conveniently placed on the face of a traveling bridge used to house the instrumentation and controls (Figure 4.19). In this case, the superstructure of the hull requires modeling. This allows variation of the steady wind load due to the motion of the model hull and may introduce a low frequency variation in load if a wind spectrum is simulated. Instead of modeling the wind speed, the steady wind load is sometimes modeled by adjusting the speed and location of the fans. It is also common to simulate the horizontal wind load by the use of an appropriate weight hung from a pulley. A stiff line (e.g., Kevlar or steel cable) is used to suspend the weight, thus avoiding any spring effect. The pulleys should be low friction type (e.g., aircraft pulleys, use of ball bearings). The line is attached at or near the deck of the structure. Special care should be taken to ensure that the system does not add any significant amount of damping in the floating system in surge and pitch. This may be verified by studying and comparing the free damped oscillation of the system with and without the wind load applied to the system.

While the wind loads on the structure may be quite important for a particular design, e.g., the floating moored structure, useful modeling of these loads is limited by the scaling problems associated with them. If the wind load alone is important, e.g., on the superstructure of an offshore platform, Reynolds number is more important and wind tunnel tests are performed on that portion of the structure model. In these cases, it is easier to achieve Reynolds number similarity.

4.11 INSTRUMENTED TOWING STAFF

For performing towing resistance tests, the wave tank's carriage is generally fitted with an instrumented staff. The staff is arranged so that any combination of roll, pitch and heave motion of a floating model may be accomodated during a test. Instrumentation is installed to measure the loads on the model in the downtank, cross-tank and vertical directions. Moments about any of these axes can also be measured. The staff is designed so that drag forces on the model can be counterbalanced by a weight on an adjustable arm. When the drag force is counterbalanced in this way, a more sensitive load cell can be used to measure the downtank force, and more accurate measurements of the effects of drag reducing modifications to a model can be made. A sketch of the towing staff is shown in Fig. 4.23.

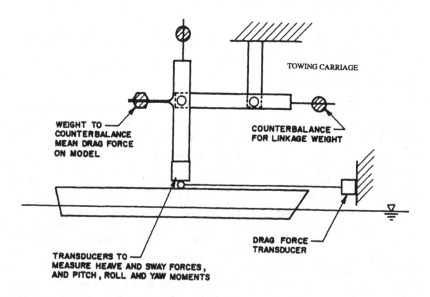

FIGURE 4.23
INSTRUMENTED TOWING STAFF

4.12 PLANAR MOTION MECHANISM

A planar motion mechanism (PMM) is used for forced oscillations of moderately heavy models. There are several variations in the design of a PMM. Some are quite complicated. In a simple design, the mechanism may be driven by two parallel hydraulic cylinders (Fig. 4.24) that can be controlled to produce pure translation, pure rotation or combined translation and rotation of the attached model. By varying its orientation in the setup, the translational axis of the mechanism can be aligned with the surge, sway, or heave axis of the model. Rotation of the model can take place about any axis that is normal to the mechanism's translational axis. The maximum displacement and velocity requirements of either hydraulic cylinder are prescribed in their design.

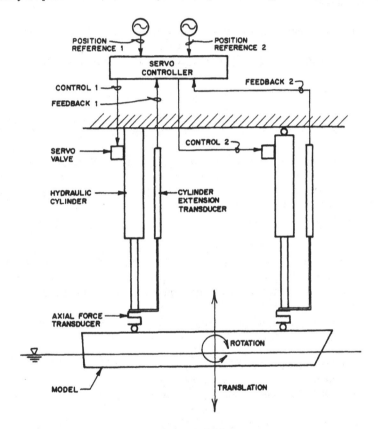

FIGURE 4.24
PLANAR MOTION MECHANISM

The extension of each cylinder (measured by a position feedback transducer) is controlled by a servo system and is driven to match a computer generated position reference signal.

If the planar motion mechanism is driven by servo-controlled hydraulic cylinders rather than cams or some other form of eccentric drive, the motions produced are not limited to regular sinusoidal motions, but can be random and may contain any number of frequency components in a prescribed range. Software is written to tailor the frequency distribution of the drive signal to match any desired spectrum model including white noise, if desired. The drive signal may be generated from a stored time series (similar to wave generation), and the motions can be duplicated for different test configurations.

The planar motion mechanism can be used to determine the stability and control characteristics and the hydrodynamic added mass and damping coefficients in all six degrees of freedom for wave tank models such as surface ships or submerged structures.

4.12.1 Single Axis Oscillator

A single axis oscillator is a simple PMM used to force models in a single direction in the wave tank (see Fig. 4.25). The oscillator is driven by a servo controlled hydraulic cylinder and is capable of generating velocities and displacements up to the design value of the mechanism. The control system of the oscillator is identical to that used in the planar motion mechanism. The position reference signal used to control the extension of the cylinder is generated by a microcomputer from a stored digital time series. The system can generate regular sinusoidal motions or random motions containing frequency components in its design range.

Figures 4.25 and 4.26 illustrate two of the tests that have been performed using the single axis oscillator. Figure 4.25 shows a test of a resonant chamber being studied as a heave damping addition to a deepwater caisson structure. The chamber was driven vertically in still water to confirm that it was properly tuned so that the water inside the chamber oscillated 180° out of phase with the motion of the chamber. The load cell at the top of the chamber was used to measure the vertical load generated by the oscillating water column. A capacitance wave staff fixed to the cylinder was used to determine the free surface elevation of the water in the chamber. The second test, shown in Fig. 4.26, used the oscillator to force horizontal motion at the top of a model riser. The riser was one component of a deepwater mooring system. The riser was instrumented with strain gauges at three elevations to measure the bending stresses induced by the forced motion. Single period sinusoidal motions and random multi-frequency motions were produced.

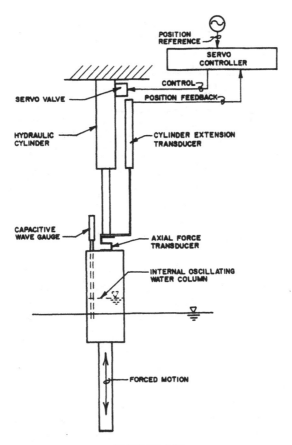

FIGURE 4.25
FORCED OSCILLATION OF A HEAVE DAMPING SYSTEM

4.13 LABORATORY WAVE ABSORBING BEACHES

It is important to simulate open ocean conditions in a laboratory environment as closely as possible. One of the factors that contaminate the generated wave at the test site is the reflection of the incident waves from the end walls of the wave basin. The coastal and offshore structure models are placed near the center of the basin. Even when the basin is long, the model experiences reflection from the end wall during the course of a test run of reasonable duration. In order to dissipate the wave energy and minimize the problem of wave reflection, wave absorbers are generally installed at the

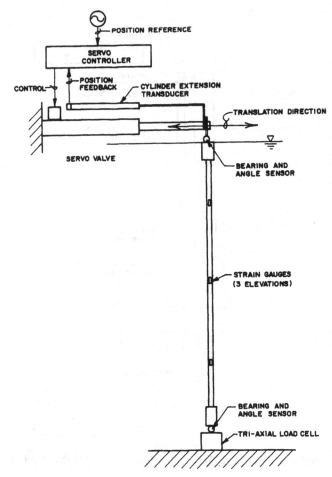

FIGURE 4.26
FORCED OSCILLATION TEST OF A DEEP WATER RISER

end opposite to wave generation. The amount of wave reflection is quantified as a wave reflection coefficient (C_r) which is defined as the ratio of reflected wave height to the incident wave height (expressed as a percentage). For a solid vertical wall, the reflection coefficient is 100 percent, i.e., the magnitude of the reflected wave is equal to that of the wave incident upon it. For an efficient beach, the reflection coefficient should be consistently less than 10 percent and preferably less than 5 percent [Jamieson and

Mansard (1987)] over the range of wave heights and periods that the basin is capable of producing. Since the wave absorber takes up valuable space from the wave basin, the design of an efficient wave absorber is always a challenge. If there is only limited space available for a wave absorber, efficient wave absorption may be very difficult to achieve.

The most commonly used wave absorbers are beaches of constant slope which extend near the bottom of the basin. It is constructed of concrete, sand, gravel or stones. The slope of these beaches must be mild for efficient wave energy absorption. Typical slopes are from 1:6 to 1:10 [Ouellet and Datta (1986)]. A beach with variable slope may reduce the total required length of the beach. A parabolic slope is sometimes used along with surface roughness and porous materials. The position of the parabolic beach can be made adjustable with the water depth in order to maintain its efficiency. Geometry of a variety of existing beaches in testing facilities is shown in Fig. 4.27.

Another concept for laboratory beaches is a progressive wave absorber. The concept of a progressive wave absorber [LeMehaute (1972)] consists of material whose porosity decreases towards the rear of the wave absorber. LeMehaute used aluminum shavings to achieve this. The shavings were more compacted (less porous) away from the wavemaker. To achieve the same efficiency as a beach with uniform porosity, the length of the beach required was reduced using variable porosity.

The same principle was adopted by Jamieson and Mansard (1987) with perforated upright wave absorber. The upright wave absorber consisted of multiple rows of perforated vertical metal sheets. The porosity of the sheets decreased towards the rear of the absorber. This provided an efficient wave absorber over a relatively short length with about a 5 percent reflection coefficient. This absorber will be discussed in further detail in Section 4.12.2.

In prototype situations, an upright caisson breakwater having a perforated wall in front of an impervious back wall [Jarlan (1961)] is occasionally used to reduce the high reflections associated with a solid wall breakwater.

4.13.1 Background on Artificial Beaches

The use of permeable material (e.g., crushed rocks, cast concrete products, wire mesh, etc.) in the sloping beaches is well known in both the field and the laboratory. If sufficient space is available, a long absorber with a low surface slope can result in very efficient wave reflections (such as the two-dimensional tank at the Oregon State University, Corvallis). However, limited space sometimes necessitates the use of an absorber with minimum length and steeper slopes. Laboratory tests have been

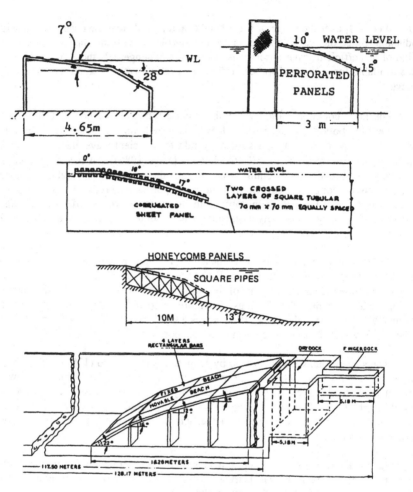

FIGURE 4.27a
BEACH GEOMETRIES

performed to determine the optimum porosity, steepness and length of the absorbers [Beach Erosion Board (1949), Straub, et al. (1956)]. It was concluded that a material with a high porosity is desirable if minimum length is needed. The best absorption was obtained with a porosity of 60 to 80 percent.

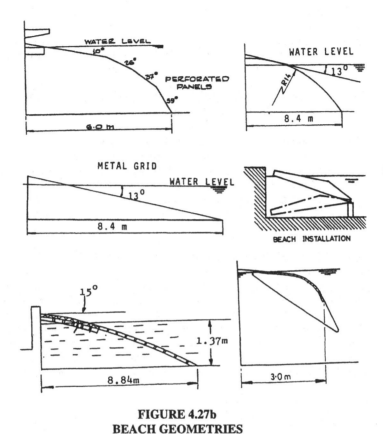

FIGURE 4.27b
BEACH GEOMETRIES

Even though mention was made earlier about perforated plates being used as a beach [Straub, et al. (1956)], Jarlan (1960) originally suggested the vertical porous wall breakwater, and presented experimental data on one. Jarlan (1961) examined the effects of mass transfer in waves through a perforated vertical wall. The theoretical development followed by Jarlan was based on acoustic theory. This theory demonstrated a simple relationship between the reflection coefficient and porosity.

4.13.2 Progressive Wave Absorbers

Progressive wave absorbers are upright wave absorbers for use in laboratory wave tanks. They require limited space compared to conventional beaches. They are efficient in variable water depths and over a wide range of wave conditions. The wave efficient in variable water depths and over a wide range of wave conditions. The wave absorber is constructed of multiple rows of vertical sheets of varying porosity and spacing between sheets. An open tubular framework supports the sheets and allows some variations in the overall geometry depending upon the test condition. They can also be used as the tank side absorbers (locally, if needed) to absorb reflected waves from large 3-D structures [Jamieson, et al. (1989), Clauss, et al. (1992)]. They are equally applicable in multi-directional basin.

These absorbers have been successfully used in the NRC, Canada multi-directional test basin. They have been tested in laboratory waves at NRC. The reflection coefficients in both regular (Fig. 4.28) and irregular waves have been limited between 3 and 7 percent over a wide frequency range. These absorbers have also been placed at the Texas A&M, College Station wave basin.

4.13.3 Active Wave Absorbers

If a wave basin with a wave generator at one end of the tank and a wave absorber at the other end is considered an "open loop" system, then there is also a closed-loop system available. In this case, a conventional active wave generator may be used both as a generator and an absorber of wave energy. A closed-loop control system is a device that will generate only the wave train required, while at the same time absorbing the unwanted reflection components returning to the wave generator after reflection from the beach. This mode of operation requires a sophisticated control system and actuating mechanism. Thus, it is feasible to generate a train of progressive waves without re-reflection at the generator of waves reflected by the structure being tested in the basin.

4.13.4 Corrected Wave Incidence

A correction to the measured wave at a point in the tank due to beach reflection can be made by using a simple reflection theory developed by Ursell, et al. (1960). The reflection effects modify the generated wave in a tank of finite length in the following way. Even if a beach exists at the far end, the reflection coefficient is not zero. Successive reflections occur between the beach and the wave generator. There is a time lag in the reflections to take place, but each reflection becomes an order of magnitude smaller, and a steady state is eventually reached. The initially generated wave train, called the primary incident wave, is partially reflected from the opposite end to form a primary reflected wave component which is then reflected from the generator surface to form a secondary reflected wave, and so on. The primary reflected wave is most significant, and the higher order terms in subsequent reflections may usually be considered insignificant. This is particularly valid if the wave generator has active absorption capability.

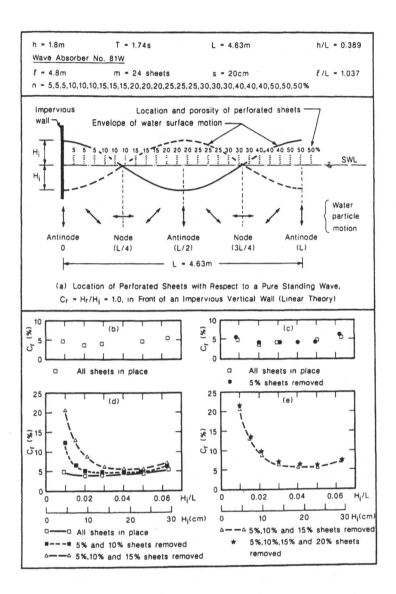

h = 1.8m   T = 1.74s   L = 4.63m   h/L = 0.389

Wave Absorber No. 81W

ℓ = 4.8m   m = 24 sheets   s = 20cm ℓ/L = 1.037

n = 5,5,5,10,10,10,15,15,15,20,20,20,25,25,25,30,30,30,40,40,40,50,50,50%

(a) Location of Perforated Sheets with Respect to a Pure Standing Wave,
$C_r = H_r/H_i = 1.0$, in Front of an Impervious Vertical Wall (Linear Theory)

FIGURE 4.28
PERFORATED SHEET LOCATION AND CORRESPONDING REFLECTION
COEFFICIENT [Jamieson and Mansard (1988)]

The primary incident wave and its partial reflection from the wave absorber will create a partial standing wave system. Ursell, et al. (1960) showed that under this condition the measured wave amplitude at a point x in the tank is given by

$$a^2 = a_i^2 \left\{ 1 + 2e \cos\varepsilon + 2e \cos(2kx + \varepsilon) + 2e^2 \cos 2kx + 2e^2 \right\}$$

(4.43)

where a_i = primary incident wave amplitude, k = wave number, e and ξ = unknown parameters due to reflection. Considering e to be small,

$$a = a_i \left[1 + e \cos(2kx + \varepsilon) + e \cos\varepsilon + 0(e^2) \right]$$

(4.44)

Thus, 'a' represents a spatially slowly-varying wave amplitude with peaks and valleys. Moreover, the envelope is a sinusoid having amplitudes between a_{max} and a_{min} where

$$a_{max} = a_i (1 + e + e \cos\varepsilon)$$

(4.45)

$$a_{min} = a_i (1 - e + e \cos\varepsilon)$$

(4.46)

Writing C_r = reflection coefficient, it may be computed from

$$C_r = \frac{a_{max} - a_{min}}{a_{max} + a_{min}}$$

(4.47)

Substituting the values of a_{max} and a_{min},

$$C_r = \frac{e}{1 + e \cos\varepsilon}$$

(4.48)

If the measured wave amplitude along the spatial coordinate x is recorded by a traversing wave gauge such that its maximum and minimum values are found, then the reflection coefficient, C_r may be computed. Moreover, at these locations, $2kx + \varepsilon = 0$ or π, respectively, so that ε may be evaluated. Then from Eq. 4.48, ε is determined and hence a_i.

4.14 REFLECTION OF REGULAR WAVES

The reflection of waves from the beach of a laboratory wave tank is an important parameter to consider in a model test. Moreover, the reflection parameter is often desired of a model being tested in a tank, particularly if the model is a two-dimensional structure, e.g., a breakwater.

There are several methods available to compute the reflection coefficient. As stated earlier, the reflection coefficient is defined as the ratio of the amplitude of the reflected wave to the corresponding amplitude of the incident wave. Note that for a regular wave, the frequency of the reflected wave is the same as that of the incident wave. As shown earlier, the reflected wave creates a standing wave system in the tank having a node and an antinode. Therefore, the reflection coefficient may be computed from the wave heights measured by a single traversing wave probe.

A second approach may use two fixed wave probes instead of a traveling wave probe. In this case, the two probes are located at distances of L/4 and L/2 from the reflecting structure, where L is the wave length. This assumes that the waves are linear and the points of node and antinode as well as the wave length are known. However, in many instances, such as perforated or sloped structures (beaches), the phase of the reflected wave is not known apriori.

In general, there are three unknown parameters due to the reflection of a regular wave: the incident wave height, the reflected wave height and the phase difference between the two. Assuming a_i $(= H/2)$ and a_r to be the amplitudes of the incident and reflected waves ($C_r = a_r/a_i$) and ε to be the phase difference, the free surface elevation, η_n, for a wave probe at a location given by x_n is

$$\eta_n = a_i \cos(kx_n - \omega t) + a_r \cos(-kx_n - \omega t + \varepsilon), \quad n = 1, 2, \ldots \tag{4.49}$$

in which k and ω are the wave number and wave frequency given by the linear wave theory for a given wave period, T, and a water depth, d. Assuming the location of the first probe to be x_1 and writing the distance between the nth probe and first probe to be δ_n, we have

$$x_n = x_1 + \delta_n \tag{4.50}$$

or

$$kx_n = kx_1 + \Delta_n \tag{4.51}$$

where $\Delta_n = k\delta_n$. Then, the profile for the nth probe is given by

$$\eta_n = a_i \cos(kx_1 + \Delta_n - \omega t) + a_r \cos(kx_1 + \Delta_n + \omega t - \varepsilon) \tag{4.52}$$

In complex notation,

$$\eta_n = \{a_i \exp (ikx_n) + a_r \exp[-i(kx_n - \varepsilon)]\} \exp(-i\omega t)$$
$$= \{a_i \exp[i(kx_1 + \Delta_n)] + a_r \exp[-i(kx_1 + \Delta_n - \varepsilon)]\} \exp(-i\omega t) \tag{4.53}$$

The recorded elevation at the nth probe may be written in terms of an amplitude, A_n, and a phase, δ_n, relative to the first wave probe record (i.e., $\delta_1^{(m)} = 0$) and is given as

$$\eta_n^{(m)} = A_n \cos\left(\omega t - \varepsilon_1 - \delta_n^{(m)}\right) \tag{4.54}$$

where ε_1 is the phase angle corresponding to the first wave probe and the superscript m refers to the measured quantities. An equivalent expression may be written in the complex domain as in Eq. 4.53.

4.14.1 Two Fixed Probes

Using the expressions for the assumed and actual elevations in Eqs. 4.53 and 4.54 and using the complex equivalent form, the following set of equations may be written [Goda and Suzuki (1976), Isaacson (1991]

$$a_i \exp(ikx_n) + a_r \exp[-i(kx_n - \varepsilon)] = A_n \exp[i(\varepsilon_1 + \delta_n)], \ n = 1,2 \tag{4.54}$$

The incident and reflected wave amplitudes are derived from this set of equations.

$$a_i = \frac{1}{2|\sin \Delta_2|}\left[A_1^2 + A_2^2 - 2A_1 A_2 \cos(\Delta_2 + \delta_2)\right]^{\frac{1}{2}} \tag{4.55}$$

$$a_r = \frac{1}{2|\sin \Delta_2|}\left[A_1^2 + A_2^2 - 2A_1 A_2 \cos(\Delta_2 - \delta_2)\right]^{\frac{1}{2}} \tag{4.56}$$

The phase difference between the two is written in terms of $\lambda = 2kx_1 - \varepsilon$,

$$\cos\lambda = \frac{A_1^2 - a_i^2 - a_r^2}{2a_i a_r} \tag{4.57}$$

From Eqs. 4.55 and 4.56 the method fails for $\Delta_2 = 0$, π, 2π, ... etc. which corresponds to a probe spacing of multiples of half wave lengths. Therefore, λ_2 should be outside the range of $0.4L < \lambda_2 < 0.6L$ [Goda and Suzuki (1976)].

4.14.2 Three Fixed Probes

The derivation of the incident and reflected wave is based on the waves being represented by the sine function (linear theory). Any nonlinearity in the wave will produce error in the measured amplitudes and phases for the two probe arrangement. This error may be reduced if additional probes are used so that a least square estimation is possible. If a three probe arrangement [Mansard and Funke (1980)] is used, then five measured quantities, namely, three wave amplitudes (A_n, n = 1,2,3) and two phase angles (δ_2 and δ_3) may be used to derive the three unknowns.

For convenience, let us write Eqs. 4.53 and 4.54 as follows:

$$\eta_n = \left[b_i \exp(i\Delta_n) + b_r \exp(-i\Delta_n) \right] \exp(-i\omega t) \tag{4.58}$$

$$\eta_n^{(m)} = B_n \exp(-i\omega t) \quad n = 1,2,3 \tag{4.59}$$

where

$$b_i = a_i \exp(ikx_1) \tag{4.60}$$

$$b_r = a_r \exp\left[-i\left(kx_1 - \varepsilon \right) \right] \tag{4.61}$$

$$B_n = A_n \exp\left[i\left(\varepsilon_1 + \delta_n \right) \right] \tag{4.62}$$

In order to minimize the error in the estimate in the least square sense, the error term is written as

$$E^2 = \sum_{n=1}^{3} \left[b_i \exp(i\Delta_n) + b_r \exp(-i\Delta_n) - B_n \right]^2 \tag{4.63}$$

The quantity E^2 is minimized in the usual least-squares sense giving rise to the following two equations in the unknown complex quantities b_i and b_r.

$$\sum_{n=1}^{3} \exp(i\Delta_n) \left[b_i \exp(i\Delta_n) + b_r \exp(-i\Delta_n) - B_n \right] = 0 \tag{4.64}$$

$$\sum_{n=1}^{3} \exp(-i\Delta_n) \left[b_i \exp(i\Delta_n) + b_r \exp(-i\Delta_n) - B_n \right] = 0 \tag{4.65}$$

Once b_i and b_r are known, a_i, a_r and λ may be computed.

$$a_i = |X_i| \tag{4.66}$$
$$a_r = |X_r| \tag{4.67}$$
$$\lambda = \text{Arg}(X_i) - \text{Arg}(X_r) \tag{4.68}$$

where

$$X_i = \frac{S_2 S_3 - 3 S_4}{S_5} \tag{4.69}$$

$$X_r = \frac{S_1 S_4 - 3 S_3}{S_5} \tag{4.70}$$

and

$$S_1 = \sum_{n=1}^{3} \exp(i 2 \Delta_n) \tag{4.71}$$

$$S_2 = \sum_{n=1}^{3} \exp(-i 2 \Delta_n) \tag{4.72}$$

$$S_3 = \sum_{n=1}^{3} A_n \exp\left[i(\delta_n + \Delta_n)\right] \tag{4.73}$$

$$S_4 = \sum_{n=1}^{3} A_n \exp\left[i(\delta_n - \Delta_n)\right] \tag{4.74}$$

$$S_5 = S_1 S_2 - 9 \tag{4.75}$$

The method will fail when $S_5 = 0$. For equal probe spacing Δ, this occurs when $\Delta = 0$, π, 2π.... For unequal probe spacing such that $\mu = \Delta_2/\Delta_3$, the condition $S_5 = 0$ corresponds to

$$\sin^2 \Delta_3 + \sin^2(\mu \Delta_3) + \sin^2(\Delta_3 - \mu \Delta_3) = 0 \tag{4.76}$$

which gives

$$\sin \Delta_3 = \sin(\mu \Delta_3) = 0 \tag{4.77}$$

This occurs where $\Delta 3 = 0$, π, 2π,... and $\Delta 2 = 0$, π, 2π....

Since in a three probe arrangement the number of measurements available is more than the number of unknowns, one may use fewer measurements to solve for these unknowns. Since only three unknowns are involved, the three wave heights at the three probes are sufficient [Isaacson (1991)]. This eliminates the phase angles in the computation.

In terms of earlier notation using Eq. 4.55, the measured wave amplitudes are related to the unknowns as

$$A_n^2 = a_i^2 + a_r^2 + 2a_i a_r \cos(\lambda + 2\Delta_n), n = 1,2,3 \tag{4.78}$$

where $\lambda = 2x_1 - \varepsilon$ and $\Delta_1 = 0$ as before. Solving for λ from Eq. 4.78

$$\cos \lambda = f_1 \tag{4.79}$$

$$\sin \lambda = \frac{f_1 \cos(2\Delta_n) - f_n}{\sin(2\Delta_n)} \tag{4.80}$$

where

$$f_n = \frac{A_n^2 - a_i^2 - a_r^2}{2a_i a_r} \tag{4.81}$$

Eliminating λ from Eqs. 4.79 and 4.80 and after some algerbraic manipulation,

$$a_i^2 + a_r^2 = \Lambda \tag{4.82}$$

$$2a_i a_r = \Gamma \tag{4.83}$$

where

$$\Lambda = \frac{A_1^2 \sin[2(\Delta_3 - \Delta_2)] - A_2^2 \sin(2\Delta_3) + A_3^2 \sin(2\Delta_2)}{\sin[2(\Delta_3 - \Delta_2)] + \sin(2\Delta_2) - \sin(2\Delta_3)} \tag{4.84}$$

$$\Gamma = \frac{1}{2} \left\{ \left[\frac{A_1^2 + A_3^2 - 2\Lambda}{\cos \Delta_3}\right]^2 + \left[\frac{A_1^2 - A_3^2}{\sin \Delta_3}\right]^2 \right\}^{1/2} \tag{4.85}$$

The solution for a_i and a_r in terms of Λ and Γ is

$$a_i = \frac{1}{2}\left(\sqrt{\Lambda+\Gamma}+\sqrt{\Lambda-\Gamma}\right) \tag{4.86}$$

$$a_r = \frac{1}{2}\left(\sqrt{\Lambda+\Gamma}-\sqrt{\Lambda-\Gamma}\right) \tag{4.87}$$

Once a_i and a_r are known, the incident wave height, the reflection coefficient and the phase angle for the reflected wave may be computed.

This method fails when the denominator in Eq. 4.84 becomes zero. Then, in terms of Δ_3

$$\sin(2\mu\Delta_3) - \sin(2\Delta_3) + \sin(2\Delta_3 - 2\mu\Delta_3) = 0 \tag{4.88}$$

which gives $\Delta_3 = n\pi$ or $n\pi/\mu$ or $n\pi/(1-\mu)$ for an integer n [Isaacson (1991)]. For example, when $\mu = 0.4$ or 0.6, the method fails when $\Delta_3 = 0$, π, $5\pi/3$, 2π, $5\pi/2$.... Moreover, in certain areas, the solution becomes imaginary. For example, when k is near unity and the height measurement is not accurate, the method may fail.

As can be expected, the least-squares method which uses the maximum number of measured variables in the computation is the most accurate. The recommended spacing for this method is $\mu = 0.45$ or 0.65 [Isaacson (1991)].

4.15   REFLECTION OF IRREGULAR WAVES

Similar to regular waves, it is important to know the reflection characteristics of a random wave in a wave tank [Mansard and Funke (1987)]. A simplified approach is to obtain the reflection characteristics of the wave tank over the range of its frequency generation by studying regular waves. Waves generated by the wavemaker get reflected back and forth between the beach and the wavemaker. Therefore, the wave system when steady state is reached is a superposition of positive and negative wave trains of the same frequency. For a wave of amplitude a and frequency ω, the wave profile of the multi-wave reflection system [Goda and Suzuki (1976)] is given by the infinite series

$$\eta = a\{\cos(kx-\omega t)+C_r\cos(kx-2kl+\omega t)+C_R C_r\cos(kx+2kl-\omega t)+$$
$$C_R C_r^2\cos(kx-4kl+\omega t)+C_R^2 C_r^2\cos(kx+4kl-\omega t)+...\} \tag{4.89}$$

where l = length of the wave tank, and C_r and C_R = reflection coefficients of the beach and wavemaker respectively. This reduces to a closed form as

$$\eta = \frac{a}{\left[1 - 2C_r C_R \cos 2kl + C_r^2 C_R^2\right]^{1/2}} \{\cos(kx - \omega t - \varepsilon)$$

$$+ C_r \cos(kx - 2kl + \omega t + \varepsilon)\} \tag{4.90}$$

and

$$\varepsilon = \tan^{-1}\left\{\frac{C_r C_R \sin 2kl}{1 - C_r C_R \cos 2kl}\right\} \tag{4.91}$$

Thus, the two wave trains propagate in the opposite directions due to multi-reflection.

In order to solve for the incident and reflected wave amplitudes, a_i and a_r respectively, wave probes are placed at a known distance, δ, apart. The observed profiles of composite waves for a given frequency at these stations (x_1 and x_2) are given by

$$\eta_1 = (\eta_i + \eta_r)_{x=x_1} = A_1 \cos \omega t + B_1 \sin \omega t \tag{4.92}$$

$$\eta_2 = (\eta_i + \eta_r)_{x=x_2} = A_2 \cos \omega t + B_2 \sin \omega t \tag{4.93}$$

where

$$A_1 = a_i \cos \Phi_i + a_r \cos \Phi_r \tag{4.94}$$

$$B_1 = a_i \sin \Phi_i - a_r \sin \Phi_r \tag{4.95}$$

$$A_2 = a_i \cos(k\delta + \Phi_i) + a_r \cos(k\delta + \Phi_r) \tag{4.96}$$

$$B_2 = a_i \sin(k\delta + \Phi_i) - a_r \cos(k\delta + \Phi_r) \tag{4.97}$$

$$\Phi_i = kx_1 + \varepsilon_i \tag{4.98}$$

$$\Phi_r = kx_1 + \varepsilon_r \tag{4.99}$$

and ε_i and ε_r are the phase angles associated with the incident and reflected waves. From the above equations, the amplitudes of the incident and reflected waves may be estimated.

$$a_i = \frac{1}{2|\sin k\delta|}\left\{\left(A_2 - A_1\cos k\delta - B_1\sin k\delta\right)^2 + \left(B_2 + A_1\sin k\delta - B_1\cos k\delta\right)^2\right\}^{1/2} \qquad (4.99)$$

$$a_r = \frac{1}{2|\sin k\delta|}\left\{\left(A_2 - A_1\cos k\delta + B_1\sin k\delta\right)^2 + \left(B_2 - A_1\sin k\delta - B_1\cos k\delta\right)^2\right\}^{1/2} \qquad (4.100)$$

For the irregular waves, the principle of superposition is applied. The maximum number of components that can be analyzed is limited by the Nyquist frequency (i.e. one-half the number of data sampling). Fast Fourier Transform technique is used to compute the quantities A_1 thru B_2 for all component waves. The spectra of incident and reflected waves are obtained by smoothing the periodograms based on the estimates of a_i and a_r.

An example of reflection coefficient due to a submerged, upright breakwater is taken from Goda and Suzuki (1976). The crest of the breakwater was 60 mm below the mean water level (MWL) in a water depth of 410 mm. The wave gauges were located 4.8 and 5.0 m from the breakwater face. The significant wave height of the irregular wave was $H_S = 82$ mm with a period $T_S = 1.44$ sec. The duration of the wave record was 68.3 sec at a sampling rate of 1/15 sec. The results are shown in Fig. 4.29. The spectra for the incident and reflected waves show the divergence of spectral density near $f = 0$ and $f = 1.97$ Hz as may be expected. Therefore, the resolution is effective within a certain range of wave frequency, as demonstrated below.

$$f_{min} : \delta/L_{max} = 0.05 \qquad (4.101)$$

$$f_{max} : \delta/L_{min} = 0.45 \qquad (4.102)$$

where L_{min} and L_{max} are wave lengths corresponding to f_{min} and f_{max}.

4.16   LIMITED TANK WIDTH

Wave tank model tests may exhibit large experimental scatter due to wall interaction effects. Therefore, in model testing, it is important to know the effect of the tank walls on the measurements. For a long narrow tank, the side walls may have an influence on the measured forces on a fixed structure placed on the center line of the tank. The interaction becomes significant if the structure transverse dimension becomes

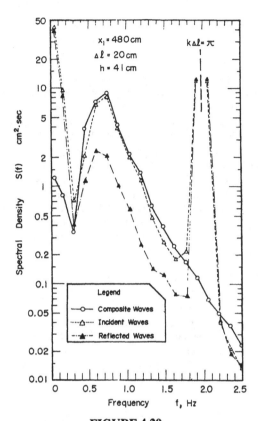

FIGURE 4.29
REFLECTION FOR IRREGULAR WAVE ON A SUBMERGED, UPRIGHT
BREAKWATER [Goda and Suzuki (1976)]

the same order of magnitude as the width of the tank. The common rule of thumb used in reducing this effect on the measurement of the inline force is to use a model such that the ratio of tank width to the model transverse dimension is at least 5:1. For example, for a vertical cylinder, the walls should be at least 2.5 diameters from the center lines of the cylinder. In towing tank hull-resistance tests, the ship bow waves should not reflect from the tank walls back onto the model. This rule limits the possible model length for a given tank width. The interference effect of the tank wall for a ship model towed in a tank is given [Goodrich (1969)] in Fig. 4.30. According to this figure, for a given tank breadth to model length ratio (B/ℓ), any combination of the wave length/model length ratio and Froude number that lies above a horizontal line through minimum β value will cause a wall effect. The figure, however, does not give any indication of the degree of wall effect.

In the experiments with floating vessels in a wave tank, the motion, in particular heave motion, produces radiated wave which will invariably radiate from the tank walls. In general, the radiated waves decrease in magnitude with distance from the model. This factor determines the optimum size of a model that may be tested in a given tank without significant adverse effect. Numerical computations have been carried out by Calisal and Sabuncu (1989) with floating vertical cylinders to determine the effect of the tank wall on the hydrodynamic coefficients. The numerical results show that the added mass coefficients exhibit peak values at resonant frequencies. These frequencies correspond to frequencies of waves with a wave length equal to the tank width and increase with the decrease in tank width. Several analytical and experimental data on a variety of submerged objects [Vasques and Williams (1992)], [Yeung and Sphaier (1989)] have been presented to quantify this effect. The forced oscillation test of a floating cylinder in still water will also experience similar interaction from the radiated waves reflected back from the tank walls.

4.17   TESTING FACILITIES IN THE WORLD

There are numerous testing facilities that exist in various countries in the world. Many of these are used in the commercial and defense related testing of marine structures. A few of these are summarized here. Note that the larger facilities used for contract activities are mainly included. Many of the smaller research facilities that exist in universities have not been included here. One may refer to the International Towing Tank Conference (ITTC) manual for their description.

4.17.1 *Institute of Marine Dynamics Towing Tank, St. John's, Newfoundland, Canada*

Deep Water Wave Tank

Tank Size: 200 m Long, 12 m Wide, 7 m Deep

Carriage Speed : 10 m/s

Waves: Regular and Irregular;1 m

Wavemaker: The wavemaker is of dual-flap, dry back construction. The lower board operates at up to 0.1 Hz, while the upper board can operate to 1.8 Hz frequency. Each wave board is powered by a single actuator in the center of the board. Two additional actuators on either side of the center provide hydrostatic support and compensation for the wave load. The sides are sealed by pneumatically pressurized sealing gaskets.

Beach: The beach consists of corrugated plates bolted to a rigid framework. The surface has a large circular profile of 30 m radius and extends 20 m into the tank at the toe. Transverse wooden slats are placed near the water surface. A portion of the beach can be lowered for access to the dock area.

4.17.2 Offshore Model Basin, Escondido, California

Tank Size: 90 m Long, 14.6 m Wide, 4.6 m Deep

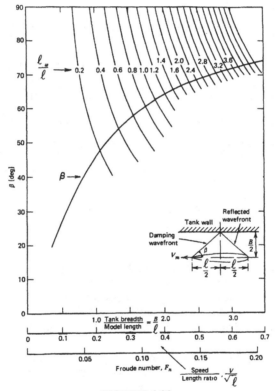

FIGURE 4.30
TANK WALL INTERFERENCE IN SHIP RESISTANCE TESTS
[Goodrich (1969)]

Deep Section: Circular Pit 9 m Deep

Carriage Speed: 6 m/s (20 ft/sec.)
                 Transverse Rotating Subcarriage

Waves: Regular and Irregular; 0.74 m (29 in.)

Wavemaker: The wavemaker is a servo-controlled, hydraulically driven, bottom pivoted, single flap wave board. Waves are generated by a pre-recorded tape created on a minicomputer.

Beach: Metal Shavings

4.17.3 Offshore Technology Research Center, Texas A&M, College Station, Texas

Tank Size: 45.7 m (150 ft) Long x 30.5 m (100 ft) Wide x 5.8 m (19 ft) Deep

Deep Section: 16.7 m (55 ft) Deep Pit With Adjustable Floor

Wavemaker: Programmable, Hydraulically Driven Hinged Flap

Waves: Regular and Irregular Seas

Maximum Wave Height: 0.8m (34 in.)

Frequency Range: 0.5-4.0 sec

Beach: Progressive Expanded Metal Panels

4.17.4 David Taylor Research Center, Bethesda, Maryland

Maneuvering and Seakeeping Facilities (MASK)

Tank Size: 79.3 m (260 ft) Long x 73.2 m (240 ft) Wide x 6.1 m (20 ft) Deep

Wavemakers: 8 Pneumatic Wavemakers on One Side
             13 Pneumatic Wavemakers on adjacent side

Waves: Regular and Irregular Waves
        Multi-Directional Waves Programmed on Magnetic Tape

Maximum Height 0.6 m (2 ft)
Wave Length 0.9 - 12.2 m (3 - 40 ft)

Beaches: Concrete Wave Absorbers With Fixed Bars

Carriage Speed: 7.7 m/s (15 knots)

Deep Water Basin

Tank Size: 846 m (2,775 ft) Long x 15.5 m (51 ft) Wide x 6.7 m (22 ft) Deep

Waves: Maximum Height 0.6 m (2 ft)
Wave Length 1.5 - 12.2 m (5 - 40 ft)

Carriage Speed: 10.2 m/s (20 knots)

High Speed Basin

Tank Size: 905 m (2,968 ft) Long x 6.4 m (21 ft) Wide x 3 m (10 ft) (for
1/3 ℓ), 4.9 m (16 ft) (for 2/3 ℓ) Deep

Waves: Maximum Height 0.6 m (2 ft)
Wave Length 0.9 - 12.2 m (3 - 40 ft)

Carriage Speed: 35.8 - 51.2 m/s (70 - 100 knots)

4.17.5 *Maritime Research Institute, Netherlands (MARIN)*

Seakeeping Basin

Tank Size: 100 m Long x 24.5 m Wide x 2.5 m Deep

Deep Section: Pit Depth 6 m

Waves: Regular and Irregular Waves 0.3 m significant; 0.7-3 sec

Carriage Speed: 4.5 m/s

Wave and Current Basin

Tank Size: 60 m Long x 40 m Wide x 1.2 m Deep

Deep Section: Pit Depth 3 m

Waves: Regular and Irregular Waves

Carriage Speed: 3 m/s

Current Speed: 0.1 - 0.6 m/s

Deep Water Towing Tank

Tank Size: 252 m Long x 10.5 m Wide x 5.5 m Deep

Carriage Speed: 9 m/s

High Speed Towing Tank

Tank Size: 220 m Long x 4 m Wide x 4 m Deep

Wavemaker: Hydraulic flap-type

Waves: Regular and Irregular Wave
          Maximum height: 0.4 m sig.
          Period range: 0.3 - 5 sec.

Carriage: Manned, motor driven; Unmanned, jet-driven

Carriage Speed: 15 m/s and 30 m/s

Beach: Lattice on circular arc plates

4.17.6 *Danish Maritime Institute, Lyngby, Denmark*

Tank Size: 240 m Long x 12 m Wide x 5.5 m Deep

Wavemaker: Numerically Controlled Hydraulic Double Flap Wavemaker

Waves: Regular and Irregular Waves
        Maximum Height: 0.9 m
        Period Range: 0.5 - 7 sec

Carriage Speed: 0 - 11 m/s(accuracy $\pm$ 0.2 percent)

4.17.7 *Danish Hydraulic Institute, Horsholm, Denmark [Aage and Sand (1984)]*

Tank Size: 30 m Long x 20 m Wide x 3 m Deep

Deep Section: 12m Deep in Center

Wavemaker: 60 Flap Type Hydraulically Driven Wavemakers on One Side Controlled by a Minicomputer

Waves: Maximum Height $\approx$ 0.6 m
Period Range $\approx$ 0.5 - 4 sec

4.17.8 *Norwegian Hydrodynamic Laboratory, Trondheim, Norway (MARINTEK)*

The Ocean Basin [Eggestad (1981)]

Tank Size: 80 m long x 50 m wide x 10 m deep

Wavemaker: Hinged double flap, 144 individually controlled; Hydraulically driven hinged type

Waves: Regular and Irregular; 0.9 m

Current: Max speed 0.2m/s

4.18 REFERENCES

1.   Aage, C. and Sand, S.E., "Design and Construction of the DHI 3-D Wave Basin", Symposium on Description and Modelling of Directional Seas, Copenhagen, Denmark, 1984.

2.   Beach Erosion Board, "Reflection of Solitary Waves", BEB Technical Report No. 11, 1949.

3.   Biesel, F. and Suquet, F., "Les Appareils Generateurs de Houle en Laboratoire", La Houille Blanche, Nos. 2, 4 and 5, 1953 St. A. Falls, No. 6, 1953.

4.   Biesel, F., et al., "Laboratory Wave Generating Apparatus", St. Anthony Falls Hydraulic Laboratory Report No. 39, Minneapolis, MN., March, 1954.

5.   Biesel, F., "Wave Machines", Proceedings of First Conference on Ships and Waves, Stevens Institute of Technology, Hoboken, New Jersey, 1954.

6.   Bullock, G.N. and Murton, G.J., "Performance of a Wedge-Type Absorbing Wave Maker", Journal of Waterway, Port, Coastal and Ocean Engineering, Vol. 115, No. 1, Jan., 1989, pp. 1-17.

7.   Calisal, S.M. and Sabuncu, T., "A Study of a Heaving Vertical Cylinder in a Towing Tank", Journal of Ship Research, Vol. 33, No. 2, June 1989, pp. 107-114.

8.   Clauss, D., Riekert, T. and Chen, Y., "Improvements of Seakeeping Model Tests by Using the Wave Packet Technique and Side Wall Wave Absorber", Proceedings on Eleventh International Offshore Mechanics and Arctic Engineering Symposium, ASME, Calgary, Canada, 1992.

9.   Eggestad, I., "NHL Ocean Laboratory: Engineering and Construction of Building and Sub-Systems", Symposium on Hydrodynamics in Ocean Engineering, Norwegian Hydraulics Laboratory, Trondheim, Norway, 1981.

10. Galvin, C.J., "Heights of Waves Generated by a Flap-Type Wave Generator", CERC Bulletin, Corps of Engineers, Department of Army, Vol. II, 1966, pp. 54-59.

11. Gilbert, G., Thompson, D.M. and Brewer, A.J., "Design Curves for Regular and Random Wave Generators", Journal of Hydraulic Research, ASCE, Vol. 9, No. 2, 1971.

12. Goda, Y. and Suzuki, Y., "Estimation of Incident and Reflected Waves in Random Wave Experiments", Proceedings of Fifteenth Coastal Engineering Conference, Vol. 1, 1976, pp. 828-845.

13. Goodrich, G.J., "Proposed Standards of Seakeeping Experiments in Head and Following Seas", Proceedings on Twelfth International Towing Tank Conference, 1969.

14. Hudspeth, R.T., Jones, D.F. and Nath, J.H., "Analyses of Hinged Wavemakers for Random Waves", Proceedings of Tenth Coastal Engineering Conference, ASCE, 1978, pp. 372-387.

15. Hudspeth, R.T., Leonard, J.W., and Chen, M-C.,"Design Curves for Hinged Wavemakers: Experiments", Journal of the Hydraulics Division, ASCE, Vol. 107, No. HY5,May., 1981, pp. 553-574.

16. Hyun, J.M., "Theory for Hinged Wavemakers of Finite Draft in Water of Constant Depth", Journal of Hydronautics, Vol. 10, No. 1, 1976, pp. 2-7.

17. International Towing Tank Conference Catalogue of Facilities, Sixteenth ITTC Information Committee. Annapolis, MD, 1979.

18. Isaacson, M., "Measurement of Regular Wave Reflection", Journal of Waterway, Port, Coastal and Ocean Engineering, ASCE, Vol. 117, No. 6, 1991, pp. 553-569.

19. Jamieson, W.W. and Mansard, E.P.D., "An Efficient Upright Wave Absorber", Proceedings on Coastal Hydrodynamics, University of Delaware, ASCE, 1987, pp. 124-139.

20. Jamieson, W.W., Mogridge, G.R. and Brabrook, M.G., "Side Absorbers for Laboratory Wave Tanks", International Association for Hydraulic Research Proc. XXIII Congress, Ottawa, Canada, August 1989.

21. Jarlan, G.E., "Note on the Possible Use of a Perforated, Vertical Wall Breakwater", Proceedings of Seventh Conference on Coastal Engineering, Aug., 1960.

22. Jarlan, G.E., "A Perforated Vertical Wall Breakwater - An Examination of Mass-Transfer Effects in Gravitational Waves", The Dock and Harbour Authority, Apr., 1961, pp. 394-398.

23. LeMehaute, B., "Progressive Wave Absorbers", Journal of Hydraulic Research, IAHR, Vol. 10, No. 2, 1972, pp. 153-169.

24. Mansard, E.P.D. and Funke, E.R., "The Measurement of Incident and Reflected Spectra Using a Least Squares Method", Proceedings of Seventh International Conference on Coastal Engineering, Sidney, Australia, ASCE, Vol. 1, 1980, pp. 154-172.

25. Mansard, E.P.D. and Funke, E.R., "On the Reflection Analysis of Irregular Waves", National Research Council of Canada, Hydraulic Laboratory Technical Report TR-HY-017, 1987.

26. Miles, M.D., Laurich, P.H. and Funke, E.R., "A Multimode Segmented Wave Generator for the NRC Hydraulics Laboratory", Proceedings of Twenty-First American Towing Tank Conference, Washington, D.C., 1986.

27.   Ouellet, Y. and Datta, I., "A Survey of Wave Absorbers", Journal of Hydraulic Research, IAHR, Vol. 24, No. 4, 1986, pp. 265-280.

28.   Patel, M.H. and Ionnau, P.A., "Comparative Performance Study of Paddle-and Wedge-Type Wave Generators", Journal of Hydronautics, AIAA, Vol. 14, No. 1, Jan., 1980, pp. 5-9.

29.   Ploeg, J. and Funke, E.R., "A Survey of Random Wave Generation Techniques", International Conference on Coastal Engineering, ASCE, 1980, pp. 135-153.

30.   Salter, S.H., "Absorbing Wave-Makers and Wide Tanks", Conference on Directional Wave Spectrum Applications, Berkeley, California, ASCE, 1981.

31.   Straub, L.G., Bowers, C.E. and Herbich, J.E., "Laboratory Tests of Permeable Wave Absorbers", Proceedings of Fifth Conference on Coastal Engineering, 1956, pp. 729-742.

32.   Ursell, F., Dean, R.G. and Yu, Y.S., "Forced Small Amplitude Water Waves: A Comparison of Theory and Experiment", Journal of Fluid Mechanics, Vol. 7, Part 1, 1960, pp. 33-52.

33.   Wang, S., "Plunger-Type Wavemakers: Theory and Experiment", Journal of Hydraulic Research, Vol. 12, No. 3, 1974, pp. 357-388.

34.   Vasquez, J.H. and Williams, A.N., "Hydrodynamic Loads on a Three-Dimensional Body in a Narrow Tank," Proceedings of Offshore Mechanics and Arctic Engineering Conference, ASME, Calgary, Canada, 1992, pp. 369-376.

35.   Yeung, R.W. and Sphaier, S.H., "Wave Interference Effects on a Truncated Cylinder in a Channel", Journal of Engineering Mathematics, Vol. 23, 1989.

CHAPTER 5

MODELING OF ENVIRONMENT

5.1 WAVE GENERATION

In the testing of an offshore structure model, the environment experienced by the structure requires proper simulation in the laboratory. This environment generally consists of waves, wind, current and possibly earthquake. This last item is not covered in this book. The excitation due to earthquake loading is generally simulated by placing the model on a shaker table. For example, a shaker table may be placed at the bottom of the wave tank, and the instrumented structure may be attached to it. The shaker table may then simulate the earthquake motion (similar to the planar motion mechanism described in Section 4.12). The energy and frequency content of the earthquake may be simulated by one of the methods to be described for wave generation.

There are many different methods that are employed in generating irregular waves [Kimura and Iwagaki (1976), Ploeg and Funke (1980), Johnson (1981), Mansard and Funke (1988)]. It should be recognized that the irregular wave may not have the statistical features possessed by the random wave but some control may have been exercised on either their frequency or time domain characteristics. The following are some of the common methods of irregular or random wave generation:

- Harmonic Synthesis — These waves may be generated by a mechanical gear system, special purpose electronic device or by computer. It uses a range of discrete frequencies producing irregular wave form.

- Frequency Sweep, by varying the speed of the actuator drive motor.

- Pseudo Random Noise, which is generated by computer or an electronic device.

- True Random White Noise and Analog Shaping Filters [Loukakis (1968)].

- Synthesis by Fourier Transform Technique by on or off-line computer.

- Reproduction of Prototype Wave Train in time domain. This may be accomplished by magnetic tape, whether analog or digital, and on-line computer [Gravesen, et al. (1974)].

- Wind alone or in combination of one of the above methods.

A compensation technique is often used to account for the transfer function of the wave board or the servo. If a discrete time driving signal is used involving a digital to analog converter, a smoothing routine, such as analog low-pass filter or straight line interpolator, is recommended.

The type of waves that are generated in a wave tank depends on the type of wave generator and the method of wave generation. There are several forms of waves generated in the laboratory. What waves are generated in a given test depends on the application and location of the offshore structure being tested.

5.1.1 Harmonic Waves

Waves are generated in the tank by scaling down the sea wave parameters by the scaling laws described in Chapter 2. Regular harmonic waves are very useful tools in understanding the physics of the fluid-structure interaction, even though such waves of significance are seldom found in nature. The generation of regular sinusoidal waves is quite straightforward once the height and period of the wave are scaled down. Then, knowing the transfer function of the given wave generator, the proper setting for the wave height may be established.

5.1.2 Non-Harmonic Waves

There are several techniques used in wave synthesizing through the Fourier transform. For a non-harmonic wave synthesizer, a large frequency band in the waves is generally simulated. The harmonic relationship between contributing frequency components is avoided. In one method, the frequencies are located at the center of the adjacent frequency bands having equal areas. Thus, the components within the individual bands have the same amplitude [Borgman (1969), Medina, et al. (1985), Toki (1981)]. The Fourier amplitude coefficients are considered proportional to the square root of the desired spectral density. Phases are obtained from a selected random number generator [Goda (1970)]. A more realistic simulation of a Gaussian stochastic process uses a "white noise" complex spectrum. The time series is generated using statistically independent Gaussian random numbers. They are then "filtered" by the square root of the desired spectral density by cross-multiplication [Goda (1977)]. In a slightly different approach [Medina and Diez (1984)], random chi-squared distributed numbers with two degrees of freedom are used. After cross-multiplication with the desired spectral density, the square root provides the amplitudes which are paired with the uniformly distributed random phase. The complex spectrum in polar coordinates is inversely Fourier transformed to yield the desired time series. Alternatively, an iterative Fourier technique [Funke and Mansard (1979)] is used to achieve a specified distribution of energy in the frequency as well as the time domain.

Sometimes, a white noise having "equal" amplitudes at all frequencies within a band is generated. Filtered white noise generators are random noise sources which never repeat. They may be used in a model test in obtaining a response transfer function from one single run.

There are many energy spectrum models that have been proposed. Some of these are Pierson-Moskowitz (PM), Bretschneider, JONSWAP and Ochi and Hubble [Chakrabarti (1993)]. While generation of waves in the tank to match a given energy density spectrum model is a common occurrence, it is sometimes desirable to duplicate the natural sea waves. In this case, a record of the short term wave activity at a specific site is available. This record is scaled down and reproduced point by point in the wave tank. The digital data is converted to the analog signal, rectified to account for the wave generator transfer function and then is used to drive the wave generator.

Sometimes, an extreme wave height in a time series is achieved by experimenting with different random number seeds until the desired zero-crossing wave height is found in the time series. The same technique is applicable to achieve a particular degree of wave grouping once the energy density is matched.

No matter how perfect the simulation technique is, the physical model is handicapped by several factors causing departure from the theoretical model:

- Imperfect knowledge of the dynamic transfer characteristics of the wave generator;
- Reflection from side walls, beach and generator;
- Distortion by mechanical device of the wavemaker in contact with water;
- Wave-wave energy transfer;
- Resonance in oscillation from inline or lateral standing wave; and
- Recirculation due to a net mass transport in the tank.

For the generation of waves, a digital input signal is computed from the target spectrum. The digital signal uses a wave machine transfer function. The transfer function generally accounts for the relationship between the mechanical displacement to the water displacement, the hydraulic servo control system, and the dynamic effect of the low-pass filter. This compensation can be effectively implemented through a Fourier transform technique.

However, since this is a linear process, it does not account for the nonlinear effect in the wave generation. The discrepancy between the desired and output signal can, however, be compensated in an iterative method by generating and recording the wave a few times until a satisfactory recording of the wave at the test site is achieved. Unfortunately, this technique would not be applicable if any reflected wave is generated at the test site (for example, between a structure and the tank wall). Hence this iterative technique is generally adopted when calibrating the waves before placing the test structure in position.

5.1.3 Imperfect Waves

An efficient wavemaker can reach a wave steepness of about 0.1 tanh kd which is about 70 percent of the theoretical limit. The excess energy fed into an inefficient wavemaker not used in the production of waves leads to standing, cross-mode waves and wave breaking. The pneumatic wavemaker described in Chapter 4 is capable of producing waves of period, $T > 1.0$ sec. If one attempts to generate waves below 1 second with this wavemaker, large transverse waves are produced in the tank which propagate in the tank in cross-mode. The period of this wave corresponds to the standing wave oscillation based on the width of the tank. A similar observation can also be found when generating shallow water waves by flapper motion of a conventional type of wave board or deep water waves by translatory motion.

Another area of concern is spurious long waves created while simulating nonlinear waves by linear wave generation theory. This aspect will be discussed in further details in Section 5.4. In some studies, the nonlinear wave components have a large influence on the outcome, e.g., in the mooring tests of a vessel or in overtopping of structures.

Nonlinear waves are known to contain, according to second order theory, high harmonics which are locked to the fundamental waves and therefore, move at the same celerity as the fundamental components. Traditional first order theory used in the laboratory reproduction of waves cannot take into account the locked nature of these harmonics and inadvertently produces undesirable free waves (called as spurious or parasitic waves) in model studies. Since the free waves move at their own celerities, they interact with the faster running locked harmonics along the length of the tank, thus resulting in profile changes such as the ones depicted in Fig. 5.1b. These undesired waves can however, be eliminated by a second order wave generation technique. With this technique, a stable second order wave moving down the tank without any appreciable profile change can be realized (see Fig. 5.1a). An example of the difference in the forces measured on a moored vessel model in a tank with and without this compensation is seen in Fig. 5.2. Forces are underestimated by about 15 percent without compensation of the undesired waves.

In the case of irregular waves, wave components of higher frequencies consisting of twice the fundamental (i.e. $2f_i$) as well as the sum of two fundamentals (i.e., $f_i + f_j$) are generated. These are also locked components and propagate at the same celerity as their fundamentals. Spurious, undesired free waves are also encountered in this case and their interactions with the locked waves during propagation result in substantial changes in the high frequency tails of their spectra (see for instance Fig. 5.3). It should, however, be pointed out that even if these free waves are eliminated by adopting a second order generation technique, variations in high frequency tails would still be encountered during propagation because of the following reason. The high frequency tail of a nonlinear wave spectrum is generally composed of two types of wave components: high frequency linear components which are integral parts of the spectrum and locked high harmonics produced by the interaction of components which belong to the low frequency part. Since linear and locked components travel with different propagation celerities, the shape of the high frequency tails of the spectrum can be expected to change [see Sand and Mansard (1986)].

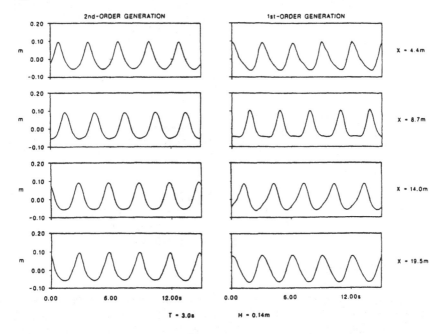

FIGURE 5.1
REGULAR WAVE PROFILES DURING PROPAGATION (a) WITH AND (b)
WITHOUT SECOND ORDER COMPENSATION
[Mansard and Funke (1988)]

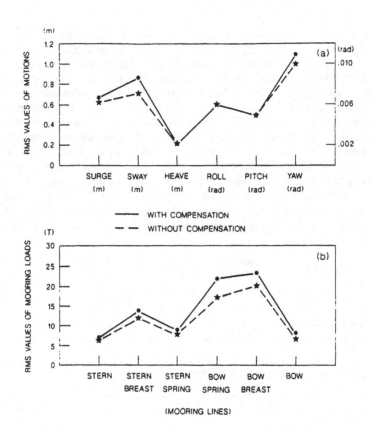

FIGURE 5.2

**EFFECT OF SECOND ORDER LONG WAVE COMPENSATION ON THE
RESPONSE OF A MOORED VESSEL**

[Mansard and Funke (1988)]

5.1.4 Shallow Water Waves

If the water depth is properly scaled in the tank, the form of the regular
shallow water waves is automatically formed during its propagation down the tank. Very
long period shallow water waves in the tank suffer from the soliton effect [Chakrabarti
(1980, 1990)]. In this case the regular wave breaks down into several peaks called
solitons, each traveling at a different speed. These waves reform as the front travels

down the tank. An example of soliton recorded in 1.53 m (5 ft) of water is shown in Figure 5.4.

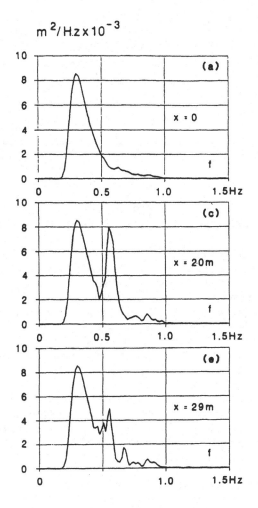

FIGURE 5.3
SPECTRA OF FIRST-ORDER WAVE GENERATION AT VARIOUS
DISTANCES X FROM PADDLE WAVEMAKER
[Sand and Mansard (1988)]

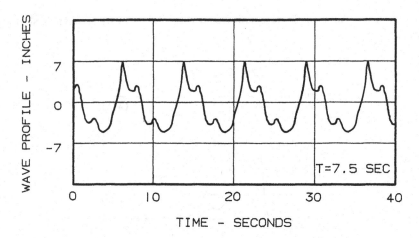

FIGURE 5.4
REGULAR LONG PERIOD WAVES IN SHALLOW WATER IN TANK

Kjeldsen and Myrhaug (1979) advanced the concepts of horizontal and vertical asymmetry parameters to characterize the nonlinear asymmetries found in natural waves. A technique was developed by Funke and Mansard (1982) to simulate these asymmetries during laboratory reproduction of waves. For this, two operations are needed. First, the wave crests are amplified and the trough is correspondingly attenuated. The duration of the trough is lengthened by a factor at the same time the crest duration is shortened by a comparable factor such that the wave period remains unchanged. The crest amplification factor is derived such that the areas under the crest and the trough are equal.

The second distortion appears in the asymmetry of the crest (called the crest front steepness). Here, the distortion in the wave crest is achieved by shifting the peak of the crest forward by a factor. The amount of distortion is limited by the amount of modification it causes in the amplitude spectrum. These distortions are simulated in the wavemaker command signal.

5.2   RANDOM WAVE SIMULATION

As we have already discussed, there exists a variety of techniques of reproducing random sea waves in the laboratory. Techniques of producing waves in the laboratory basins have evolved from the simple electro-mechanical systems in the past to the present-day sophisticated hydraulic-electric servo systems, controlled by on-line computers and capable of producing a large number of different time series of water surface elevations from a sea state defined solely by its spectrum or just by its significant wave height and peak period. Because of this possibility of variations due to different wave synthesis techniques (in addition to already existing differences in experimental setups), there is no assurance that testing the same model in different laboratories will give similar results. Therefore, an accurate definition of the required sea state has become very important. For example, matching just the significant wave height and period may not be sufficient to produce similar results. Groupiness in waves may be an important input parameter in defining a model test.

The duration of random wave record should be sufficiently large for analysis; but, in any case, should not be less than 180 times the mean wave period T_m. Usually one starts with a given wave energy density spectrum for the simulation of sea waves in a laboratory. There are several different methods for wave generation based on this energy density. Two of the most common Fourier techniques [see Rice (1945)] are the Random Phase Method (RPM) and Random Complex Spectrum Method (RCSM). The former is spectrally deterministic, while the latter is spectrally non-deterministic. The RPM method has been claimed [Tucker, et al. (1984] to be incorrect in representing nature and produces insufficient variability of wave parameters. For long non-repetitive wave records (approaching infinity), the two methods are identical. Even for short records generated numerically as well as physically, the differences in some parameters are found to be small. This area will be discussed further in Section 5.3.3.

5.2.1 *Random Phase Method*

In the so-called spectrally deterministic method, a smooth wave spectral density function is specified from which the adjacent frequency components are derived. These frequency components are not necessarily statistically independent. Therefore, in the strictest sense, the resultant wave trains are not part of a Gaussian stochastic process. The phases for each frequency component are obtained randomly; hence, it is called the random phase method. However, numerical simulation with a "random generator" may be restarted from the same initial condition (called seed) every time and is often cyclical over a finite number of entries into the generator. Such sources of random numbers are referred to as pseudo-random number generators. By using different seeds, different sequences of random phases can be generated and thus different time domain characteristics of the sea state. This technique is often employed to achieve different degrees of grouping or different wave height statistics in the simulated waves.

The sea surface is generally assumed to be Gaussian with a zero mean. Simulation of the sea surface usually consists of a finite number of Fourier components as a function of time:

$$\eta(t) = \sum_{n=1}^{N} a_n \cos(2\pi f_n t + \varepsilon_n) \tag{5.1}$$

where f_n and a_n are frequency and amplitude chosen from the wave energy spectrum and ε_n is a random phase angle. This method always reproduces the wave energy density spectrum, $S(f)$. The method, however, does not strictly model a random Gaussian surface except in the limit as $N \rightarrow \infty$ [Tucker, et al. (1984)].

The smooth spectrum is subdivided in N equal frequency increments of width Δf over the range of frequencies 0 and f_m, where f_m is the maximum generated frequency. Then, the spectrum density $S(n\Delta f)$ where n refers to the nth increment is converted into a Fourier amplitude spectrum $a(n\Delta f)$:

$$a_n = a(n\Delta f) = \sqrt{2S(n\Delta f)\Delta f}, n = 1, 2 \ldots N \tag{5.2}$$

The value of Δf (in a Finite Fourier Transform simulation) is chosen such that

$$\Delta f = \frac{1}{T_R} \tag{5.3}$$

where T_R is the length of the time series to be synthesized. The periodicity of the time series is also T_R. $S(f)$ is not generally provided at this sampling rate and, therefore, the resampling of $S(f)$ is usually carried out. The phase spectrum $\varepsilon(n\Delta f)$ is created at each frequency, $n\Delta f$, from a random number generator with a uniform probability distribution between $-\pi$ and $+\pi$. The target time series is derived by an inverse Fourier transform:

$$\eta(t) = \sum_n a(n\Delta f) \cos\left[2\pi n\Delta f t + \varepsilon(n\Delta f)\right], \ 0 \le t \le T_R \tag{5.4}$$

It is a common practice to request a mean maximum recorded elevation, η_{mR}, to be within 5 percent of the theoretical value, η_{mT}:

$$\left|\eta_{mT} - \eta_{mR}\right| / \eta_{mT} \le 0.05 \tag{5.5}$$

where

$$\eta_{mT} = \sqrt{2m_0} \sqrt{\log(T_R / T_z)} \qquad (5.6)$$

in which m_0 = area under the energy density spectrum (zeroth moment) and T_z = zero-crossing period. In practice, the amplitude/phase pairs are converted into real and imaginary components and the Fast Fourier Transform is applied to the complex spectrum. The requirement of FFT for the number of points of 2^k is satisfied by the initial choice of f_m where

$$\Delta t = \frac{1}{2 f_m} \qquad (5.7)$$

Usually, f_m is selected to be $3f_0$, where f_0 is the spectral peak frequency. Because of the method of generation of the time series, when inverted by Fourier transform it yields a close match to the target spectrum.

5.2.2 Random Complex Spectrum Method

In this method [Borgman (1969, 1979), Tuah and Hudspeth (1982)], the adjacent frequency components are made statistically independent. The specified energy spectrum is preserved within the bounds of probability only, making it a non-deterministic method (nonrepeating). It is analogous to the operation of filtering "white noise" by using a filter which has the transfer function of the desired spectrum. The filter output matches the filter transfer function on the average over very long periods of time. Over a finite duration, the filter output has spectral properties which vary randomly because of the Gaussian nature of the underlying process.

A "white noise" complex spectrum is generated in terms of the real and imaginary parts of the white noise function

$$\text{Re}\, W(n\Delta f) = RANG(n) / \sqrt{2} \qquad (5.8)$$

$$\text{Im}\, W(n\Delta f) = RANG\,(n+1) / \sqrt{2} \qquad (5.9)$$

and

$$\text{Re}\, W(0) = \text{Im}\, W(0) = 0 \qquad (5.10)$$

where RANG is random number generating function for number sequences with Gaussian probability distribution of zero mean and unit standard deviation. The white noise spectrum is filtered as

$$\mathrm{Re}(n\Delta f) = \mathrm{Re}\, W(n\Delta f) * \sqrt{S(n\Delta f)\Delta f} \qquad (5.11)$$

$$\mathrm{Im}(n\Delta f) = \mathrm{Im}\, W(n\Delta f) * \sqrt{S(n\Delta f)\Delta f} \qquad (5.12)$$

The Fourier transform of the left-hand side provides the time series having a repeat period of $1/\Delta f$. As before, the maximum frequency is chosen to make use of 2^k samples in an FFT. Resampling provides a $\Delta t = 1/(6f_0)$.

5.2.3 Random Coefficient Method

The above methods do not provide a random amplitude, a_n. An alternate method for numerical simulation is to generate a set of random numbers which have a Gaussian probability distribution. This set is passed through a linear filter with a specified frequency response to reproduce the given wave spectrum. Assuming that the sea surface is sampled at discrete times, t_m, with an equal interval of Δt, the discrete sea surface points are given by

$$\eta_m = \sum_{n=1}^{N/2}\left(b_n \cos 2\pi f_n t_m + c_n \sin 2\pi f_n t_m\right) = \sum_{n=1}^{N/2} a_n \cos\!\left(2\pi f_n t_m + \varepsilon_n\right) \qquad (5.13)$$

where $f_n = n/(N\Delta t) = n/T_R$; $a_{N/2} = 0$ and $\varepsilon_{N/2} = 0$. The Nyquist frequency is given by $f_{N/2} = 1/(2\Delta t)$. Note that the total number of Fourier coefficients is N.

This representation of η_m repeats itself after a time, $T_R = 1/\Delta f$, and is thus valid in the range $0 \le t \le T_R$. By choosing a small Δf (large N), this range can be made quite large. The computed spectrum from a sample for $0 \le t \le T_R$ is only an estimate of $S(f)$ having a large sampling variance. It will not exactly match the spectrum of the entire time history. If several different sample estimates are averaged, the average will converge on $S(f)$.

The coefficients b_n and c_n are considered to be independent random variables having Gaussian distribution with a common variance by

$$\sigma_n^{\,2} = S(f_n)\Delta f_n \qquad (5.14)$$

where Δf_n is written as a general case (as opposed to constant Δf). A variable Δf_n allows a longer nonrepeating record length (for the same N value) than T_R without repeating itself. The joint probability distribution [Chakrabarti (1990)] of b_n and c_n is

$$p(b_n, c_n) = \frac{1}{2\pi\sigma_n{}^2} \left[\exp \frac{-(b_n^2 + c_n^2)}{2\sigma_n{}^2} \right] \tag{5.15}$$

Writing $b_n = a_n \cos \varepsilon_n$ and $c_n = a_n \sin \varepsilon_n$, one obtains

$$p(a_n, \varepsilon_n) = \frac{a_n}{2\pi\sigma_n{}^2} \left[\exp \frac{-a_n^2}{2\sigma_n^2} \right] \tag{5.16}$$

Therefore, a_n has a Rayleigh distribution with an rms value of $\sqrt{2}\sigma_n \left[= \sqrt{2S(f_n)\Delta f_n} \right]$, and ε_n are independent with uniform distribution over 0 and 2π [Chakrabarti (1990)]. This method insures that the wave amplitudes follow Rayleigh distribution. In the simulation, Δf_n is often assumed constant so that $f_n = n\Delta f$.

5.3   WAVE PARAMETERS

In addition to the wave energy density, at least two other parameters may be important for the determination of the offshore structural response. These are wave grouping and wave asymmetry.

5.3.1 *Wave Groups*

High waves appear in groups. Goda (1970) demonstrated that random ocean waves are not completely randomly distributed, but rather have a tendency to appear in groups, particularly the high ones. The grouping characteristic of waves has a particular effect in the response of offshore structures, particularly those that are moored in the open ocean. Wave grouping characteristic in random seas is a significant design element which should be properly taken into account [Lonquet-Higgins (1984)]. When random wave generators are used in wave tanks, the wave groups generate second order long waves in the tanks.

The concept of groupiness may be defined in terms of the run length and total run [Goda (1970), Chakrabarti, et al. (1974)]. The run length is defined as the number of successive waves exceeding in height a given threshold value. This threshold value could be any given wave parameter, e.g., the significant wave height. The total run is defined as the number of waves in an interval between first threshold exceedance of a run to the exceedance of the next run. While run length can be shown to have an effect on structural responses, it is difficult to simulate it in a wave time series. Goda (1970) showed that they are related to a spectral peakedness parameter.

There are a variety of theories and parameters used to characterize wave groups. These are listed by Medina and Hudspeth (1990). None of the parameters that depend on the spectral shape, however, are very good at distinguishing wave groups [Medina and Hudspeth (1990)]. Three of the methods of analysis of wave groups that incorporate most of these parameters are (1) wave height function, (2) three-axes representation of run lengths, and (3) correlation coefficient between successive wave heights. Some of these and the associated probability distributions have been discussed by Chakrabarti (1990).

The initial probability estimate of Goda (1970) was consistently under the computed field and numerical simulations. The estimate for wave groups was refined by Kimura (1980) by introducing a probability density function including a correlation function, ρ. Battjes and Vledder (1984) introduced a parameter, κ:

$$\kappa = 2\rho \tag{5.17}$$

κ is derived from the variance spectrum and it characterizes wave groups. The square of the wave envelope or wave height function [Rice (1945)] may be used to analyze wave groups. The use of temporal Hilbert filters to estimate the height square function seems to be the best method [Hudspeth and Medina (1988)] to define the instantaneous variance function of a wave record.

5.3.2 Wave Asymmetry

Asymmetry in a wave profile produces a departure of the wave statistics from a Gaussian process. The horizontal asymmetry factor is defined as the ratio of the crest height to the total height. Similarly, the vertical asymmetry factor is the ratio of the crest front to the crest rear steepness. For the Gaussian wave, the former has a value of 0.5 while the latter is 1.

5.3.3 Group Statistics

Elgar, et al. (1985) simulated 100 time series for various target spectra by the random phase and random coefficient methods. The purpose was to study the difference between the two methods in the computation of certain statistical properties including properties of wave groups. Each time series was 8192 seconds long and covered frequency ranges between 0.04 and 0.3 Hz. Therefore, the frequency resolution of the target spectra was 1.22 x 10^{-4} Hz and number of random phases was 2130 (or twice the number of coefficients). The mean wave period was 10 seconds. This gives 800

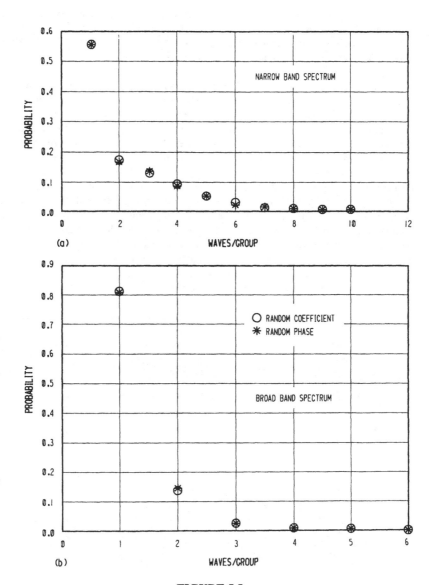

FIGURE 5.5
**EXAMPLE PROBABILITY DISTRIBUTION OF NUMBER OF WAVES PER
GROUP [Elgar, et al.(1985)]**

waves per time series and 80,000 waves per target spectrum for each simulation scheme. A group of high waves was defined as the group of the crest to trough heights exceeding the significant wave height. The number of groups in each time series was between 30 and 100 and mean length of runs was between 1 and 2.5.

The statistical quantities were averaged over the 100 realizations for each target spectrum. The mean length of runs and the variance of run lengths were found to be well correlated between the two simulation schemes. The estimated probability function of the number of waves per group produced by the random phase and random coefficient methods were also compared for each target spectrum. The results shown for a narrow-band and a broad-band spectrum in Fig. 5.5 are nearly identical. Thus, for a large number of components (e.g., 2130 random phases), the (low-order) statistical quantities discussed above have a negligible effect from either of the simulation schemes. The effect of spectral shape and the number of spectral components on the ensemble average mean length of runs produced by the two simulations is shown in Fig. 5.6. The percent difference is obtained as (random phase run length - random coefficient run length)/random phase run length * 100. For the broad-band spectrum, the mean lengths of runs by the two simulations are nearly identical even at the low number of wave components (= 66). For a narrow-band spectrum, the difference increases as the number of wave components decreases with a maximum difference of 12 percent at N = 66.

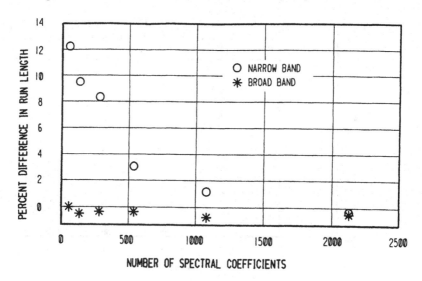

FIGURE 5.6
DIFFERENCE IN LENGTHS OF RUN FOR TWO SIMULATION SCHEMES
[Elgar, et al. (1985)]

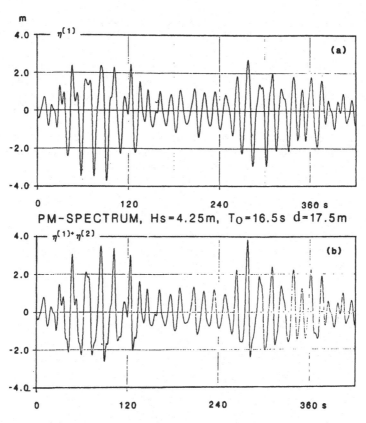

FIGURE 5.7
WAVE TRAIN SYNTHESIZED FROM PM SPECTRUM; (a) LINEAR (b)
SECOND-ORDER [Sand and Mansard (1986)]

5.4 HIGHER HARMONIC WAVES

A natural wave train is composed of numerous frequency components. In a second-order description of these waves, two of the frequency components, say f_1 and f_2, will interact to produce low and high frequency second-order components at frequencies $f_1 - f_2$ and $f_1 + f_2$, respectively. For a single sinusoid of frequency f_1, the second-order component appears at twice the frequency, $2f_1$. This component, known as bound waves, travels at the same speed as the basic wave. It modifies the regular wave by introducing a higher, sharper crest and a flatter trough. For an irregular wave train, a similar observation has been made. For example, in Fig. 5.7 taken from Sand

and Mansard (1986), the crests are sharper and troughs flatter in the second-order wave train compared to the linear wave train.

In reproducing a second-order wave train either for a numerical or a physical model study, the first-order waves must be combined with their bound (or locked) higher order components. In the case of physical models, appropriate transfer functions which can eliminate the generation of spurious waves must also be applied.

The transfer function is obtained from the solution of the Laplace equation with the second-order surface condition for the first-order input wave train. For an irregular wave of N frequencies, the first-order profile is given by

$$\eta(t) = \sum_{n=1}^{N} \left[a_n \cos(\omega_n t - k_n x) + b_n \sin(\omega_n t - k_n x) \right] \tag{5.18}$$

The higher harmonic contribution for this wave record has the form:

$$\eta^{(2)}(t) = \sum_{n=1}^{N} \sum_{m=1}^{N-n} \left[\begin{array}{l} (a_n a_m - b_n b_m) G_{nm}^+ \cos\left[(\omega_n + \omega_m)t - (k_n + k_m)x \right] \\ + (a_n b_m + b_n a_m) G_{nm}^+ \sin\left[(\omega_n + \omega_m)t - (k_n + k_m)x \right] \end{array} \right] \tag{5.19}$$

where $G^+{}_{nm}$ is the transfer function for the wave frequencies $\omega_n + \omega_m$.

$$G_{nm}^+ d = \delta \left\{ \left[2(\alpha_n + \alpha_m)^2 \left(k_n dk_m d / (\alpha_n \alpha_m) - 4\pi^2 \alpha_n \alpha_m \right) + (\alpha_n + \alpha_m) \right. \right.$$

$$\left(k_m^2 d^2 / \alpha_m + k_n^2 d^2 / \alpha_n - 4\pi^2 (\alpha_n^3 + \alpha_m^3) \right) \right] / 4\pi (\alpha_n + \alpha_m)^2$$

$$\left. -\Omega + \left[4\pi^2 \alpha_n^2 \alpha_m^2 (1 + \alpha_n / \alpha_m + \alpha_m / \alpha_n) - k_n dk_m d \right] / 4\pi \alpha_n \alpha_m \right\} \tag{5.20}$$

in which $\Omega = 2(k_n + k_m)d \tanh (k_n + k_m)d$, $\alpha_m = (d/L_0, m)^{1/2}$, $\alpha_n = (d/L_0, n)^{1/2}$, d is the water depth, and L_0 is the deep water wave length. The value of $\delta = 1$ for $n \neq m$ and $= 1/2$ for $n = m$. The transfer function has been given by Hudspeth and Chen (1979). For $n = m$, it reduces to Stokes second-order solution. The transfer function $G^+{}_{nmd}$ is plotted in Fig. 5.8 as a function of d/L_0 and the ratio ω_n/ω_m. When a regular wave of single frequency is present, $\omega_n/\omega_m = 1$ and

$$\eta^{(2)}(t) = a^2 G_{nm}^+ \cos 2\omega t \tag{5.21}$$

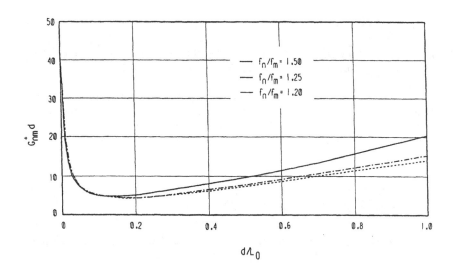

FIGURE 5.8
TRANSFER FUNCTION $G^+_{nm}d$ FOR A PAIR OF WAVE COMPONENTS AS FUNCTION OF d/L_0 [Sand, et al. (1988)]

An approximate expression for G^+_{nm} in this case [Sand, et al. (1988)] is

$$G^+_{nm} = \begin{cases} 3L_0 / (8\pi d^2) & \text{for } d / L_0 \leq 0.11 \\ 1.1 / d & \text{for } 0.11 < d / L_0 < 0.35 \\ \pi / L_0 & \text{for } d / L_0 \geq 0.35 \end{cases} \qquad (5.22)$$

For example, in shallow water where $d/L_0 = 0.05$ (e.g., T = 14 sec, d = 15.5 m) and for a regular wave of height H = 2 m, the second-order wave amplitude is

$$a^{(2)} = 1^2 \cdot \frac{3L_0}{8\pi d^2} = 0.15m \qquad (5.23)$$

Similarly, in deep water where $d/L_0 = 0.5$ (e.g., T = 8 sec, d = 50 m), and for a wave height of 2 m, the second-order amplitude becomes

$$a^{(2)} = 1^2 \cdot \pi / L_o = 0.03m \qquad (5.24)$$

Thus, in the shallow water example, the crest and trough levels are +1.15 m and -0.85 m; while in deep water, the crest and trough levels are +1.03 m and -0.97 m.

Because of numerous possible pairs resulting from N components of different frequencies, a large number of contributions are expected in the second-order wave in an irregular wave. Assuming that the components are ordered as f_1, f_2,...f_N, the number of contributions at frequency f_n is $n/2 - [1 - (-1)^n]/4$. The total number of contributions N_T for N frequencies is

$$N_T = \sum_{n=2}^{N} \left\{ n/2 - \left[1 - (-1)^n\right]/4 \right\} = N^2/4 - \left[1 + (-1)^{N-1}\right]/8 \qquad (5.25)$$

Thus, if $N = 128$, then $N_T = 4096$. In practice, the number of contributions is reduced by neglecting interactions from frequency components that contain little energy.

The energy density spectrum of the second-order wave components in the irregular wave is obtained in terms of the transfer function as follows:

$$S^{(2)}(f) = 2 \int_0^{f/2} S^{(1)}(f') S^{(1)}(f - f') \left[G^+(f, f')\right]^2 df' \qquad (5.26)$$

However, it should be noted that the total spectrum is not the sum of $S^{(1)}(f)$ and $S^{(2)}(f)$ because of the relative phase among components, and must be computed from the ensuing time series. The spectra corresponding to the first and second-order time series shown in Fig. 5.7 are shown in Fig. 5.9. A plot of the spectrum of high harmonics is also presented in this figure at an expanded scale, for easy comparison. Note that the increase in the high frequency region is mainly found near $2f_0$ where f_0 is peak frequency.

The second-order transfer functions for a JONSWAP ($\gamma = 7$) as well as a PM spectrum are shown in Fig. 5.10.

As previously shown, the second-order waves produce a high frequency and a low frequency solution. The bounded long (low-frequency) waves have a large effect on the motions of floating structures (Fig. 5.2), e.g., semisubmersibles, single-point mooring systems, tension leg platforms, etc. Similarly, the higher harmonic components can influence the tendon springing loads in a TLP. Moreover, there is a term in the high frequency wave component which decays with water depth slower than a cosh function. Thus, in a greater depth where the first order components disappear, this second-order contribution may be dominant. For a large-based fixed gravity structure or a deepwater spar the presence of this component will influence the model test results.

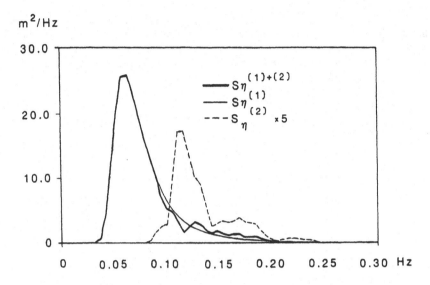

FIGURE 5.9
WAVE SPECTRA INCLUDING FIRST AND SECOND ORDER COMPONENTS
[Sand and Mansard (1986)]

For the generation of nonlinear waves in the laboratory, having a correct control signal is critical. This signal for a hinged or piston type wavemaker may be calculated from the complete boundary value problem including the wavemaker boundary condition by a perturbation technique which has been described for a linear wave in Chapter 4.

5.5 GENERATION OF MULTI-DIRECTIONAL WAVES

Wind-generated ocean waves (Fig. 5.11) can be best described by multi-directional waves. The number of laboratories capable of modeling short-crested sea states which are characteristic of the ocean environment is still limited but increased rapidly in the 1980's. Of these facilities, those most noteworthy are the facilities at the University of Edinburg in 1978, Swedish Hydrodynamic Laboratory (SSPA) in 1979, the Hydraulic Research Station, Wallingford, U.K., in 1980, the Norwegian Hydrodynamic Laboratory (MARINTEK) in 1981, the Danish Hydraulic Institute in 1984, the National Research Council of Canada Hydraulic Laboratory in 1986, Canadian Laboratory in New Foundland in 1989, Oregon State University Laboratory in

1990 and Texas A&M Laboratory in 1991. These laboratories are equipped with segmented wave generators which may be independently placed and operated to produce random multi-directional waves at the test site in the wave basin.

The generation of multi-directional waves is similar to that of 2-D waves except that the spreading of the wave components from different directions must be taken into account. Because the wave components arrive from the segmented generators at different angles, the simulation in the tank is valid only over a limited region in the wave tank.

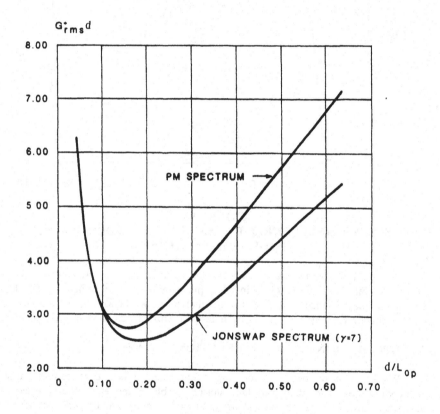

FIGURE 5.10
TRANSFER FUNCTION VERSUS RELATIVE DEPTH FOR A PM AND A
JONSWAP SPECTRUM [Sand and Mansard (1986)]

FIGURE 5.11
WIND GENERATED SEA

5.5.1 Procedure for Simulation of Sea State

The steps taken in generating a multi-directional sea [Sand (1980), Cornett and Miles (1990)] in a wave basin are as follows:

1. The sea state is described in terms of a specified frequency spectrum and a specified directional spreading function.

2. Based on the physical constraints of the wave basin, the specified frequency spectrum and spreading function are modified. For example, the wave basin generally has limitations in the generation of wave frequency ranges and a maximum wave angle at a given wave frequency. The resultant frequency spectrum and spreading function are designated as the "target" functions for the generated sea state.

3. The transfer function for the wave generator is used to compute the drive signal of the wave machine (segments) using linear theory to reproduce the target sea state at the test location.

4.   The wave field is measured at the test site using either an array of wave probes or coincident wave probe-orthogonal velocity meter combination.

5.   The measured wave data is analyzed using a Maximum Entropy Method (MEM) or a Maximum Likelihood Method (MLM) to estimate the directional wave spectrum from the test site [see Chapter 10].

6.   The measured and target directional wave spectra are compared, and an iterative process is used until the desired accuracy in the wave simulation is reached.

5.5.2 *Multi-Directional Generation Theory*

The mathematical description of the directional sea is based on waves incident from multiple directions at the point under consideration. If Eq. 5.18 is generalized to include the direction of waves, then the expression for the wave profile in the x, y plane becomes

$$\eta(x, y, t) = \mathrm{Re}[a \exp\{i(kx \cos\theta + ky \sin\theta - \omega t + \varepsilon)\}] \tag{5.27}$$

where θ is the direction of wave measured counter clockwise from the positive x axis. An irregular, short-crested wave train is obtained by linearly superimposing regular long-crested waves of different frequencies and directions.

$$\eta(x, y, t) = \mathrm{Re}\left[\sum_{j=1}^{\infty} a_j \exp\left\{i\left(k_j x \cos\theta_j + k_j y \sin\theta_j - \omega_j t + \varepsilon_j\right)\right\}\right] \tag{5.28}$$

The amplitudes a_j are complex, and the associated phases are assumed to be randomly distributed between 0 and 2π. The directional wave spectrum is given as $S(\omega, \theta)$ where the one-dimensional spectrum is derived from

$$S(\omega) = \int_{-\pi}^{\pi} S(\omega, \theta) d\theta \tag{5.29}$$

It is convenient and often customary to separate out the frequency and directional dependence of the energy spectrum as

$$S(\omega, \theta) = S(\omega) D(\omega, \theta) \tag{5.30}$$

where $D(\omega, \theta)$ is the (non-negative) directional spreading function. It is obvious from Eqs. 5.29 and 5.30 that

$$\int_{-\pi}^{\pi} D(\omega,\theta)d\theta = 1 \tag{5.31}$$

One of the most common forms of the directional spreading functions is the frequency dependent cosine power function which is considered non-zero over $\pm \pi/2$:

$$D(\omega,\theta) = C(s)\cos^{2s}(\theta-\theta_0) \quad \text{for} |\theta-\theta_0| < \pi/2 \tag{5.32}$$

where θ_0 is the principal direction of wave having a frequency ω, and s is the spreading width index. The value of the coefficient C(s) should be such that

$$C(s) = \frac{\Gamma(s+1)}{\Gamma(s+1/2)\sqrt{\pi}} \tag{5.33}$$

where Γ is the gamma function. Note that as the value of the spreading index s goes up, the spreading function becomes narrower. For $s \to \infty$, we obtain a long-crested wave. The standard deviation of wave direction, σ_θ, provides an alternate and more general measure of spreading width than s since it applies to any type of spreading function [Cornett and Miles (1990)]. For the relationship in Eq. 5.33, Fig. 5.12 shows the relationship between σ_θ and s.

Most earlier studies [Forristall (1981), Pinkster (1984)] considered the following model to simulate water surface elevation

$$\eta(x,y,t) = \sum_{i=1}^{N}\sum_{j=1}^{M} a_{ij}\cos\left[k_i\left(x\cos\theta_j + y\sin\theta_j\right) - \omega_i t + \varepsilon_{ij}\right] \tag{5.34}$$

where (a_{ij}, ε_{ij}) are amplitude-phase of wave components having frequencies, ω_i and directions, θ_j. However, for finite values of N and M, this wave form is neither ergodic nor homogeneous [Jeffreys (1987)]. Any two wave components with same ω_i but different θ_j are "phase-locked" (losing the randomness of the phase).

An alternate model uses a single direction for each frequency which produces a spatially homogeneous wave field

$$\eta(x,y,t) = \sum_{i=1}^{N} a_i\cos\left[k_i\left(x\cos\theta_i + y\sin\theta_i\right) - \omega_i t + \varepsilon_i\right] \tag{5.35}$$

where N is the total number of frequency components in the sea state.

The values of the wave amplitude, a_i, and corresponding phase angle, ε_i, may be computed by several methods as described for the unidirectional waves. In the random phase method, they are determined from

$$a_i = \sqrt{2S(\omega_i)D(\omega_i,\theta_i)\Delta\omega\Delta\theta}$$ (5.36)

$$\varepsilon_i = 2\pi U[0,1]$$ (5.37)

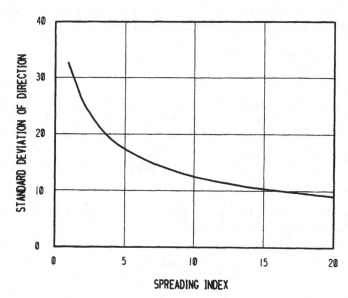

FIGURE 5.12
RELATIONSHIP BETWEEN σ_θ AND s
[Cornett and Miles (1990)]

where U is a uniform distribution function between 0 and 1. The chosen frequency range of the spectral energy density $S(\omega_i)$ is divided into N equal intervals of $\Delta\omega$. In each frequency band $\Delta\omega$, the directional spreading function is computed at M wave angles. Thus, there are M sub-frequencies within each frequency bandwidth $\Delta\omega$, each having a unique direction. The wave frequencies are therefore obtained from

$$\omega_i = i\frac{\Delta\omega}{M}$$ (5.38)

The angles for each frequency component may be random. As an alternative, they may be computed in a linear variation as

$$\theta_i = \theta_0 + (i-1)\Delta\theta - \theta_{max} \qquad (5.39)$$

where θ_{max} (e.g., $= 65°$) is the maximum angle of wave propagation relative to θ_0 and

$$\Delta\theta = \frac{2\theta_{max}}{(M-1)} \qquad (5.40)$$

If N and M are chosen to be powers of 2, the wave elevation time series may be obtained by an inverse FFT technique. The record length of the simulated time series is $T_R = N\Delta t$.

Equation 5.34 may be generalized by replacing ω_i by ω_{ij} and k_i by the corresponding k_{ij}. The wave field may be represented by L frequency bands with M wave components in each band. Considering also M uniformly spaced wave angles within each band, this equation may be written in a more convenient manner as

$$\eta(x,y,t) = \sum_{i=1}^{L}\sum_{j=1}^{M} a_{ij}\cos\left[\omega_{ij}t - k_{ij}\left(x\cos\theta_{ij} + y\sin\theta_{ij}\right) + \varepsilon_{ij}\right] \qquad (5.41)$$

where

$$\omega_{ij} = 2\pi f_{ij} = 2\pi\left[M(i-1)+j\right]\Delta f \qquad (5.42)$$

$$a_{ij} = \left[2S\left(f_{ij}\right)D\left(f_{ij},\theta_{ij}\right)M\Delta f\Delta\theta\right]^{1/2} \qquad (5.43)$$

$$L = N/M \qquad (5.44)$$

The wave angles increase linearly with frequence between the values $-\theta_{max}$ and $+\theta_{max}$.

Once the time history of the wave elevation is known, the water particle kinematics may be derived using appropriate transfer functions from linear wave theory.

For the simulation and generation of a multi-directional sea state, a sea state representing JONSWAP spectrum having $H_s = 12.2$ m (40 ft), $T_0 = 15$ sec and $\gamma = 2$ is chosen. The spreading width for T_0 (the peak period) is taken as $s = 4.5$. It is assumed to decrease linearly with frequency for $T < T_0$ until $s(f) = 0.2$. At lower periods, it is held constant at 0.2. The mean wave direction at T_0 is taken as $\theta_0 = 0$ and assumed to increase linearly with frequency at the rate $\partial\theta/\partial f$ = 423 deg/Hz. The specified directional sea is given in Fig. 5.13. This wave was simulated [Cornett and Miles (1990)] at the NRC Hydraulics Laboratory in a water depth of 1.55 m. The wave field was simulated using the random phase method and assuming one direction for each frequency. The wave elevation is given by the single summation model of Eq. 5.35.

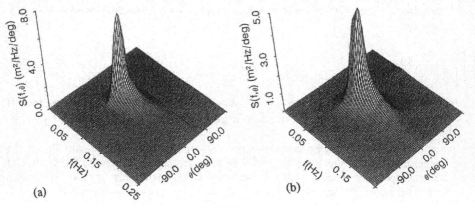

FIGURE 5.13
DIRECTIONAL SEA SPECTRUM [(a) TARGET (b) MEASURED] GENERATED IN A BASIN [Cornett and Miles (1990)]

In the mathematical simulation of the example spectrum, N = 2048, M = 32, L = 64, Δf = 0.0002778 Hz and θ_{max} = 65° which is limited by the wave generation capability of the basin. This gave a Δt = 0.8789 sec and the duration of 1 hr. full scale at a scale of 1:28. Each major frequency band, $M\Delta f$ = 0.008889 Hz.

5.6   GENERATION OF STEEP WAVES

The generation of extreme waves is important in the study of wave impact loads on fixed structures and the capsize tests of floating structures. For either a stationary or a moving model, it is frequently desired to make a steep wave break precisely at the location of the structure [Anderson and Johnson (1977)]. The method utilizing sweep frequency modulation to generate transient waves may be used for this purpose [Kjeldsen and Myrhaug (1979, 1981)].

Extreme waves may also be generated at a particular location through the use of the inverse Fourier Transform technique with predetermined phase control rather than randomly selecting the phases. In the simplest form of inverse FFT generation, the spectral densities of each component are specified. The group velocity of each component is computed in order to predict the appropriate initial phase relationships to produce an additive superposition of all frequency components at a particular location in the wave tank.

Davis and Zarnick (1964) developed a linear sweep technique to generate a transient wave train of decreasing wave frequency. Since the longer wave components have higher celerity, the wave train concentrates at a predetermined location of the wave tank and converges to its highest amplitude. This method was later improved by Takezawa, et al. (1976) who started with a desired wave spectrum. The input signal of the wavemaker was derived from the transfer function of the wave generator using inverse Fourier transformation. Therefore, the shape of the spectrum is well controlled under this technique.

One of the simplest methods for generating extreme waves [Longuet-Higgins (1976)] in deep water considers the frequency increment for the wave components to be given by

$$\Delta \omega = -\left(\frac{g}{2x}\right) t_s \qquad (5.45)$$

where t_s represents the time interval assigned to the particular wave component to arrive at the location x in the tank.. Using a sampling rate of Δt, the wave profile for a wave at a frequency ω_i is given by

$$\eta_i(t) = \frac{H}{2} \sin(\omega_i t + \varepsilon_i) \qquad (5.46)$$

where H is the wave height, t varies over the particular t_s and ω_i varies between an initial and a final frequency, ω_1 and ω_N, respectively. The phase angle ε_i is chosen such that the continuity in profile between the components is maintained.

$$\varepsilon_1 = 0 \qquad (5.47)$$

and

$$\varepsilon_{i+1} = \sin^{-1}\left(\frac{2\eta_i(t_i)}{H}\right) - \omega_{i+1} t_i \qquad (5.48)$$

Thus, the frequency of the harmonic wave components varies from ω_1 and ω_N while the wave amplitudes remain the same.

Mathematically, the wave profile may be expressed

$$\eta(x,t) = \text{Re}\left[iBG(\tau)\exp\left[i\left(\omega t - \omega^2 \, x/g\right)\right]\right]$$

(5.49)

where B is a constant (independent of time) and depends on the type of wavemaker used in generating waves.   For example, for a wedge-type wavemaker, B has the following form:

$$B = H\left[1 - \exp\left(-\omega^2 \, D/g\right)\right]\sin\beta$$

(5.50)

where the quantities D and β are the depth of the wedge and its angle at the bottom. The quantity τ and function G are given by

$$\tau = \left(\frac{g}{2\pi x}\right)^{1/2}\left(t - \frac{2\omega x}{g}\right)$$

(5.51)

$$G(\tau) = \frac{1}{2} + \frac{1}{\sqrt{2i}}\int_0^\tau \exp\left(i\pi\lambda^2/2\right)d\lambda$$

(5.52)

Note that when t = 0,

$$\tau = -\omega\left(\frac{2x}{\pi g}\right)^{\frac{1}{2}}$$

(5.53)

Instead of using a uniform wave amplitude having decreasing frequency, another technique [Salsich, et al. (1983)] consists of generating a drive signal which combines decreasing wave frequency with an exponentially increasing wave amplitude of specified form. As before, the resulting wave energy is focused at some point downstream of the wavemaker and results in a nonlinear, asymmetric and, possibly, breaking wave.

The group velocity of a wave in deep water is one half the phase velocity and is the speed at which the wave energy travels down the tank

$$c_{g_o} = \frac{c_0}{2} = \frac{g}{2\omega}$$

(5.54)

If there are two wave components present in a wave front with frequencies close to each other, then the two will meet at a distance x from the wave generator if

$$d\omega = -\frac{g}{2x}dt \tag{5.55}$$

where $d\omega$ is the frequency difference between the wave components and dt is the time lag between them at the generator. Assuming ω varies from ω_1 to ω_N at an increment of $d\omega$, the wave profile is given by

$$\eta_i(t) = \left(\frac{\omega_N}{\omega_i}\right)^{P_1} \exp\left[P_2 \frac{\omega_N - \omega_i}{\omega_N}\right] \sin \omega_i t_i \tag{5.56}$$

and

$$t_{i+1} = \frac{\omega_i}{\omega_{i+1}} t_i + \Delta t \tag{5.57}$$

where Δt is the sampling rate, P_1 and P_2 are the polynomial and exponential multipliers, respectively. Values of P_1 and P_2 have been recommended to be 5 and 3 respectively. The profile is created such that the maximum amplitude corresponds to the maximum generator voltage.

5.7 GAUSSIAN WAVE PACKET FOR TRANSIENT WAVES

Unlike random waves whose duration is generally several minutes (e.g., 5-10 mins) transient waves are necessarily short. Transfer functions may be equally derived from these transient waves. They have the added advantage of being less contaminated with (beach) reflection due to short duration. One such transient wave system is called Gaussian wave packet.

The Gaussian wave packet allows control of both the shape of the wave train and its spectrum. It is composed of an infinite number of harmonic components superimposed on each other. The wave profile is written [Kinsman (1965)] as

$$\eta(x,t) = \int_{-\infty}^{\infty} a(k) \exp[i(kx - \omega t)]dk \tag{5.58}$$

where the wave frequency ω is given by $\omega(k) = \sqrt{gk \tanh kd}$. The amplitude is wave number dependent, and has a Gaussian-shaped spectrum described by

$$a(k) = \frac{a_0}{\sigma_a \sqrt{2\pi}} \exp\left[-(k - k_0)^2 / 2\sigma_a^2\right] \tag{5.59}$$

where a_0 is the maximum wave amplitude and k_0 is the corresponding wave number. The quantity σ_a is the standard deviation of the amplitude spectrum, and it is called the form factor of the Gaussian wave packet. On substituting the value of a(k), the wave profile becomes

$$\eta(x,t) = \int_{-\infty}^{\infty} \frac{a_0}{\sigma_a\sqrt{2\pi}} \exp\left[-(k-k_0)^2/2\sigma_a^2\right] \exp\left[i(kx-\omega t)\right] dk \qquad (5.60)$$

This integral is evaluated by expanding $\omega(k)$ into a three-term Taylor series about $k = k_0$:

$$\omega(k) = \omega_0 + A(k-k_0) + \frac{1}{2}B(k-k_0)^2 \qquad (5.61)$$

where

$$\omega_0 = \omega(k_0), A = \frac{d\omega}{dt}\bigg|_{k=k_0}, \text{ and } B = \frac{d^2\omega}{dt^2}\bigg|_{k=k_0} \qquad (5.62)$$

On substitution of this expression for $\omega(k)$ and algebraic manipulation, the real and imaginary parts of $\eta(x, t)$ take the form [Clauss and Bergman (1986)]:

$$\begin{matrix} \text{Re} \\ \text{Im} \end{matrix}\{\eta(x,t)\} = a_0 S^{\frac{1}{4}} \exp\left[-\frac{\sigma_a^2 S}{2}(x-At)^2\right]$$

$$\begin{matrix} \cos \\ \sin \end{matrix}\left\{k_0 x - \omega_0 t + \frac{\tan^{-1}(-\sigma_a^2 Bt)}{2} + \frac{\sigma_a^4 BtS}{2}(x-At)^2\right\} \qquad (5.63)$$

in which the quantities A and B are obtained from the differentiation of the dispersion relation:

$$A = \frac{c_0}{2}\left[1 + \frac{2k_0 d}{\sinh(2k_0 d)}\right] \qquad (5.64)$$

$$B = c_0 d\left[\frac{1 - 2k_0 d \coth(2k_0 d)}{\sinh(2k_0 d)} - \frac{1}{2\sinh(2k_0 d)}\left(\frac{\sinh(2k_0 d)}{2k_0 d} - \frac{2k_0 d}{\sinh(2k_0 d)}\right)\right] \qquad (5.65)$$

and

$$S = \frac{1}{1 + \sigma_a{}^4 B^2 t^2} \tag{5.66}$$

$$c_0 = \left[\frac{g}{k_0} \tanh(k_0 d) \right]^{\frac{1}{2}} \tag{5.67}$$

Thus, the Gaussian wave packet is characterized by three parameters: the dominant wave period T_0 (corresponding to ω_0), the water depth d and the form factor σ_a. Note that as the value of σ_a decreases, the amplitude spectrum (Eq. 5.59) becomes sharper. A large form factor (standard deviation) produces a wide spectrum with quick dispersion of the wave packet.

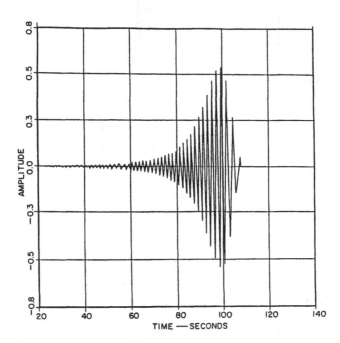

FIGURE 5.14
COMPUTER GENERATED INPUT SIGNAL TO WAVEMAKER

Equation 5.63 has three distinct parts. The exponential term is called a modulation function while the trigonometric function is an oscillation. The modulation function travels with the velocity A in the positive x-direction.   Therefore, A is interpreted as the group velocity of the Gaussian wave packet. The quantity B is referred to as the damping factor because of its role in the damping term which has a maximum value (of unity) at the wave concentration point t = 0. The amplitude increases up to this   point as the long wave components overtake the shorter ones. The amplitude decreases ·again beyond this point because of the passing of the long waves. The product of the damping term and the modulation function gives the envelope of the Gaussian wave packet.

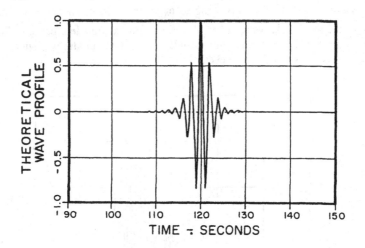

FIGURE 5.15
THEORETICAL WAVE TIME HISTORY AT DESIRED LOCATION

The generation of the actual wave groups requires modification of the input signal using the transfer function of the wave generator. Generation of a wave in a wave tank using this wave packet [Chakrabarti and Libby (1988)] is illustrated here. The middle of the tank (of 76 m length) is taken as the spatial reference point (x = 0). The water depth in the tank was set at 2.1 m. The central wave number (k_0) was chosen as 1.0 (1/m) and the form factor (σ_a) was taken as 0.1. The input signal to the wavemaker is shown in Fig. 5.14. The wavemaker in this case was a pneumatic type. Therefore, the transfer function between the analog input signal and the wave generated in the tank is expected to be nonlinear. Figure 5.15 shows the theoretical wave time history at the reference point in the tank. The actual recorded wave at this location in the tank is shown in Figure 5.16. Note that the measured wave is not as symmetric as the theoretical signal. The energy density spectra of the theoretical wave time history and the wavemaker input signal match quite well (Fig. 5.17). The energy density spectrum of the generated wave is given in Fig. 5.18.

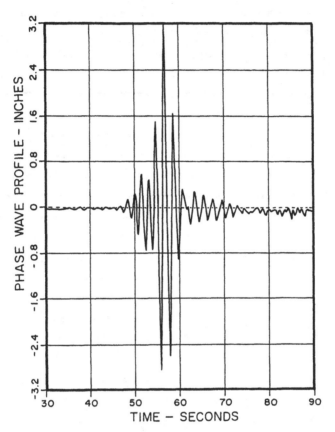

FIGURE 5.16
ACTUAL WAVE PROFILE RECORDED AT DESIRED LOCATION

5.8 CURRENT GENERATION

Techniques of current generation in a wave tank have already been described in Chapter 4. In the generation of current, it is generally assumed to be steady. The knowledge of any unsteadiness of the generated current is required and must be confirmed in order to establish its acceptability as a steady current. The generated current is assumed to be steady if the high frequency turbulence as well as low frequency oscillation are within acceptable limits. For acceptable steady current, the standard deviation is generally less than 10 percent of the mean value. Plots of mean current generated in a tank have been illustrated in Fig. 4.16. The standard deviations about the mean are also shown. Note that the coefficient of variations is generally less than 10 percent. Note, however, that if a current meter with a built-in low-pass filter is used, the filter eliminates the high frequency turbulence from the recording. If it is desired to quantify the high frequency turbulence level, velocity probes without built-in filters should be employed.

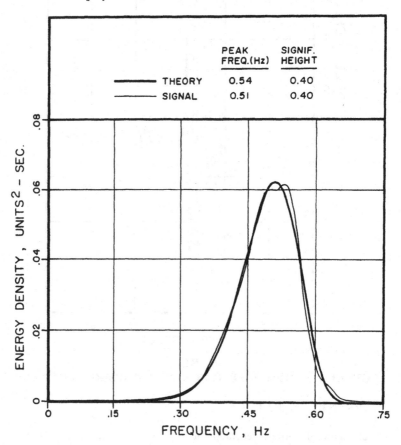

FIGURE 5.17
COMPARISON OF THEORY VS. SIGNAL SPECTRA

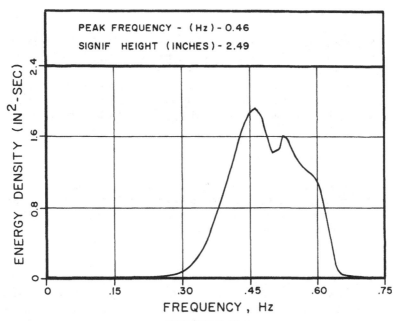

FIGURE 5.18

MEASURED WAVE ENERGY DENSITY SPECTRUM AT TEST SITE

The generation of current in a large wave tank is usually limited to studying the effect of waves and current together. If the effect of current alone on a submerged structure is desired, then facilities that can generate high speed uniform current should be sought. Several such facilities are available commercially. One facility that has many years of experience in studying current effect on structures in a natural flow is the hydraulic testing facility located at St. Anthony's Falls, Minnesota.

Steady current effect may also be simulated by towing a structure with the overhead carriage in a towing tank. It is assumed that the effect of flow past a stationary object is equivalent to towing the object at a steady speed through calm water except for the nature of turbulence present in the flow. In case of towing, the control and temporal variation of speed are better than those for current. However, the vibration of the carriage and the towing staff is introduced into the model, and should be carefully examined. Such vibrations may cause dynamic amplification if they coincide with the vortex shedding frequencies generated at the model.

5.8.1 Wave Current Interaction

When a steady current is superimposed on waves, the characteristics of the wave change. In particular, in the presence of current, the wave height and the wave length experience modification. For example, if the current is in the direction of wave propagation, the wave amplitude decreases and its length increases. On the other hand, if the current opposes the wave, the wave steepens with an increase in magnitude and decrease in length. These changes take place due to the interaction between the waves and current.

5.8.1.1 Interaction Theory

The subject of uniform steady current on waves in a constant depth has been investigated for many decades. The interaction theory of current on waves was first derived by Longuet-Higgins and Stewart (1961). The encounter frequency, in this case for waves colinear with current, is given by

$$\omega_e = \omega + Uk \tag{5.67}$$

where ω_e = encounter frequency, and U = current velocity. Note that the encounter frequency ω_e is greater than the wave frequency ω for positive value of current, U. This is the well-known Doppler's effect. A similar phenomenon is encountered by a ship moving at a constant speed in a seaway.

In deep water, certain simplifications of results are possible. In the presence of a uniform current, the wave number, k_e, is related to the wave frequency, ω, [see Mei (1983) for details] by the generalized dispersion relationship:

$$k_e = \frac{2\omega^2/g}{\left[1 + \left(1 + 4U\,\omega/g\right)^{1/2}\right]} \tag{5.68}$$

where U may be positive (in the direction of wave propagation) or negative (opposite to the wave direction). Note that the expression in Eq. 5.68 reduces to the deep water dispersion relation $(k = \omega^2/g)$ in the absence of current $(U = 0)$. When U is positive, the value of k_e is smaller so that the wave length is larger. Likewise, when U is negative, the value of k_e increases, and the wave length is smaller than when no current is present.

The wave amplitudes are also modified in the presence of current. For colinear currents in deep water, a rather simple expression may be derived. If a_e is the modified amplitude, then

$$a_e = \frac{2a}{\left(1+4U\omega/g\right)^{1/4}\left[1+\left(1+4U\omega/g\right)^{1/2}\right]}$$ (5.69)

where a is the deep water wave amplitude in the absence of current. Note that for U = 0, the above expression reduces to a_e = a. The change in the wave length and amplitude has been provided graphically by Brevik and Aas (1980).

From Eqs. 5.68-5.69, the effect of current on deep water random waves may be determined. In particular, a simple expression may be derived for the wave energy density spectrum in terms of the square of the transfer function shown in Eq. 5.69. It has been shown by Huang, et al. (1972) that the modified wave energy density spectrum in the presence of steady current in deep water may be written as

$$S^*(\omega) = \frac{4S(\omega)}{\left(1+\dfrac{4U\omega}{g}\right)^{1/2}\left[1+\left(1+\dfrac{4U\omega}{g}\right)^{1/2}\right]^2}$$ (5.70)

where S(ω) is the wave energy density spectrum in the absence of current and asterisk denotes the modified form of the spectrum in current. The spectra of fluid particle velocity and acceleration at an elevation y measured negatively below the still water level (y = 0) are given by

$$S_u^*(\omega) = \omega^2 e^{key} S^*(\omega)$$ (5.71)

$$S_{\overset{*}{u}}^{\cdot}(\omega) = \omega^4 e^{key} S^*(\omega)$$ (5.72)

Once the modified spectra of wave kinematics are known, the fluid forces on an offshore structure, e.g., through the application of the Morison equation, may be determined [Tung and Huang (1973)].

As in the case of regular waves, Eq. 5.70 suggests that the energy S*(ω) will tend to decrease in the presence of in-line current and will increase in magnitude with opposing current. Equation 5.70 also shows that S*(ω) will become infinite when ω = -g/(4U). However, the wave components which propagate with current may also be influenced by breaking in this range. Hedges, et al. (1979) showed that an equilibrium

range of spectrum will be achieved and developed an expression for the constraint
for limiting the values of the energy density spectra given by

$$S_c^*(\omega) = \frac{A^* g^2}{\omega^5} \frac{1}{\left(1 - \frac{\omega U}{g}\right)^5 \left[1 + \frac{2\omega U}{g}\left(1 - \frac{\omega U}{g}\right)\right]}$$
(5.73)

where A* = constant. This expression reduces to Phillip's equilibrium range form for
the no-current case for which the range of A* is 0.008 - 0.015. Hedges, et al. suggested
that Eq. 5.73 will replace Eq. 5.70 whenever $S_c^*(\omega)$ is less than $S^*(\omega)$.

Hedges et al. (1985) used this modified form in correlating wave and water
particle velocity data measured in a wave flume. The value of A* was derived by a
least-squares fit to the observed data. They found that the equilibrium range constraint
provides a good fit to the data, provided a suitable value of A* is chosen.

5.8.1.2 Combined Wave-Current Tests

While theory on wave-current interaction including random waves is shown, the
corresponding experimental data are limited. Brevik and Aas (1980) performed
experiments in the laboratory with regular waves which provided good correlation
between the theory and experiment. However, test data on random waves in the
presence of current and the corresponding correlation of wave kinematics are limited.
A few such correlations from the data generated on random wave-current interaction in
deep water are illustrated here.

Regular waves were generated at a period of 1.25 sec with and without opposing
current. The plots of these data are given in Fig. 5.19. The mean current is evident in
the horizontal velocity measurement. The wave without current has a mean amplitude
(over 10 cycles) of 70 mm (2.76 in.) while in current, the amplitude changed to a_e = 86.9
mm (3.42 in.) For a mean current speed of U = -0.15 m/s (-0.48 ft/sec) and a period of
T = 1.25 sec, the transfer function from Eq. 5.69 has a value of 1.256. Then, the
computed amplitude becomes 88.1 mm (3.47 in.) which is within 1.5 percent of the
measured value.

Random waves with or without current were generated in the tank. The current
was colinear with waves. The water depth was maintained at 3m (10 ft). Therefore,
considering the frequency range of the waves generated, it may be considered deep
water.

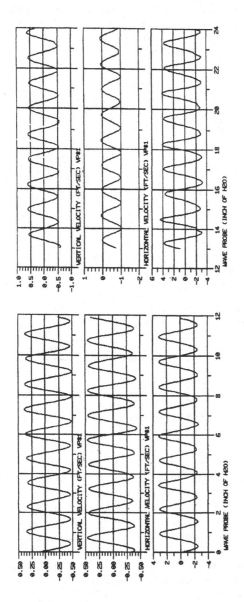

FIGURE 5.19

LINEAR WAVE ALONE AND IN THE PRESENCE OF OPPOSING CURRENT

The random wave modeled a JONSWAP spectrum with a peaked parameter, γ of 3.3. The analog driving signal was based on 100 harmonic wave components. The comparison of the JONSWAP model and the measured wave spectrum in the absence of current is shown in Fig. 5.20. The significant wave height of 84.3 mm (3.32 in.) is matched by the generated wave within 2 percent.

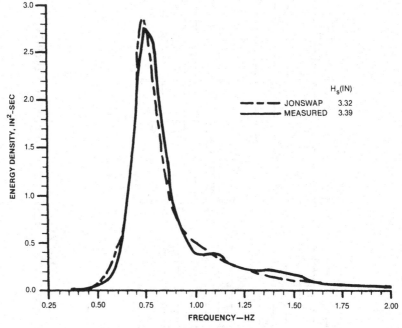

FIGURE 5.20

CORRELATION OF JONSWAP SPECTRUM WITH MEASURED SPECTRUM OF WAVES

The same random wave was repeated in the presence of current. The colinear current in the wave direction had a time-averaged speed of 0.16 m/s (0.52 ft/sec). The temporal variation from this mean value was limited to about 5 percent. The comparison of the modified wave energy density spectrum with the computed spectrum from Eq. 5.70 is shown in Fig 5.21. There is a small shift in the peak frequency, but the frequency resolution in the spectrum estimated was not fine enough. The measured significant wave height was within 5 percent of that predicted. The case of opposing current is taken from Hedges, et al. (1985). The strength of current was 0.2m/s (0.66 ft/sec) on a PM spectrum. The wave spectra with and without current are shown in Fig. 5.22.

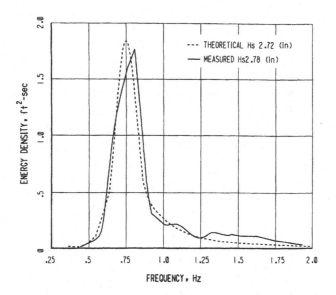

FIGURE 5.21
THEORETICAL VS. MEASURED WAVE ENERGY DENSITY SPECTRUM IN COLINEAR CURRENT

The modified spectra in the presence of current are computed based on measured spectra from Eqs. 5.70 and 5.73. The value of A* in Eq. 5.73 is taken as 0.0314. According to Eq. 5.70, the waves in opposing current continue to rise at high frequency components (dotted line in Fig. 5.22). However, measurement as well as Eq. 5.73 show that the waves at these frequency components must break and reform showing much lower energy contents. The comparison here is generally good except that the measured data reduces in magnitude faster at high frequencies.

5.9 WIND GENERATION

The wind loading on an offshore structure is best determined in a wind tunnel where the wind speed and turbulence may be properly controlled. This area has been covered briefly in Chapter 7. However, it is a common practice to simulate wind on the superstructure of a moored floating vessel subjected to waves. Common methods of generating wind in a wave tank environment have been described in Chapter 4.

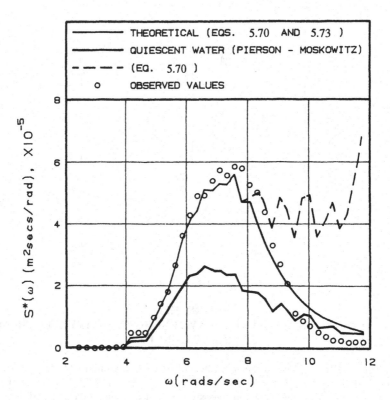

FIGURE 5.22
THEORETICAL VS. MEASURED WAVE ENERGY DENSITY SPECTRUM IN OPPOSING CURRENT [Hedges, et al. (1985)]

Often, the wind load is simulated by a fixed weight imposed on the floating structure with the use of strings and pulleys. However, several possible problems may exist in such a simulation, including friction in the system.

In the simulation of actual wind in the model, the Reynolds number for wind on the superstructure is not reproduced. Because of the nature of the wind generation, the wind at the vessel will be quite turbulent. Sometimes, the calculated mean wind load is scaled using Froude's law, and the mean load rather than the wind speed is reproduced in the model by adjusting the location and speed of fans as well as the placement of the superstructure.

The effect of the frequency range in the generated wind is investigated in model tests. The random wind covers a wide band of frequencies. This frequency band may have a large influence on the moored structures, and the combined effect of wind with wave is an important modeling criterion.

There are several empirical models available to represent wind spectra. The frequency in the spectrum may range from 0.005 to 1 Hz. The proposed models generally agree at the high frequency range (≥ 0.05 Hz). However, at the low end, the models differ substantially. A recent model proposed by Ochi and Shin (1988) considers the average of many offshore data and is given by

$$\hat{S}\left(\hat{\omega}\right) = \begin{cases} 92.8\hat{\omega} & \text{for } 0 < \hat{\omega} \leq 0.019 \\[3mm] \dfrac{116\hat{\omega}^{0.70}}{\left(1+0.53\hat{\omega}^{0.35}\right)^{11.5}} & \text{for } 0.019 \leq \hat{\omega} \leq 0.628 \\[3mm] \dfrac{133.4\hat{\omega}}{\left(1+0.53\hat{\omega}^{0.35}\right)^{11.5}} & \text{for } \hat{\omega} > 0.628 \end{cases}$$

(5.74)

in which the nondimensional quantities $\hat{\omega}$ and $\hat{S}\left(\hat{\omega}\right)$ are given by

$$\hat{\omega} = \frac{10\omega}{U_w}$$

(5.75)

and

$$\hat{S}\left(\hat{\omega}\right) = \frac{\omega S_w(\omega)}{C_w U_w}$$

(5.76)

where S_w is the wind spectrum, U_w is the mean wind speed at an elevation of 10m (33 ft) above sea level and C_w the corresponding surface drag coefficient. The spectral model is plotted in nondimensional form in Fig. 5.23. In the lower frequency range ($\omega \leq 0.019$) where the moored structure will experience the largest effects, the spectral distribution is linear, and approaches a value of zero. At the higher end of the frequency the wind may excite the vibrational frequency mode of the structure. The frequency spectrum model for wind is generally not reproduced in a wave tank test, even though for certain structure models, it may be important.

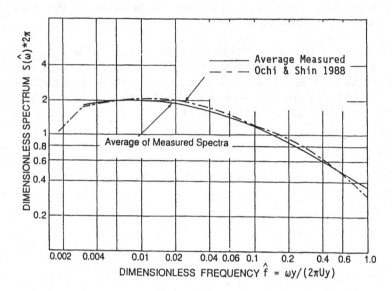

FIGURE 5.23
OFFSHORE WIND SPECTRUM MODEL
[Ochi and Shin (1988)]

5.10   REFERENCES

1.   Anderson, C.H. and Johnson, B., "An Algorithm for Predicting Breaking Waves in a Towing Tank", Proceedings of Eighteenth American Towing Tank Conference, Annapolis, Maryland, August 1977.

2.   Battjes, J.A. and Vledder, V., "Verification of Kimura's Theory for Wave Group Statistics", Proceedings on Nineteenth International Conference on Coastal Engineering, ASCE, 1984, pp. 642-648.

3.   Borgman, L.E., "Ocean Wave Simulation for Engineering Design", Journal of the Waterways and Harbors Div., ASCE, Vol. 95, No. WW4, November 1969, pp. 557-583.

4.   Borgman, L.E., "A Technique for Computer Simulation of Ocean Waves", Submitted to ASCE as part of a report by the Committee on the Reliability for Offshore Structures, 1979.

5.   Brevik, I. and Aas, B., "Flume Experiments on Waves and Currents I. Rippled Bed", Coastal Engineering, Vol. 3, 1980, pp. 149-177.

6.   Chakrabarti, S.K., Snider, R.H., and Feldhausen, P.H., "Mean Length of Runs of Ocean Waves", Journal of Geophysical Research, Vol. 79, No. 36, December 1974.

7.   Chakrabarti, S.K., "Laboratory Generated Waves and Wave Theories", Journal of the Waterway, Port, Coastal and Ocean Division, ASCE, No. WW3, August, 1980, pp. 349-367.

8.   Chakrabarti, S.K. and Libby, A.R., "Further Verification of Gaussian Wave Packets", Applied Ocean Research, Vol. 10, No. 2, 1988, pp. 106-108.

9.   Chakrabarti, S.K., Nonlinear Methods in Offshore Engineering, Elsevier Publishers, Netherlands, 1990.

10.   Chakrabarti, S.K., Hydrodynamics of Offshore Structures, 2nd Edition, Computational Mechanics Publications, Southampton, U.K., 1993.

11.   Clauss, G.F. and Bergman, J., "Gaussian Wave Packets - A New Approach to Seakeeping Tests of Ocean Structures", Applied Ocean Research, Vol. 8, No. 4, 1986, pp. 190-206.

12.   Cornett, A. and Miles, M.D., "Simulation of Hurricane Seas in a Multidirectional Wave Basin", International Conference on Offshore Mechanics and Arctic Engineering, Houston, Texas, February, 1990, pp. 17-25.

13.   Davis, M.C. and Zarnick, E.E., "Testing Ship Models in Transient Waves", Proceedings of Fifth Symposium on Naval Hydrodynamics, 1964, pp. 507-543.

14.   Elgar, S., Guza, R.T. and Seymour, R.J., "Wave Group Statistics From Numerical Simulations of a Random Sea", Applied Ocean Research, Vol. 7, No. 2, 1985, pp. 93-96.

15.   Forristall, G.Z., "Kinematics of Directionally Spread Waves", Proceedings of Conference on Directional Waves Spectra Applications, University of California, Berkeley, 1981, pp. 129-146.

16.   Funke, E.R. and Mansard, E.P.D., "On the Synthesis of Realistic Sea States", Hydraulics Laboratory Technical Report LTR-HY-66, National Research Council of Canada, 1979.

17.   Funke, E.R. and Mansard, E.P.D., "The Control of Wave Asymmetries in Random Waves", Proceedings of Eighteenth Conference on Coastal Engineering, Cape Town, South Africa, 1982.

18.   Goda, Y., "Numerical Experiments on Wave Statistics with Spectral Simulation", Report, Port and Harbour Research Institute, Vol. 9, No. 3, 1970.

19.   Goda, Y., "Numerical Experiments on Statistical Variability of Ocean Waves", Report of the Port and Harbour Research Institute, Yokusuka, Japan, Vol. 16, No. 2, June 1977, pp. 3-26.

20.   Gravesen, H., Frederiksen, E. and Kirkegaard, J., "Model Tests With Directly Reproduced Nature Wave Trains", Proceedings of Fourteenth Coastal Engineering Conference, Copenhagen, Denmark, June, 1974, pp. 372-385.

21.   Hedges, T.S., Burrows, R. and Mason, W.G., "Wave-Current Interaction and Its Effect on Fluid Loading", Report No. MCE/3/79, Department of Civil Engineering, University of Liverpool, December, 1979.

22.   Hedges, T.S., Anastasiou, K. and Gabriel, D., "Interaction of Random Waves and Currents", Journal of Waterway, Port, Coastal and Ocean Engineering, ASCE, Vol. 111, No. 2, March, 1985, pp. 275-288.

23.   Huang, N.E., Chen, D.T., and Tung, C.C., "Interactions Between Steady Non-Uniform Currents and Gravity Waves with Applications for Current Measurements", Journal of Physical Oceanography, Vol. 2, 1972, pp. 420-431.

24.   Hudspeth, T.R. and Chen, M.C., "Digital Simulation of Non-Linear Random Waves", Journal of the Waterway, Port, Coastal and Ocean Division, ASCE, Vol. 105, WW1, February, 1979, pp. 67-85.

25.   Hudspeth, R.T. and Medina, J.R., "Wave Groups Analyses by the Hilbert Transform", Proceedings on Twenty-First International Conference on Coastal Engineering, ASCE, 1988, pp. 884-898.

26.   Jeffreys, E.R., "Directional Seas Should Be Ergodic", Applied Ocean Research, Vol. 9, No. 4, 1987, pp. 186-191.

27. Johnson, B., "A State-of-the-Art Review of Irregular Wave Generation and Analysis", Proceedings of Sixteenth International Towing Tank Conference, Leningrad, USSR, U.S. Naval Academy Report No. EW-9-82, September, 1981.

28. Kimura, A. and Iwagaki, Y., "Random Wave Simulation in a Laboratory Wave Tank", Proceedings of Fifteenth Coastal Engineering Conference, Honolulu, Hawaii, July, 1976, pp. 368-387.

29. Kimura, A., "Statistical Properties of Random Wave Groups", Proceedings on Seventeenth International Conference on Coastal Engineering, ASCE, 1980, pp. 2955-2973.

30. Kinsman, B., <u>Wind Waves, Their Generation and Propagation on the Ocean Surface</u>, Prentice-Hall, Inc., Englewood Cliffs, New Jersey, 1965.

31. Kjeldsen, S.P. and Myrhaug, D., "Breaking Waves in Deep Water and Resulting Wave Forces", Proceedings of the Eleventh Offshore Technology Conference, OTC 3646, Vol. 4, 1979, pp. 2515-2522.

32. Kjeldsen, S.P., "Methods, Difficulties and Standards for Producing Breaking Waves in Model Basins", Contribution to the Seakeeping Committee Report, Sixteenth International Towing Tank Conference, Leningrad, USSR, 1981.

33. Longuet-Higgins, M.S., "Breaking Waves in Deep and Shallow Water", Proceedings of Tenth Conference on Naval Hydrodynamics, MIT, Boston, Massachusetts, June, 1976, pp. 597-695.

34. Longuet-Higgins, M.S., "Statistical Properties of Wave Groups in a Random Sea State", Philosophical Transactions of Royal Society of London, Series A, Vol. 312, 1984, pp. 219-250.

35. Longuet-Higgins, M.S., and Stewart, R.W., "The Changes of Amplitude of Short Gravity Waves on Steady Non-Uniform Currents", Journal of Fluid Mechanics, Cambridge, England, Vol. 10, No. 4, June 1961, pp. 529-549.

36. Loukakis, T.A., "Random Sea Generation by the 'White' Noise Technique", Fifteenth American Towing Tank Conference, Ottawa, Canada, 1968.

37. Mansard, E.P.D. and Funke, E.R., "Selected Papers on Two-Dimensional Wave Generation and Analysis", Report NRCC No. 28750, Hydraulics Laboratory, National Research Council of Canada, Ottawa, Canada, 1988.

38.   Medina, J.R. and Diez, J.J., "Comparisons of Numerical Random Sea Simulations", Discussion, Journal of Waterway, Port, Coastal and Ocean Division, ASCE, Vol. 110, No. 1, 1984, pp. 114-116.

39.   Medina, J.R., Aguilar, J. and Diez, J.J., "Distortions Associated With Random Sea Simulators", Journal of Waterway, Port, Coastal and Ocean Division, ASCE, July, 1985, pp. 603-628.

40.   Medina, J.R. and Hudspeth, R.T., "A Review of the Analyses of Ocean Wave Groups", Coastal Engineering, Vol. 14, 1990, pp. 515-542.

41.   Mei, C.C., The Applied Dynamics of Ocean Surface Waves, John Wiley & Sons, New York, New York, 1983.

42.   Ochi, M.K., and Shin, Y.S., "Wind Turbulent Spectra for Design Consideration of Offshore Structures", Proceedings of the Twentieth Annual Offshore Technology Conference, OTC 5736, Houston, Texas, May 1988, pp. 461-467.

43.   Pinkster, J.A., "Numerical Modelling of Directional Seas", Proceedings of Symposium on Description and Modelling of Directional Seas, Copenhagen, Denmark, Paper C-1, 1984, 19 pages.

44.   Ploeg, J. and Funke, E.R., "A Survey of Random Wave Generation Techniques", Proceedings of Seventeenth International Conference on Coastal Engineering, Sydney, Australia, 1980, pp. 135-153.

45.   Rice, S.O., "Mathematical Analysis of Random Noise", Bell System Technical Journal, Vol. 23, 1944 and Vol. 24, 1945.

46.   Salsich, J.O., Johnson, B. and Holton, C., "A Transient Wave Generation Technique and Some Engineering Applications", Proceedings of Twentieth American Towing Tank Conference, Hoboken, New Jersey, Vol. 2, 1983, pp. 949-969.

47.   Sand, S.E., "Accurate Reproduction of Short Crested Seas in Physical Models: Experience From a DHI Research Facility", Danish Hydraulics Institute, Denmark, Internal Report, July, 1980.

48.   Sand, S.E., Mansard, E.P.D., and P. Klinting, "Sub and Super Harmonics in Natural Waves", Journal of Offshore Mechanics and Arctic Engineering, Volume 110, August, 1988.

49. Takezawa, S. and Takekawa, M., "Advanced Experimental Techniques for Testing Ship Models in Transient Water Waves: Part I, The Transient Test Technique on Ship Motions in Waves", Proceedings of Eleventh Symposium on Naval Hydrodynamics, London, United Kingdom, 1976.

50. Takezawa, S. and Hirayama, T., "Advanced Experimental Techniques for Testing Ship Models in Transient Water Waves: Part II, The Controlled Transient Water Waves for Using in Ship Motion Tests", Proceedings of Eleventh Symposium on Naval Hydrodynamics, London, United Kingdom, 1976.

51. Toki, N., "On the Generation of Irregular Wave to be Used for the Design Studies of Floating Structures: Preliminary Study on the Concept of Design Irregular Wave", Proceedings of Sixty-Fourth Japan Towing Tank Conference, June, 1981.

52. Tuah, H., and Hudspeth, R.T., "Comparisons of Numerical Random Sea Simulations", Journal of the Waterway, Port, Coastal and Ocean Division, ASCE, Vol. 108, No. WW4, 1982, pp. 569-584.

53. Tucker, M.J., Challenor, P.G. and Carter, D.J.T., "Numerical Simulation of Random Seas: A Common Error and Its Effect Upon Wave Group Statistics", Applied Ocean Research, Vol. 6, No. 2, 1984, pp. 118-122.

54. Tung, C.C., and Huang, N.E., "Statistical Properties of Wave-Current Force", Journal of Waterways, Harbours and Coastal Engineering Division, ASCE, Vol. 99, No. WW3, August 1973, pp. 341-354.

CHAPTER 6

INSTRUMENTATION AND SIGNAL CONTROL

6.1 DATA ACQUISITION

In offshore structure model testing, we are generally interested in characterizing the model environment and the responses of the model structure to that environment. The responses of the prototype structure are then predicted for a given environment by scaling up the model response. Usually, the test environment is intended to scale a specified ocean environment. Instruments such as wave height probes and current meters are used to confirm that the sea state has been properly modeled during the laboratory test. Structure responses of interest might include motions of the model, foundation loads for bottom mounted structures and stress levels in individual members. Also, it might be desirable to quantify the interaction of the structure with the environment. For example. wave reflection or the runup of waves on the face of the structure can be important consideration in the design of a breakwater or a structure platform. In all model testing, there are processes occurring during the model test that we want to measure and record. Various specialty instruments [for example, Goldstein (1983)] are used in making the required measurements.

The transducer is connected to an automatic data acquisition system so that the transducer signal may be automatically recorded. In the past, strip charts were used for this purpose. Strip charts are capable of recording analog signals. With the fast analog-to-digital systems available today, the signal can be demultiplexed (i.e., arranged sequentially by data channel), converted to digital data and recorded in engineering unit.

A simple schematic of a data acquisition system is shown in Fig. 6.1. The transducer signal is conditioned first and transferred to the computer memory where data is stored for later analysis. The signal conditioner amplifies the signal from the microvolts level to volts.

6.1.1 Transducer

A transducer is a device that receives energy from one system, converts it to a different form and transmits it to another system. As used in a testing facility, transducers receive a physical input such as force, displacement, acceleration, pressure, etc. and then produce an electrical output (generally an analog voltage) that is proportional to the magnitude of the input signal. A few common means of measuring an input signal include a bonded strain gauge, a linear variable differential transformer (LVDT) and a capacitance probe. These component transducer systems are designed to measure the response and send out a corresponding voltage as the output.

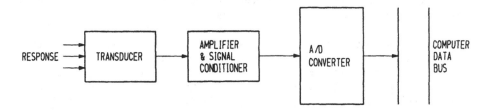

FIGURE 6.1
A TYPICAL FLOW OF DATA FROM THE MODEL RESPONSE TO THE COMPUTER

6.1.2 Signal Conditioning

The output of a transducer is in the form of a voltage. Since the output of different types of transducers varies in both magnitude (microvolts to volts) and form (DC or AC voltage), they must be conditioned or normalized in such a way as to be compatible with the device being used to record output. Typical functions performed by signal conditioners include AC to DC demodulation, gain multiplication, DC offset removal and filtering of noise or unwanted data.

Many transducers work on AC voltage. Since the data is normally required as a DC output, the signal conditioner demodulates the AC signal to a DC output. One method is to rectify the AC signal and then compute the rms value of the rectified signal as the equivalent DC output of the transducer.

Transducer outputs (such as, from a strain gauge) are often at a very low voltage (microvolts). In order to obtain a good resolution, this output is multiplied by a predetermined gain at the signal conditioner (through electrical circuitry) during the calibration of the transducer. This magnification of the data permits the data to be recorded in a common voltage range (e.g., ± 5 volts).

Often, the transducer output drifts with time. If this drift is large (such as, due to temperature change), then the transducer is unusable. The drift due to temperature may often be compensated by inserting appropriate additional resistors in the circuit. However, a small drift over time is quite common. This drift may be removed by analog or digital means. It is often desired to record the drift and remove it digitally. A zero reading is taken before the actual test run is started. This zero is saved for future study of the transducer behavior, but subtracted from actual data before data is recorded on computer. Thus, the transducer records only the variation due to responses.

The purpose of an analog filter circuit is to selectively pass or reject signals based on their frequency. There are many types of filter circuits, but we are mainly concerned here with lowpass type filters (Chapter 10 discusses digital filters). As the name suggests, a lowpass filter passes any signal with a frequency lower than a specified cutoff or corner frequency. Signals with frequencies higher than the cutoff frequency also pass through the filter but at a reduced amplitude. Filters can also amplify signals with frequencies within the pass band. Generally this amplification is limited to only a few percentages.

In addition to unwanted data, filtering is also done to prevent aliasing of the data. If a sinusoid is sampled at a frequency that is less than twice its oscillation frequency, then the sinusoid will appear to occur at a lower frequency. This shifting of frequency is called aliasing. Data is usually contaminated with the 50/60 Hz electrical noise. Aliasing can cause this energy to shift into the range of frequencies that we are typically interested in and may corrupt the data. Filters can prevent this ambiguity by eliminating all frequencies in the signal that are 1/2 the sampling frequency or greater. Analog filtering must be done prior to sampling, and sampling must occur at a rate that is greater than twice the highest frequency in the signal if aliasing is to be avoided.

Once the data has been recorded, digital filters may be used to separate frequency components or to remove noise from the data. Digital filters are mathematical algorithms that simulate the action of analog filters. However, aliasing cannot be prevented through digital filtering. This type of filtering occurs after sampling, and aliasing has already occurred at the time the samples are taken.

6.1.3 Data Recorder

Data can be recorded in many different ways including strip charts, magnetic tape and digital systems. Most wave tank data today are recorded digitally. Each analog data channel is terminated with a sample and hold circuit. With a common sample command controlling all of the channels, simultaneous samples of the inputs are taken. The sampled analog voltages are multiplexed into an analog to digital converter which generates a binary digital representation of each sample. The binary numbers are converted to floating point format and written to data files for storage.

6.2   BONDED STRAIN GAUGE

The term strain refers to the change in any linear dimension of a body, generally due to the application of external forces. A device commonly used to measure strain and to indirectly measure force is the bonded strain gauge. A bonded strain gauge consists of a grid or filament of very small diameter wire or thin metallic foil mounted on paper or plastic. The filament material has the property of linear variation of electrical resistance with strain. In order to measure the strain in an active member of an instrument, one or more gauges are bonded directly to the surface of the member. When the gauge is connected to an electrical instrument which will indicate small changes in resistance, such as a Wheatstone bridge, it can be used to record any strains occurring in the test section in the direction of the gauge axis (i.e., parallel to the gauge filaments). A Wheatstone Bridge configuration using one strain gauge and three fixed value resistors is shown in Fig. 6.2. With equal resistance in each leg of the bridge, the voltage levels +SIG and -SIG will be identical, and the differential bridge output will be zero volts. As the strain gauge resistance changes due to strain, the bridge will become unbalanced and a measurable voltage will occur between +SIG and -SIG. The excitation to the bridge can be either AC or DC. The output will be AC or DC, accordingly.

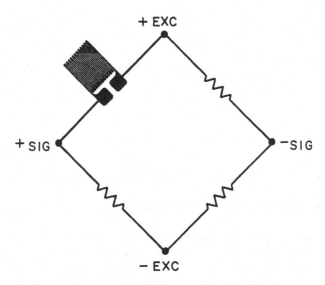

FIGURE 6.2
LAYOUT OF A WHEATSTONE BRIDGE CIRCUIT

6.3   POTENTIOMETER

Like a strain gauge, a potentiometer is a variable resistor. It can be used in a Wheatstone Bridge in the same manner as a strain gauge. It can also be used as shown in Fig. 6.3 if its resistance is large enough for the current through the potentiometer to be within the capabilities of the excitation power supply. The output voltage varies between +EXC and -EXC depending on the instrument's position.

Measurements of linear displacements with amplitudes greater than 152 mm (6 in.) are normally made with linear potentiometers.   A potentiometer consists of a strip of conductive plastic with an electrical resistance and a metallic contact that is coupled to the moving object.   As the contact moves along the conductive strip, the resistance between the end of the strip and the contact point varies linearly with the displacement.   When a voltage is applied between the two ends of the conductive strip, an output voltage results between the end of the strip and contact point which is proportional to the displacement.

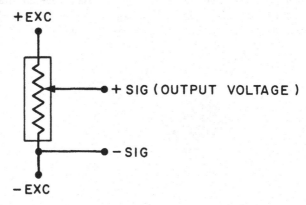

FIGURE 6.3
VARIABLE POTENTIOMETER CIRCUIT

6.4   DISPLACEMENT AND ROTATIONAL TRANSDUCERS

Linear displacements in the range of $\pm$ 0.25 mm (0.01 in.) to $\pm$ 152 mm (6.0 in.) are measured   with LVDTs. Differential transformers are electromagnetic devices commonly used for translating the displacement of a magnetic armature into an AC voltage. An   LVDT produces an electrical   output which   is proportional to   the mechanical displacement of a separate moveable core. It consists of a primary coil and two secondary coils symmetrically spaced on a cylindrical form. A free-moving rod-shaped magnetic core inside the coil assembly provides a magnetic flux linkage between the coils.

When the primary coil is energized by an external AC source, voltages are induced in the two secondary coils (Fig. 6.4). These coils are connected in series so that the two voltages are opposite in polarity. Therefore, the net output of the transducer is the difference between these voltages and the net output is zero when the core is at the center or null position. When the core is moved from the null position, the induced voltage in one coil increases, while the induced voltage in the opposite coil decreases. This action produces a differential voltage output that varies linearly with the change in the core position. The phase of this output voltage changes abruptly by 180° providing negative values as the core is moved from one side of null to the other. No physical connection exists between the core and the coils. Therefore, the device is totally friction free when the core assembly is properly guided.

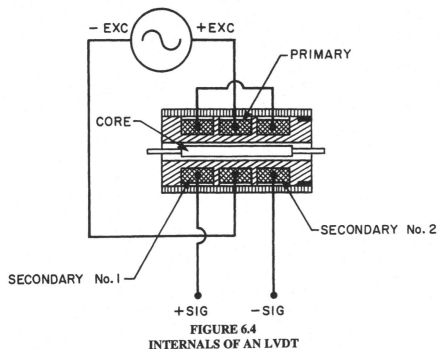

FIGURE 6.4
INTERNALS OF AN LVDT

Angular displacements are measured with Rotary Variable Differential Transformers (RVDTs) and rotary potentiometers. LVDTs and RVDTs are similar in that they both consist of a primary coil, two secondary coils wound in series opposition, and a core. Although mechanically arranged to measure either linear or rotational displacements, the means of operation of the two types of differential transformers is identical. These transducers have their components arranged to measure angular displacements rather than linear displacements.

6.5   VELOCITY TRANSDUCERS

Current velocities in the wave tank may be measured by electromagnetic current probe, an ultrasonic current probe, and a laser-based anemometer among other systems [Chang, et al. (1993)].   The principle of ultrasonic current probe consists in measuring the transit time of a sound probe between two small crystals. It is widely used in oceanographic measurements.   This probe is also used for current measurement in multiple directions in a model basin.

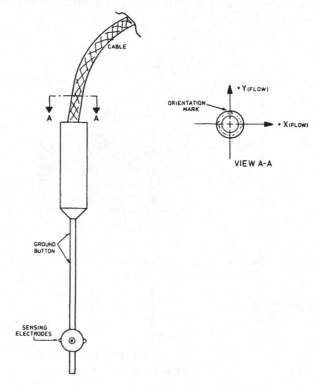

FIGURE 6.5
ELECTROMAGNETIC CURRENT PROBE

The laser-based anemometers are not yet used in standard model testing. They are expensive, but may be very useful for more special flow studies. What makes them advantageous over other probes is their inherent ability to measure very locally without introducing a probe in the flow field.

The electromagnetic current probes (Fig. 6.5) are most widely used in the laboratory. They are also used in the field. The sensing head of the probe is generally made smaller for laboratory use to avoid averaging over a large area. However, it also introduces additional noise in the measurement. The probe is usually equipped with a built-in filter to reduce this high frequency noise level.

The operation of these current meters is based upon the Faraday principle of electromagnetic induction. Simply stated, a conductor (in this case, water) moving in a magnetic field (generated from within the probe) produces a voltage that is proportional to its velocity. Because the flux of the permanent magnet is constant, the rate of change of flux in each coil is a linear function of the rate of change of displacement (i.e., velocity) of the magnetic core.

Therefore, the magnitude of the output voltage is linearly proportional to core velocity. However, if both poles of the magnet core entered a coil, the net voltage in that coil would be zero, because the opposite magnetic poles would produce equal but opposite polarity voltages. To prevent this condition, the two coils are connected in a series-opposing configuration. The resulting output voltage is then the sum of the differential voltages in the two coils. The polarity of the output voltage depends on the direction of core motion and the orientation of the core's magnetic poles.

The voltages present across the electrode pairs are of the same frequency as the magnet drive frequency and have amplitudes that are proportional to the flow component detected by the electrodes. This AC signal which is generally small is amplified, synchronously demodulated (phase relation of the drive signal and electrode signal determines numeric sign of output), and filtered to yield two DC voltages that represent the velocity vectors of the flow in two orthogonal directions.

6.6   ONE DIMENSIONAL CURRENT PROBE

Commercially available current meters suitable for measuring water particle velocity in the wave tank are often limited by their size, sensitivity and cost. These current meters utilize various physical principles in the measurement of current velocity [Hardies (1975), McNamee, et al. (1983)], but those that meet the criterion of size and sensitivity have an associated high cost. A simple, inexpensive current probe as well as wave particle velocity probe may be designed using the principle of wave force measurement. Such designs, along with their inherent features and shortcomings, are described in the following.

A current probe [Sharp (1964)] was developed on the principle of measuring hydrodynamic force. The probe is shown in Fig. 6.6. It consists of a strain gauged cantilever with a length of thread spanned between the tip of the cantilever and a second support. The flow of water past the thread will introduce a (mainly) drag-type hydrodynamic force on the thread. The thread is placed horizontally and normal to the flow. The force is recorded by the strain-gauged cantilever. The thread in the example is 51 mm (2 in.) long. The force is directly related to the velocity of flow. The drag-type force makes the relation parabolic.

6.7   TWO DIMENSIONAL CURRENT PROBE

The concept of the current meter design described in the above section may be extended to include two component measurements by strain gauges. This allows the measurement of magnitude and direction of the horizontal current velocity. It also could measure, through suitable orientation, the horizontal and vertical components of the water particle velocity.

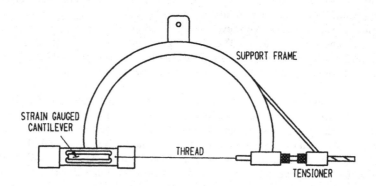

FIGURE 6.6
A SIMPLE ONE-DIMENSIONAL CURRENT PROBE [Sharp (1964)]

6.7.1 Basic Properties

Wave tank usage of a current meter poses several problems that should be considered in the design of a suitable prototype. Two such problems are the (primarily) two-dimensional time-varying nature of the fluid flow and the stability of calibrations of the meter during extended use in the tank. A device similar to that shown in Fig. 6.7

may be designed for this purpose. The perforated ball [Bishop (1979)] is attached to a cantilever arm whose movements are measured with LVDTs. The motion of the cantilever arm is proportional to the hydrodynamic force on the ball. The fluid velocity is determined from this force by laboratory calibrations.

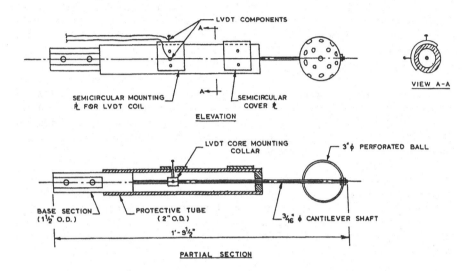

FIGURE 6.7
LVDT TYPE WATER PARTICLE VELOCITY METER

The physical principle utilized in this design is that the cantilever arm is of small section, and the hydrodynamic force on the ball is proportional to the square of the velocity. This proportionality is only valid for a constant drag coefficient which should be realized in the wave tank for a small range of velocity. Any variation will become evident during calibration in any case. For the water particle velocity in waves, inertia can be a problem. The inertial effect can be reduced in magnitude by making the ball of a light, perforated material. The perforation also allows the flow to be locally turbulent, reducing the laminar flow effect and rapid change of the drag coefficient. The square law causes problems in accuracy at low fluid velocity; hence, there is a limit on the minimal fluid velocity that can be accurately measured. Conversely, the squared relationship gives greater sensitivity at higher velocities and this can be used to our advantage if the ball and/or cantilever is matched to the desired range of measurement.

6.7.2 Design Information

Once the most appropriate configuration of the current meter had been determined, the next stage is the selection of material and actual design of component pieces. Due to the chlorinated environment of the wave tank, a corrosion resistant aluminum alloy 6061-T6, or stainless steel 304 may be suitable. The actual shape of the device is determined after consideration of a number of factors. The spherical shape of the sensing head eliminates directional problems. The perforations in the sensing head reduce the size of any vortices formed on the downstream side of the device which likewise reduce the amount of vortex noise created.

Sizing of the cantilever shaft begins with the determination of the allowable measuring range available on the LVDTs. With this range in mind, coupled with the estimated drag force on the perforated ball, deflections are calculated for several different lengths and diameters of cantilever shafts. A round cross section may be appropriate primarily due to ease in construction.

Several features are incorporated into the design to make it as flexible as possible. All connections between component pieces are either screwed or friction fit. The cantilever shaft and LVDT core collar may have friction connections using stainless steel set screws. A protective tube, semicircular pieces and stop collars may be fastened in place using brass screws. Such fastenings allow for easy breakdown and adjustment as needed.

Several methods for varying the sensitivity of the device can be used. For example, the location of the LVDT measured from the fixed end of the cantilever or the length of the cantilever may be changed. The size of the perforated ball could also be altered. Moving the LVDT apparatus would not necessitate rewiring the electrical connections as the core of the LVDT is mounted on a machine screw which is easily removable. The coil of the LVDT is mounted on the semicircular piece interchangeable with the cover plate. However, once the LVDT core is moved, it should be relocated at its electronic neutral position within the coil.

Each current meter is capable of measuring forces on the sensing head in two directions by using two LVDTs mounted orthogonally to each other on the cantilever shaft. In order to determine the remaining component of a three dimensional velocity, two current meters may be deployed rather than incorporating the capability of measuring a third degree of freedom from one current meter. To prevent damage to the LVDTs from excess motion in any direction, a stop is designed to limit movement of the cantilever shaft past a maximum allowable deflection. Such a stop is necessary because by nature of the orthogonal mounting, as one LVDT core moves longitudinally, the other core moves laterally.

Consideration should also be given to the possibility of vibration-induced problems of the probe in waves. For the probe in Fig. 6.7, theoretical calculations showed a natural period of 0.4 sec assuming the sensing head filled with water. This is below the 1 to 8 sec period of waves typically generated in the tank. Once the probe is built, its natural period may be established by a pluck test under water and collecting unfiltered data.

6.7.3 Calibration and Testing

Prior to calibration of the current meter, the LVDT cores are electrically balanced within their corresponding coils using a sine wave generator and an oscilloscope. The core is adjusted manually with the machine screw until the scope indicates a minimum voltage output.

The current meter is checked for accuracy on a calibration stand. The prototype is set up so that only one direction (parallel to the longitudinal axis of the LVDT coil in question) is loaded at a time with known weights in increments, and the cross-talk in the transverse LVDT is recorded along with the inline LVDT. Ideally, a linear scale factor will be expected in the inline direction while the cross-talk in the transverse direction should be small (typically less than 5%). This calibration shows the relationship between applied force and corresponding voltage output. While it verifies the construction technique and the accuracy of the prototype, it is not useful in the current measurement in the tank. In the latter case, a scaling curve is desired from a towing test with known towing speeds.

For this purpose, the current meter is mounted on a traveling bridge and a towing test is performed within a velocity range (e.g., 0.3 m/s or 1 ft/sec to 0.6 m/s or 2 ft/sec at intervals of 0.03 m/s or 0.1 ft/sec). This provides a relationship between speed and voltage.

Combining the two calibrations, a direct relationship between the force and the towing velocity may be derived. Theory states that the relationship between force and velocity is of the second order in velocity. The actual calibration of the prototype design is shown in Fig. 6.8. The results of a least squares fit of the data are as follows:

$$F = -0.000926 + 0.00536U + 0.0329U^2, \qquad \text{for } 0 \le F \le 0.14 \qquad (6.1)$$

Alternately,

$$U = -0.0815 + [0.0348 + 30.395F]^{1/2} \qquad (6.2)$$

where F is the applied force in pounds, and U is the tow velocity in feet per second. Alternatively, the velocity may be directly related to the output voltage of the LVDT. The constant in Eq. 6.1 probably reflects errors in the original data. The linear term is the friction or shear drag on the immersed object possibly coupled with data errors. The   dominant coefficient   on the U^2 term   is the form drag.   This coefficient incorporates the drag coefficient, C_D, cross-sectional area of the immersed object and the density of the fluid into one constant for the towing speeds involved. Figure 6.8 is a plot of the original towing test data overlaid with the curve generated by Eq. 6.1.

A prototype current meter of the above design, if subjected to regular waves, will show that the current meter output leads the wave probe profile at the same location. A

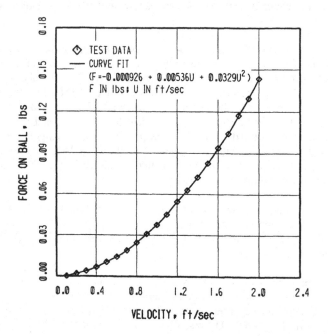

FIGURE 6.8
PROBE CALIBRATION CURVE BY TOWING

closer look at the make-up of the hydrodynamic force on the current meter is helpful in determining how this phase lead developed.

The force on the current meter in waves has two components, one due to fluid dynamic drag and the other due to fluid inertia on the sensing head. Morison's equation (Chapter 7) expresses the relationship between the total force and the parameters that govern each of these force components. For a given set of values for C_M and C_D (see Eq. 7.1) a relationship can be developed for the phase shift in the total measured force. A simple calculation is carried out using $C_M = 1.5$ and $C_D = 0.75$. The theoretical phase shift and component forces are compared to the experimental values from the prototype design in Table 6.1. Considering that the inertia and drag coefficients are approximate, the comparison in Table 6.1 shows that the phase lead is due primarily to the presence of an inertial component in the total force.

Therefore, there are limitations in the use of this kind of probe in oscillatory flow. It can be shown that if the component of total force due to drag is at least 4 times as large as the inertial component, the error in the sum of these components is within 5%; the magnitude of the corresponding phase shift is approximately 10 degrees.

TABLE 6.1
PHASE AND FORCE RELATIONSHIPS FOR A PERFORATED BALL TYPE CURRENT PROBE

Wave		Phase Lead		Force	
Regular Amplitude (ft)	Wave Period (sec)	Experimental (deg)	Theoretical (deg)	Experimental (lbs)	Theoretical (lbs)
0.168	3.5	50.6	62.6	0.0136	0.0141
0.327	3.5	36.0	41.6	0.0293	0.0307
0.265	3.5	42.4	50.6	0.0222	0.0223
0.151	4.0	51.0	63.1	0.0099	0.0116
0.233	4.0	43.4	51.7	0.0175	0.0180
0.297	4.0	38.2	44.8	0.0230	0.0243

1ft = 0.3m;   1 lb = 4.45 N

6.8   AN ALTERNATE CURRENT PROBE

The original design appears promising, but the LVDTs as installed might have accuracy problems due to the mechanical motions involved. Accordingly, an alternate design using strain gauges may be used rather than LVDTs as the output devices. Figure 6.9 is a sketch of the base section of the strain gauged current meter. The reduced gauge section is machined from a 25 mm (1 in.) diameter rod which experiences the current force. The difficulty with this strain gauge design is the placement and waterproofing of the gauges at the appropriate locations.

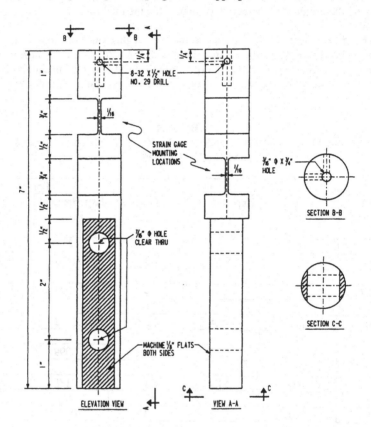

FIGURE 6.9
STRAIN GAUGED CURRENT PROBE ELEMENT

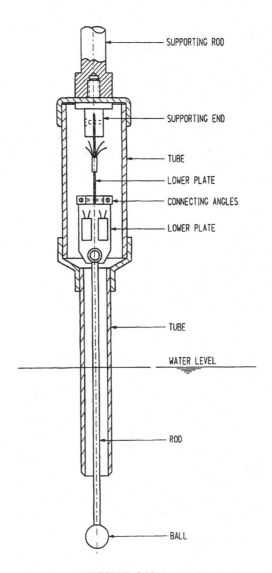

FIGURE 6.10
STRAIN GAUGED CURRENT METER ASSEMBLY
[Abdel-Aziz & Abuelnaga (1983)]

A modified version of this current meter design [Abdel-Aziz and Abuelnaga (1983)] is shown in Fig. 6.10. It also incorporates strain gauges as sensing elements. By placing strain gauges on two plates at right angles to each other, two directional components are measured. The current forces on a ball acting as a cantilever transfers the strains to the gauges in terms of a bending moment. The rod connecting the ball to the plates at the top is covered by a protective tube so that essentially only the sphere experiences the fluid flow.

The ball is submerged to the depth at which the current velocity measurement is desired. The ball and a short part of the rod are subjected to the current force. The majority of the force is experienced by the ball since the rod diameter is small. In order to minimize the inertial effects, the ball and the rod are made of light material such as plastic or aluminum.

The plates contain four strain gauges each; two on each face forming a Wheatstone bridge. The gauges in the circuit are arranged such that the neighboring gauges in the Wheatstone arms are subject to different signs. This provides maximum possible sensitivity. The size of the plates is chosen based on the size of the sphere and the moment arm to provide allowable design stress in the plate. Usually, this stress is taken as about 30 percent of the yield stress to assure linearity within the range of strain. The output of the bridge is sensitive only to the bending moment caused by the current force on the ball in a direction perpendicular to the corresponding plate. The symmetric placement of the gauges on the plate provides temperature compensation so that the bridge remains balanced with changes in its environment.

The calibration of the current meter may be achieved by using a towing carriage. Since the steady current force on the ball is mainly drag, the output voltage for different towing speed will provide a quadratic relationship of the form:

$$U = Cv^n \tag{6.3}$$

where v is the output voltage due to the current (or towing) velocity U, n is the slope (~ 0.5) and C is a constant. Thus, a linear relationship is expected on a log-log scale, e.g., shown in Fig. 6.11. In this particular calibration of the prototype design, the values of C and n were found to be 7 and 0.51, respectively. By waterproofing the strain gauge area, the current meter may be made to work under water.

6.9   ACCELEROMETERS

Accelerometers used in wave tank applications are generally commercially available. There are two types of accelerometers: closed-loop and open-loop. In open-loop accelerometers, a mass is displaced against a spring by the acceleration, and the resulting load or displacement is measured. In a closed-loop accelerometer, the force required to prevent displacement of the mass is applied by a servo-loop. These are pendulous devices that operate on a torque balance principle. The applied linear acceleration acting on an unbalanced, bearing-mounted mass develops a torque. An equilibrium is maintained by arranging the mass so that any resultant motion develops an electrical signal. This signal is properly amplified and supplied to an electrical torque generator acting on the mass. The net result is a current proportional to the acceleration, accompanied by an infinitesimal displacement of the pendulous mass. By permitting the current to pass through a stable resistor, a voltage proportional to acceleration is developed.

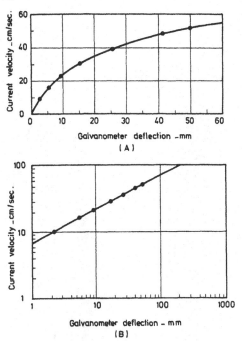

FIGURE 6.11
CALIBRATION CURVE, CURRENT VELOCITY VERSUS GALVANOMETER DEFLECTION (A) LINEAR SCALE; (B) LOG-LOG SCALE [Abdel-Aziz and Abuelnaga (1983)]

A closed-loop system has several possible advantages. For example, two main sources of inaccuracy in the open-loop accelerometer, namely, the mechanical spring and the displacement-to-voltage transducer, are eliminated. Thus, problems of nonlinear springs, mechanical hysteresis, changes of the modulus of elasticity with temperature, spring deflection due to temperature and fatigue are non-existent with torque balance devices. Therefore, closed-loop systems achieve levels of accuracy, stability and reliability   that are comparatively much greater than   open-loop accelerometers.

6.10   PRESSURE TRANSDUCER

Pressure sensing can be accomplished by measuring the deflection or strain of a diaphragm element exposed to the pressure source (Fig. 6.12). The deflection of the

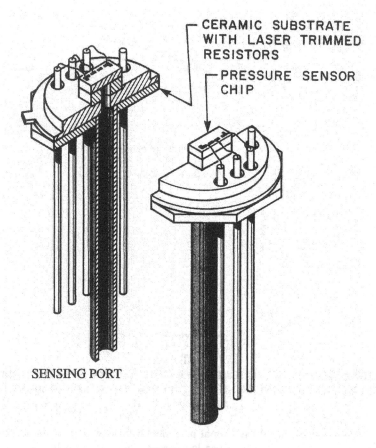

CERAMIC SUBSTRATE
WITH LASER TRIMMED
RESISTORS

PRESSURE SENSOR
CHIP

SENSING PORT

FIGURE 6.12
A SINGLE PORT PRESSURE TRANSDUCER DESIGN

sensing diaphragm can be measured with an LVDT or a capacitive gauge. In general, strain gauged pressure transducers are more accurate than other types. If the sensing element is large enough, bonded strain gauges may be attached to the element's surface to measure its strain. Where size and stability are of primary importance, as in wave tank tests, silicon pressure sensors are typically used. The sensing element in a silicon-diaphragm pressure transducer is an integrated circuit silicon chip containing a micro-machined or ground diaphragm. Four piezoresistive strain gauge resistors are diffused into the diaphragm to form a fully active Wheatstone Bridge. The single crystal nature of the diaphragm assures an essentially unmeasurable hysteresis in the calibration curve. In some cases, integrated circuit signal conditioning amplifiers are included in the pressure transducer package, eliminating the need for external conditioning.

Commercially available pressure sensors are of two different kinds: absolute and differential. The absolute pressure gauges have one sensing port exposed to water and are good for hydrostatic heads (Fig. 6.12). They may be used in the measurement of dynamic pressures by setting the static head to a zero reference before starting the test run. This method may not be satisfactory if the static head is too high compared to the dynamic amplitude, or if the expected dynamic amplitude becomes higher than the static head.

The differential pressure gauges are more accurate in measuring dynamic pressure head, such as pressure due to wave motion. In this case, one sensing end of the transducer is exposed to waves while the other end only experiences the hydrostatic head. These gauges are quite suitable for mounting on the surface of an offshore structure model flush with the outer shell surface. They may also be placed submerged in the waves if one end is vented to the atmosphere. In this case, the vent tube is filled with water to the still water level to match the hydrostatic head. The pressure transducer may be calibrated by varying the static head on either end of the gauges in sequence. A special purpose calibration stand may be built for this purpose. These transducers are generally not commercially available in a waterproof condition. They may be potted for underwater use with one of many potting compounds to embed the electrical connections, leaving the sensing ends open.

6.11 WAVE ELEVATION

The water surface elevation is measured by a resistive (or, better yet, conductive) or a capacitive type wave probe. The capacitive type wave probe is considered more reliable and has more widespread application. In either case, the sensing element changes the amount of conductance or capacitance as the water rises or lowers at the probe, thus causing a change in voltage output. The success of these probes depends on the quick response in the output due to change in the water elevation. Also, any splashing or wetness of the probe should not affect the reading. The capacitance probe is superior in this respect.

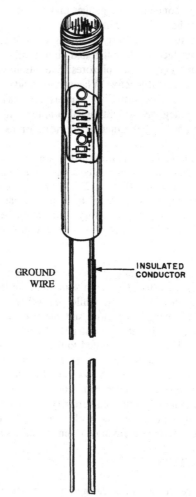

GROUND
WIRE

INSULATED
CONDUCTOR

FIGURE 6.13
A TYPICAL CAPACITANCE WAVE PROBE

The sensing element of a capacitance wave probe approximates a co-axial capacitor (Fig. 6.13). It consists of an insulated (e.g., teflon-coated) metal (e.g., stainless steel) rod (about 1.6 mm or 1/16 in. in diameter). The operating principle of this probe [Chang, et al (1993)] is that the wire acts as a capacitor. The wire core and the water act as the two plates of the capacitor and the insulation acts as the dielectric. A voltage is generated by the difference between that at the top of the capacitance wire and an immersed ground wire. The wave height is obtained by measuring the time that it takes the capacitor to reach a certain percentage of its fully charged value. The time is then related back to the capacitance and indirectly (through use of calibration constants) to its change in immersion. The capacitor (wave probe) is excited with an AC voltage. An AC output is produced which is linear with the instantaneous water level. The AC output is rectified and passed through a low pass filter to produce a final DC voltage level.

These probes have a superior linear response, and the full scale error is typically much less than 1 percent. An oscillation of 6 mm (1/4 in.) amplitude of water column over the range of gravity wave frequencies is easily detectable by the probe. One problem with these probes is that surfactants (primarily hydraulic fluids) stick to the wires causing them to lose their hydrophobic properties. Water may slide off the probe slowly, thus lowering its frequency response and attenuating signals. Therefore, they need daily cleaning with a dry cloth.

A sonic probe may be placed near the water surface pointing towards the surface at a distance of 457 - 610 mm (18-24 in.), depending on the wave amplitude. The time of reflection of sound from the water surface as it travels to and from the probe is converted to voltage. By the nature of the probe, the reflection takes place over a large area of the water surface depending on its distance from the surface.

Unlike a sonic wave probe which takes an average reading over a finite area of the water surface, the capacitance probes have sizes that are able to provide readings over a very small area on the surface.

A design similar to the capacitance probe, but using a small strip of metal or foil, may be used against a structure to measure wave elevation at the submerged vertical member of the structure. This allows the measurement of wave run up at the structure which is useful in the design of deck clearance, boat bumper height, etc.

Chang, et al (1993) described two new wave height measurement systems: a surface sensing mechanical probe and a laser wave height probe which uses a linear array of charged couple devices. The mechanical probe is minimally intrusive. A senso-motor electrode in contact with water moves up and down via a cantilevered arm oscillating with the water surface. For the laser probe, fluid is seeded with florescene dye and a laser beam is shot down through the free surface. The change in radiation intensity on the array is used to compute wave height.

6.12   FREE-BODY MOTIONS

There are numerous methods of detecting and recording motion, but many have the undesirable tendency to affect the motion that they are meant to measure. Some methods measure displacement (e.g., linear transducers); other methods measure rotation (e.g., gyroscope). Many of these transducers are commercially available, with or without built-in amplifiers.

In the past, the motions of a body were often measured by attaching displacement transducers, such as potentiometers, to the body. In many cases, however, the attachment of these sensors affects the motion of the body. For example, the friction in such a system may inhibit the motion. An example of this will be given in Chapter 9. In order to measure the motions of a body without physical connection to the body, an optoelectronic remote motion monitoring system, such as SELSPOT or OPTOPOS, can be used.

The optoelectronic movement monitoring system uses small light-emitting diodes (LEDs) to identify selected points on the moving body. An infrared sensitive camera detects the position of the diodes for registration and analysis of static and dynamic motion in real time. Infrared beams from the LEDs are focused on the detector surface through a lens system, and the position (velocity, and acceleration) of each diode used can be studied in real time. The detector can measure the position of several diodes simultaneously. The use of one camera will monitor positions in two dimensions, whereas, two cameras properly positioned permit three dimensional monitoring.

When three LEDs are attached to the body, the three dimensional displacements of all three lights are recorded by the system. Computer software determines the six degrees of freedom motion – surge, sway, heave, roll, pitch, and yaw – of a floating vessel from the displacements of these lights.

In operation, tiny infrared LEDs are positioned and attached to a model in such a way that they are visible to two infrared sensitive cameras (Fig. 6.14). The LEDs emit infrared light sequentially, and the output from each camera is demultiplexed, sampled and stored to provide two analog voltage outputs per camera. These outputs represent the tangent of the vertical and horizontal displacement angles taken from the camera line of sight axis. Using two cameras and a minimum of three LEDs, it is possible to obtain data that allows the calculation of the six motions of a model. A fourth light is generally added to the system so that the calculated motions at this light may be overlaid on the measured displacement for verification of the systems' accuracy. Typically, the accuracy is better than 5 percent.

There are limitations to the use of this system. The LEDs work in the dry so that they must be placed over the waves. The two cameras have a field of vision which, when combined, form a cubic region in space. If the light emitting diodes move out of this cubic region, the cameras can not record their position. As the lights are placed farther away from the cameras, this region becomes larger, but the measurement accuracy reduces. Sunlight and reflections of the lights off the water or other reflective surfaces can create spurious readings. The lights must be positioned on the body such that they are never shielded from the camera by the body or by one another.

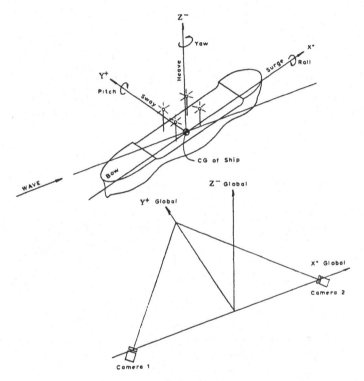

FIGURE 6.14
PLACEMENT OF OPTOELECTRONIC SYSTEM IN THE WAVE TANK

These are not serious drawbacks to the optoelectronic system in a controlled laboratory environment. The field of vision and the resulting accuracy are easily calculated from the calibration table. The ambient light is controlled, and all reflections can be negated by camera placement and shielding techniques. The location of each light is selected to maintain lines of sight with both cameras.

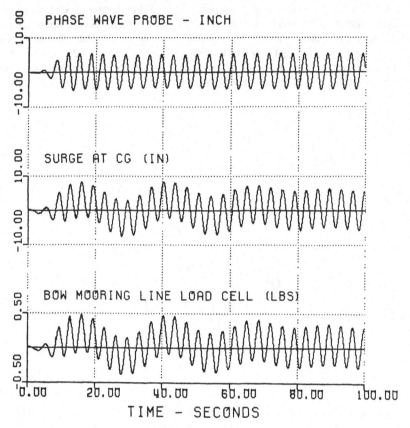

FIGURE 6.15
MEASUREMENT OF SURGE AND MOORING LINE LOAD ON A BARGE MODEL IN REGULAR WAVES

This system has been employed quite successfully in wave tanks. For example, a barge tested in waves with this system is considered here. It was moored in head seas by similar linear springs fore and aft such that it had a natural period in surge of approximately 30 seconds. Figure 6.15 demonstrates typical results from the regular wave test. The bow mooring line load was measured by a ring gauge. The wave and surge were measured by a capacitance wave probe and the SELSPOT system. A load was predicted from the surge measurement by multiplying the instantaneous surge value by the linear spring constant. This predicted load was compared to the measured load by overlaying the plots of each load versus time in Fig. 6.16. The two measurements are within about 10 percent of each other. The measured load in Fig. 6.16 correspond to the loads measured as shown in Fig. 6.15.

6.13 FORCE MEASUREMENT

There are two types of load cells one can build. One is a direct force measuring device using strain gauges. The other utilizes displacement transducers. In a direct force type cell, a material is chosen whose stress-strain properties are known to be linear in its elastic range. The design of the force block is made in such a way that under full load, about 30 percent of the elastic range is reached. The force is then measured with strain gauges. This type of force cell is typically quite accurate; it could, however, be expensive, depending on its requirement for waterproofing.

In the displacement type load cell, the deflection of a flat spring under the action of a force as a cantilever is measured by a position transducer. The system is designed so that the deflection is linear within the range of forces. If properly designed, this type of system can be quite accurate. It is also relatively inexpensive, since LVDTs are factory sealed.

Forces are measured using instrumented load cells arranged so that the measured force is transmitted through the load cell to the model's support structure. The load cells are designed so that the deflection or strain in their active member is proportional to the load being supported by the cell. The deflection of the cell is measured with an LVDT; the strain is measured by a strain gauge.

When the load is required in a single direction, a strain type ring gauge may be employed. This is typical of mooring lines attached to a floating body. These load cells consist of a machined aluminum ring that is necked down at two diametrically opposite locations where it is instrumented with strain gauges to measure the deformation of the ring under load.

The calibration of these load cells is accomplished by loading the transducer with a series of NIST traceable weights and recording the circuitry output through the data acquisition system. A least squares best fit calculation produces a calibration scale factor that directly relates the transducers' electrical output to the applied load. A plot is generated which shows the data points and the best-fit line. In addition to the scale factor, the correlation coefficient, the root mean squared (rms) error, and the percent full scale output (%FSO) error are also important parameters. The percent full scale output error is defined as the maximum difference between the transducer response estimated by utilizing the calculated calibration scale factor and the measured response divided by the full scale output range of the transducer.   The accuracy of these load rings is typically found to be better than 0.50 %FSO during calibration.

Since strain gauge type load cells are used under water, they are protected by waterproofing. Provisions are made in the design for the application of waterproofing materials. Since the strain gauged area is often reduced in size for appropriate sensitivity it provides room for the waterproof. It is important to clean the surrounding area of dust and grease. It allows the waterproof material to adhere to the surface. If the strain gauge is placed on a diaphragm inside a machined pocket, the pocket may be partially filled with paraffin first. Typically, waterproofing consists of a coat of M-coat D and M-coat G (available from Micro-measurement) onto the strain gauge area.   It is then covered by generously applying a sticky tar-like compound called A-MP sealing and dielectric compound.

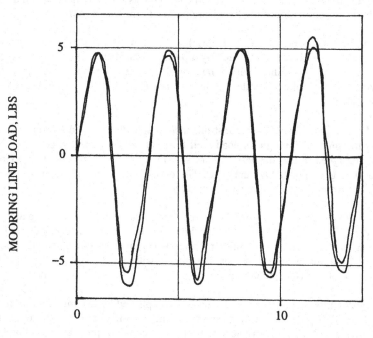

TIME, SEC.

FIGURE 6.16
PREDICTED LOAD FROM SURGE VERSUS MEASURED LOAD

The global loads and moments on a fixed structure are measured by XYZ load cells that use the tank floor as the supporting structure. These transducers consist of machined aluminum plates which are assembled in such a way as to permit shear loading in three orthogonal directions. Each axis is instrumented with an LVDT or strain gauge which measures the deformation of the plates under load. For a submersible load cell used to measure six component forces and moments, it simultaneously requires rigidity, sensitivity and response. Use of strain gauges require waterproofing and Wheatstone bridge construction.

Alternately, piezoelectric load cells may be used. Such cells are commercially available already in a sealed construction. They have inherent rigidity, high sensitivity and a wide dynamic range. In a typical piezoelectric cell design, the transducers are typically placed between the base plate and a loading platform. For three component measurement, a total of eight piezoelectric cells may be used [Bazergui, et al. (1987)] -- two each in the x and z directions and four in the y (vertical) direction. The cells are fitted with threaded rods at each end used to secure to the base plate and loading platform, respectively (Fig. 6.17). Thus, the cell assembly is rigid in tension, but soft (minor stiffness) in bending. In an example cell, the ratio of bending to axial stiffness was 0.0032/1.

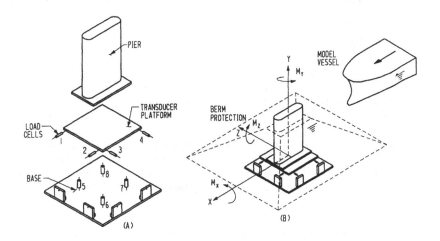

FIGURE 6.17
GENERAL ARRANGEMENT OF A 3-D LOAD TRANSDUCER PLATFORM
AND TEST PIER INCLUDING AN EXPLODED VIEW
[Bazergui, et al. (1987)]

Eight cells for three-load and three-moment measurement provide two redundant measurements -- one in the horizontal and one in the vertical direction. The extra load cells add stiffness to the platform and provide symmetry to the system. The relationship between the forces F_1 to F_8 (compression assumed positive) measured by the individual cells and the three forces given by F_x, F_y and F_z and the three moments given by M_x, M_y and M_z acting at the center of the top surface of the loading plate may be given by the following equations:

$$F_x = F_1 + F_2 \tag{6.4}$$

$$F_z = -(F_3 + F_4) \tag{6.5}$$

$$F_y = -(F_5 + F_6 + F_7 + F_8) \tag{6.6}$$

$$M_x = (F_5 - F_6 + F_7 - F_8)l/2 - bF_y \tag{6.7}$$

$$M_z = -(F_5 + F_6 - F_7 - F_8)l/2 - bF_x \tag{6.8}$$

$$M_y = (F_1 - F_2 + F_3 - F_4)l/2 \tag{6.9}$$

where l = horizontal distance between the axes of adjoining load cells and b = vertical distance between the top of the loading plate and the lateral load cell axes.

6.14   XY INSTRUMENTED SECTION

It is often desirable to measure wave loads on a small section of an offshore structure member. A typical example is the members of a jacket structure or a section of an exposed riser. The load is measured by an instrument placed inside the section. This instrument is often used to measure inline and transverse loads on the member in waves. Because of the small size of the member compared to wave length, drag is generally important in the total measured force. The hydrodynamic coefficients are computed from the measured load. Since the hydrodynamic load may vary with depth, it is important that the forces are measured on as small a section of the member as possible without sacrificing accuracy. A typical XY load cell design is described below.

In this design, the load measuring element is chosen to have an outer dimension of 0.3m (1 ft) in length x 64 mm (2-1/2 in.) in diameter. These dimensions are particularly chosen by the space available to provide relative ease in manufacturing the load cell while at the same time maintaining the accuracy in the measurement of load in the test range. The length of the section relative to the wave periods of interest is chosen for the tests such that the distribution of loads on the section is reasonably uniform so that the variation of the relevant parameters (e.g., KC and Re) over this length is minimal. While a smaller diameter would provide larger possible range of KC within the limits of the testing facility, diameters smaller than 64 mm (2-1/2 in.) is difficult to build and to maintain an acceptable accuracy of measurement.

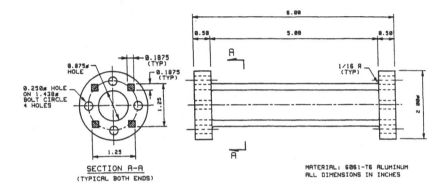

FIGURE 6.18
MACHINED INSTRUMENTATION FOR TWO-COMPONENT LOAD MEASUREMENT

The load cell was designed as a strain gauge device. The strain gauges were mounted on four beams (Fig 6.18) of square cross-section. The cross-section was chosen to limit the maximum stress to 10,000 psi. The four beams provided two active axes orthogonal to each other. The strain gauges were designed so that the load cell was sensitive to normal loads only, but insensitive to the location of load, or to moments and torsion. The load cell was fabricated from a single solid aluminum block to avoid problems with welds or mechanical connections. Tolerances on the dimensions and locations of the square sections were carefully maintained. The circular discs at the ends provided continuity of the member. The discs were designed so that essentially all deflections take place in the beams. A shroud of 64 mm (2-1/2 in.) in diameter was placed on the cell which determined the outer dimension of the test member. The strain gauges were waterproofed for underwater operation.

The load cell was calibrated along the two active axes on a calibration stand using NIST traceable weights. The rms error of the linear scale factor was about 0.07 percent in either direction. The cross-talk which measures the erroneous reading along the second (perpendicular) axis while loading along one axis was less than 1 percent.

6.14.1 Experimental Setup

The load cell was made a part of a 1.8 m (6 ft) continuous 64 mm (2-1/2 in.) diameter cylinder. The instrumented section was placed 0.3m (1 ft) below the still water level. This allowed the maximum possible load on the load cell in waves without exposing the cell by the changing free surface. The cylinder extended 1.2 m (4 ft) below the instrumented section to maintain continuity and to avoid end effects.

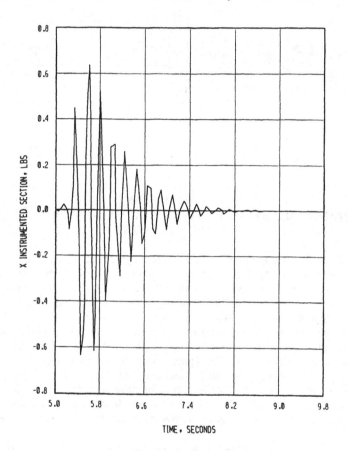

FIGURE 6.19
NATURAL VIBRATION TEST OF THE INSTRUMENTATION SETUP

In order to determine the natural vibration frequency of the setup, the cylinder was allowed to resonate by pulling on it and then releasing it. The recorded free vibration of the cylinder on the instrumented load is shown in Fig. 6.19. The frequency of vibration was found to be about 4 Hz. An analog filter of 5 Hz cut-off frequency was used in each channel during testing. This eliminated most of the noise from the recorded data.

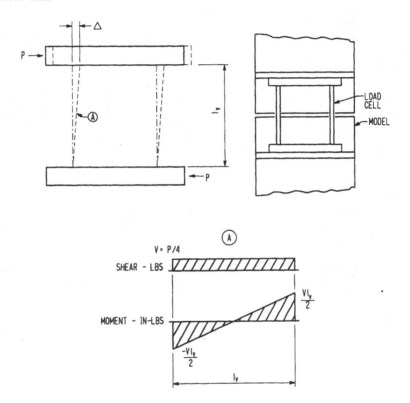

FIGURE 6.20
PRINCIPLE OF SHEAR MEASUREMENT AND USE OF CELL IN A TYPICAL MODEL

6.15   DESIGN OF A LOAD CELL

The purpose of this section is to describe the feasibility of designing and building a load cell that would be equally responsive to shear and bending moment in any direction. The cell is to be designed to use a minimum number of analog channels and preferably read shear and bending directly upon calibration.

6.15.1 Shear Force and Bending Moment

The XY bending moments and XY horizontal shearing forces on a structure are measured with moment shear load cells. The basis of design for this type of load cell is a fixed-end beam, with only end deflections (i.e., no end rotation). The load cell consists of four beams of square section (for ease of placement of strain gauges) spaced equally between two virtually fixed ends of circular sections. The entire load cell is machined out of one solid aluminum block.

By placing strain gauges near the end of a beam section on opposite sides and wiring as a half Wheatstone Bridge, the bending through the section can be measured and any axial loads in the beam will automatically be cancelled out. See Fig. 6.20 for a schematic of the cell as a frame with a shear applied. The ends of the beams being fixed against rotations, the shear is proportional to the bending through a beam section.

The present load cell was designed to measure load on a submerged structural member of an offshore structure based on a maximum bending moment of 1.93×10^8 N-m (142,000 ft-kip) and shear of 4.27×10^5 m. ton (942,000 kips) with a scale factor of 50. Therefore, the maximum bending and shear for a model should be 31 N-m (273 in. lbs) and 33.8 N (7.6 lbs), respectively.

To have equal resolution of shear and bending, the cell is designed with the stresses of the same magnitude. The shear and bending stresses are obtained [Timoshenko (1955)] as follows:

$$\sigma_v = \frac{F}{8}(l_v - 2x)\frac{6}{D^3} \tag{6.10}$$

and

$$\sigma_b = \frac{M_B}{2l_b D^2} \tag{6.11}$$

respectively where M_B = maximum bending moment; F = maximum shear load to be read; D = size of cell beams; l_v = length of beams; x = location of strain gauges from end of beams for measuring shear loads (assume x = D); and l_b = spacing between beams of cell.

By equating the bending due to shear and the axial/compression bending stresses, l_b can be determined.

$$l_b = \frac{2M_B D}{3F(l_v - 2x)}$$ (6.12)

By assuming various combinations of beam size and length, l_b can be calculated. By computing the maximum stresses in the cell, a final size configuration can be determined. The following dimensions were chosen for the present instrument: l_v = 50.8 mm (2.0 in.); D = 4.76 mm (0.19 in.); and l_b = 56.6 mm (2.23 in.)

6.15.2 Layout of Strain Gauges

To measure shear in a leg of the load cell, strain gauges must be placed near an end of a load cell leg and must be on opposite sides of the leg. The gauges are wired into two legs of a Wheatstone Bridge as shown in Fig. 6.21. In this manner, axial tension and compression in the leg, either from axial loads, weight, or bending, is automatically compensated for, while at the same time amplifying the effect of bending through the section.

In the present design, it was decided to perform an initial rough calibration with only one set of shear measuring gauges for the x and z directions. From this initial calibration attempt, it was discovered that the correct way to measure shear was to provide sets of shear measuring strain gauges on opposite legs of the cell. In this manner, in addition to measuring two shears to take an average, the unbalanced shear could be used to compute any torque loads that would be applied.

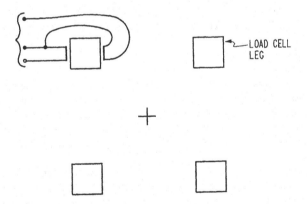

FIGURE 6.21
WIRING OF STRAIN GAUGES TO MEASURE SHEAR; ONE SET SHOWN

The strain gauges for the measurement of bending across the load cell are located at the mid-height on the load cell columns. This is where bending due to shear is a minimum and requires the least compensation. Additionally, the gauges are placed in series on opposite sides of two legs of the Wheatstone Bridge circuit as shown in Fig. 6.22. Any bending through the section is compensated for by the series opposing gauges.

Although this load cell was not specifically designed to measure axial loads, this option can be easily added. Axial load can be measured by adding eight gauges on opposite faces of all four legs. These gauges would all be in series and would be used as a quarter bridge.

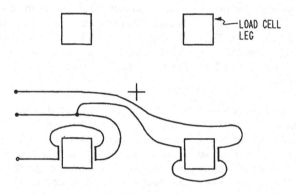

FIGURE 6.22
WIRING OF STRAIN GAGES TO MEASURE BENDING; ONE SET SHOWN

6.15.3 Calibration of Cell

The shear and bending moment cell was calibrated by being attached to an indexing table and applying various loads at a known distance from the cell. As a preferred practice, an initial check calibration was performed after attaching strain gauges, but before waterproofing. The final calibration of the load cell was done with additional gauges to measure unbalanced torques and with waterproofing in place.

If load cells of this type are to be used for measurement of loads on wave tank models, load cells of multiple elements could be considered. Standard end flanges could be machined, and various sizes of legs used to give the required stiffness and strength. By using this approach, load cells could be modified by substituting a different pair of legs and the initial cost would be greatly reduced because most of the intricate machining would be eliminated.

6.16   TWO-FORCE DYNAMOMETER

A dynamometer is used when there is a requirement to measure forces in specified directions while allowing the model to move in other degrees of freedom. Here, the design of a simple two-component force balance is described [Kowalski and Brown (1983)].

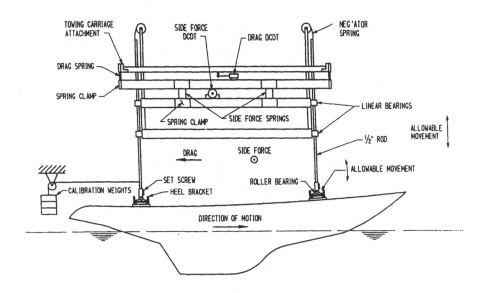

FIGURE 6.23
TWO-FORCE DYNAMOMETER (SIDE VIEW)
[Kowalski and Brown (1983)]

Let us consider a dynamometer which is capable of measuring forces in the inline and transverse directions. The arrangement of the springs is made as shown in Fig. 6.23. This allows all deflections in the respective directions to be accommodated by the appropriate springs. Thus, with careful alignment of springs, the interaction effect (cross-talk) is minimized.

The displacements may be measured by an LVDT or a direct current displacement transducer (DCDT). Typical error of a DCDT is about 1/2 percent. In the current design [Kowalski and Brown (1983)], a 12.7 mm (1/2 in.) by 76.2 mm (3 in.) stock (actual displacement length of 25.4 mm or 1 in.) was chosen for the force gauges.   The arrangement of the force cells for the side and inline forces is shown in Fig. 6.24. This arrangement allows measuring the towing (drag) load and side (lift) load on a towed body. The arrangement allows the model to heel, yaw as well as to trim during towing.

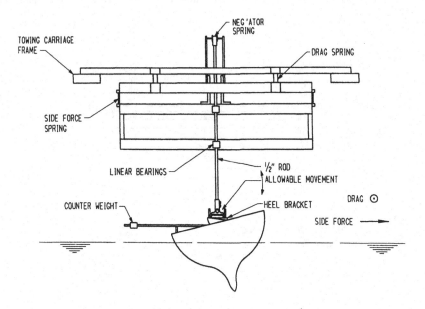

FIGURE 6.24
TWO-FORCE DYNAMOMETER (FRONT VIEW)
[Kowalski and Brown (1983)]

6.17   TOWING STAFF INSTRUMENTATION

A towing staff may be built in many different ways. The staff to tow the model is generally fixed such that the model is allowed to move in roll, heave and pitch. The staff is instrumented to measure tow speed, resistance, induced trim and sinkage of the model   (Fig. 6.25). The towing speed is measured using the moveable bridge's speed encoder. This encoder records the RPM of a wheel that is attached to the bridge as the wheel rolls along the top of the tank's side rails.

A load cell is attached at the point where the tow staff attaches to the model to measure the model's towing resistance. The induced trim angle of the model is measured with a resistive potentiometer mounted at the bearing that allows free pitch motion. The model sinkage is measured with a similar potentiometer that is mounted to measure the extension of a line attached to the bottom of the tow staff as the line unwinds from a pulley. The pulley is balanced with a constant tension spring. The mean drag force as well as some of the moments may be counterbalanced by weights mounted on adjustable arms (Fig. 6.25).

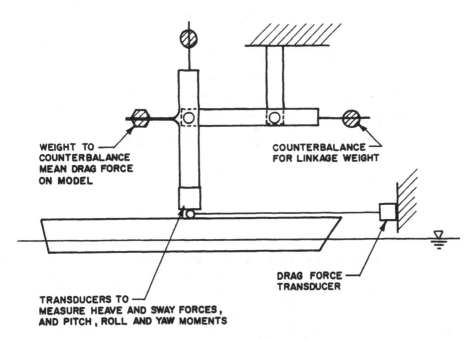

WEIGHT TO
COUNTERBALANCE
MEAN DRAG FORCE
ON MODEL

COUNTERBALANCE
FOR LINKAGE WEIGHT

DRAG FORCE
TRANSDUCER

TRANSDUCERS TO
MEASURE HEAVE AND SWAY FORCES,
AND PITCH, ROLL AND YAW MOMENTS

FIGURE 6.25
INSTRUMENTED TOWING STAFF

6.18 MECHANICAL OSCILLATION OF A FLOATING BUOY

It is often desired to force a structure to oscillate in prescribed manner in still water or in waves and/or current. The purpose of the oscillator is to induce scaled oscillation to the model to duplicate prototype behavior. Accordingly, the mechanical oscillator is often custom-made to tailor the need of the testing. Many testing facilities have a six degrees of freedom oscillator, such as PMM described in Chapter 4. Here we provide a specific example highlighting the requirements of instrumentation for such a system.

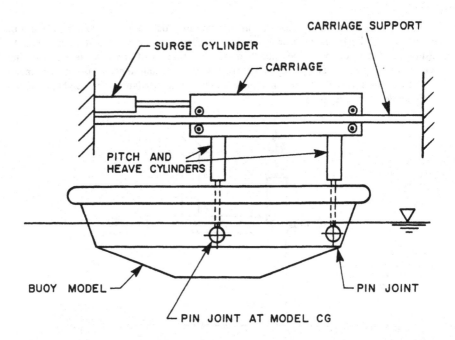

FIGURE 6.26
SCHEMATIC OF MECHANICAL OSCILLATION SYSTEM

An example of a buoy undergoing forced oscillation in several planes is shown in Fig. 6.26. The oscillator for the forced motion tests consists of three hydraulic cylinders and a horizontal carriage. Two of the cylinders are oriented vertically to produce heave or pitch of the model, depending on the relative motions of the cylinders. The third cylinder is positioned horizontally to produce the surge motion by moving the carriage structure back and forth. The two vertical cylinders are attached to the carriage and ride with it. The motion of each of the cylinders is controlled using a position servo system. Each servo loop operates by comparing the position of the cylinder (measured by a position feedback transducer) to its desired position and modulating the hydraulic pressure to the cylinder with a control valve, causing the cylinder to move until the feedback signal matches the command signal. A desired combination of heave, surge and pitch is possible if the three cylinders are driven simultaneously.

The horizontal and vertical loads on the model buoy during the forced oscillation tests are measured using load cells. The load cells are single axis devices and are mounted in pairs at the two hydraulic cylinder to buoy attachment points. Each pair is arranged to measure the horizontal and vertical loads applied to it. From this information, the total horizontal and vertical load and overturning moment are calculated.

The motions of the drive cylinders during the forced oscillation tests are measured by linear potentiometers mounted to the cylinders. These transducers are simultaneously used to measure the cylinder extension and to provide a feedback signal to the servo-system controlling the cylinder motion.

6.19 DATA QUALITY ASSURANCE

Since data acquisition is generally automated, the thrust of the quality assurance measures is aimed at the instrumentation. What follows is an outline of the routine steps usually taken to insure quality data.

Consider a test requiring the use of four types of instrumentation; load cells, wave probes, current meters and pressure transducers. Each load measuring unit may be comprised of three individual load cells assembled to measure loads in three orthogonal directions. Before the structure is placed in the wave tank each load cell assembly is calibrated in the three directions. A separate scale factor is determined for each cell. National Institute of Standard Testing (NIST) traceable weights are applied to one of the three directions to calibrate the active axis and to determine the amount of cross talk between cells. If the cross talk is small but linear, summing amplifiers are employed to remove the trend. Acceptable cross talk is typically considered to be about 2.0%.

After the structure is mounted on load cells in the tank, an in-place calibration is performed. Both static and dynamic calibrations can be performed. During a static calibration, a wire rope is attached to the structure and then slung over a pulley system. NIST traceable weights are placed in the wire and the resulting loads are measured on the load cell. The sum of all load cell readings in one direction must equal the applied load. The dynamic calibration employs a spring mass system. Knowing the mass, the spring constant and extension, the applied load is calculated and compared to the measured load.

The current is measured with a commercially obtained electromagnetic current meter. Typical published accuracies for these probes is $\pm$ 2% full scale output. The published scale factors of the manufacturer should be verified by a check calibration. The meter is "check calibrated" by attaching the probe to the moveable bridge and towing it through the water. Typically, several points are used for "check" calibrations. The entire usable range of the meter is usually covered to include the maximum value that is anticipated. The moveable bridge usually has a tachometer with a digital readout and a direct link to the data acquisition system. The tachometer is calibrated by integrating the output and comparing the results with the known travel distance of the bridge.

The manufacturer's published scale factors for pressure transducers are checked by submerging the transducers known distances in still water. Transducer scale factors are checked at a chosen increment over the usable range of the transducer.

All pieces of instrumentation are checked periodically during the test. At the beginning of each test day, the wave probes are "check calibrated". This consists of submerging and raising the probes known distances and comparing with the scaled electrical output. A static calibration of the model/load cell assembly can be performed each morning, if erratic behavior is discovered. Again, known NIST traceable weights are applied to the structure and this load is compared to the load cell outputs. The pressure transducers are also checked daily by running a benchmark wave over the structure and monitoring the difference in the transducer output between runs.

After each run, certain statistics are displayed on the control terminal to assure data quality, These may include for each data acquisition channel: maximum and minimum values, percent offset and percent usage. This allows the operator to immediately spot dead channels, drifting amplifiers or instrumentation, non-optimum amplifier or computer gains. Dead channels are easily spotted as they will show the same value for minimum and maximum values and zero percent usage. Non-optimum amplifier or computer gains are recognized from the small percentage of usable A/D capacity output. A drifting amplifier can be discovered by monitoring the percent offset value. This value will reflect the initial zero value which would vary as the amplifier drifts. In addition, a number of priority data channels are plotted on a CRT after each run and checked for magnitudes and trends.

6.20 REFERENCES

1.   Abdel-Aziz, H.A. and Abuelnaga, A., "A Two-Dimensional Strain Gauge Current Meter", Proceedings on Second Offshore Mechanics and Arctic Engineering Conference, Houston, Texas, ASME, 1983, pp. 771-777.

2.   Bazergui, A., Eryzlu, N.E. and Saucet, J.P., "A Submersible 3-D Load Transducer Platform", Experimental Techniques, September 1987, pp. 25-29.

3.   Bishop, J.P., "Measurements of Wave Particle Motion at the Christchurch Bay Tower Using a Perforated Ball Instrument", National Maritime Institute, NMI Report OT-R-7942, Feltham, U.K., July 1979, 55 pages.

4.   Chang, P.A., Ratcliffe, T.J., Rice, J. and McGuigan, S., "Two Novel Free Surface Measurement Techniques Developed at DTMB," Proceedings of the Twenty-Third American Towing Tank Conference, New Orleans, Louisiana, National Academic Press, Washingotn D.C., 1993, pp. 263-277.

5.   Goldstein, R.J., <u>Fluid Mechanics Measurements</u>, Hemisphere Publishing Corporation, Washington, D.C., 1983.

6.   Hardies, C.E., "An Advanced Two Axis Acoustic Current Meter", Proceedings on Seventh Annual Offshore Technology Conference, Houston, Texas, OTC 2293, May 1975.

7.   Kowalski, T. and Brown, D., "An Inexpensive Two-Component Force Balance for Towing Tank and Flumes", Proceedings of Twentieth American Towing Tank Conference, Hoboken, New Jersey, 1983, pp. 1207-1218.

8.   McNamee, B.P., Sharp, B.B. and Stevens, L.K., "Measurement of Water Particle Velocities in Waves", Applied Ocean Research, Vol. 5, No. 1, 1983, pp. 49-53.

9.   Sharp, B.B., "Flow Measurement With a Suspension Wire", Journal of Hydraulic Division, ASCE, Vol. 90, HY92, 1964, pp. 37-53.

10. Timoshenko, S.P., <u>Strength of Materials</u>, Third Edition, D. Van Nostrand Co., Inc., Princeton, New Jersey, 1955.

CHAPTER 7

MODELING OF FIXED OFFSHORE STRUCTURES

7.1 DESIGN LOAD COMPUTATIONS

Since the installation of the first piled platform in 30 ft. of water in the Gulf of Mexico about 40 years ago, the offshore industry has seen many innovative structures placed in deeper waters and more   hostile environments. Fixed structures have been placed in over 300 m (1000 ft) of water depth. For example, the Cognac platform was installed by Shell in the Gulf of Mexico in the eighties.   The design of such large offshore structures requires complex and sophisticated design tools. One of the most important aspects of the design method is the accurate determination of in-place wave forces on these structures. In the last 15 to 20 years the progress in the research on wave loading has been fast and steady. In this regard, measurement of wave loads on structure models has been necessary for the computation of design loads and verification of many analytical design tools.

Many of these design techniques for the wave force computation have been reviewed by Sarpkaya and Isaacson (1981), and Chakrabarti (1990, 1993). Wave forces on offshore structures are generally calculated in two different ways: (1) the Morison equation, and (2) diffraction theory.

The Morison equation arises from the fact that the accelerated flow separates from the surface of the submerged structural member forming a wake (low pressure region) "behind" the member.   It assumes that the force derived from this effect is composed of inertia and drag forces added together. These components require the knowledge of an inertia (or mass) coefficient and a drag coefficient which   are determined experimentally. Similarly, a transverse force exists on these members which is dependent on a lift coefficient. The Morison equation is usually applicable to structures whose dimensions are small compared to wave length when the drag and lift forces are significant.

When the size of the structure is comparable to the wave length, the presence of the structure is felt by the waves and is expected to alter substantially the wave field in the vicinity of the structure. In this case the diffraction of the waves from the surface of the structure must be taken into account in the evaluation of forces. While a closed form solution is possible for a few simple cases, the solution generally involves a numerical technique which solves the Laplace equation with associated boundary conditions.

A simple dimensional analysis shows that the nondimensional force is a function of depth parameter (d/D), Keulegan-Carpenter parameter (KC = $u_0 T/D$), Reynolds number (Re = $u_0 D/v$) and diffraction parameter ($\pi D/L$) where u_0 = maximum water particle velocity, D = structure dimension, d = water depth, T = wave period, L = wave length, and v = kinematic viscosity of water. The KC number is a measure of the importance of the drag force effect while the diffraction parameter determines the importance of the diffraction effect. Note from their definitions that when the KC number is large, the diffraction parameter is small and vice versa. Thus, a large diffraction effect necessarily indicates a small drag effect and, inversely, when the drag effect is large, the diffraction is generally negligible. These quantities may be used to determine whether the diffraction theory or the Morison equation should be used. For example, for $\pi D/L$ > 0.5, the structure is large enough to fall into the diffraction theory category. There are, however, structures with members that fall in both categories. These are called hybrid structures, and must consider both diffraction and drag effects.

The model testing of these different structure types will be discussed individually in the subsequent section.

7.2 SMALL-MEMBERED FIXED STRUCTURES

Small diameter circular cylinders are an integral part of many offshore structures. Numerous tests have been performed in the past thirty years to determine the hydrodynamic coefficients on cylinders due to oscillatory waves. Similarly, tests have been carried out in steady current, with harmonically oscillating cylinders or in oscillating flow. These tests differ in the range of the Keulegan-Carpenter number, Reynolds number, water depth, experimental set-up, measurement technique and often in the values of the inertia and drag coefficients. For a comprehensive and up-to-date review, the reader may refer to Sarpkaya and Isaacson (1981) and Chakrabarti (1993).

The data on vertical structural members are extensive. Inclined tubular members have received limited attentions in this respect [Cotter and Chakrabarti (1984), Chakrabarti (1985)]. In more recent tests on cylinders fixed in steady flow, the applicability of the independence principle has been investigated. The independence principle states that the inline forces on a yawed cylinder may be expressed in terms of the normal pressure forces, and the tangential velocity component can be neglected. It has been found that the independence principle applies to both subcritical and supercritical flow, but is not applicable to flow in the critical region. It has similarly been shown that the independence principle applies to a yawed cylinder in wave flow as well. Data are reduced in terms of normal kinematic components on the assumption of the validity of the independence principle.

The purpose of these tests with vertical and inclined tubular models is to determine the values of hydrodynamic coefficients. The hydrodynamic coefficients are expected to be functions of the KC and Re number. Therefore, the tests are designed to vary these parameters systematically. In waves, these parameters may change in value from one section of a structural member to another because of variation of particle velocities with depth. For this reason, a small submerged section of the member is instrumented to measure two orthogonal force components (inline and transverse). An instrument similar to what has been described in Section 6.14 may be used for this purpose. Regular waves past the fixed cylinder are generated. The cylinder may also be placed at an angle to the waves to include inclined tubular members of a jacket structure, for example. The instrumented section is generally placed near the free surface to cover the largest KC and Re range, but sufficiently away from it so that the free surface variation does not expose part of the instrumented section. The test cylinder is chosen to be long enough so that the three-dimensional end effects on the instrumented section may be minimized. In addition to the two-dimensional forces on the instrumented section, the total forces on the cylinder may be measured as reactions at its two ends. The measured inline forces from the instrumented section in conjunction with the Morison equation can be used to compute the inertia and drag coefficient. The velocity and acceleration of the water particle at the instrumented section are computed, by such means as the irregular stream function theory [Dean (1965)] using the wave profile above the center of the instrumented section.

The oscillatory water particle velocity is sometimes measured by using a velocity probe. This is desirable over the computed values. However, if Marsh McBirney current probe is used for this purpose, then one should be cautious about the inherent filter in the system and the associated phase shift. In fact, in any of these measurements the phase relationship between the measured quantities should be carefully maintained. Any small error in the phase angles may produce a large error in the computed coefficient values. The method of computation of these coefficients is described in Chapter 10. The mean values of the hydrodynamic coefficients are plotted versus the KC and Re number. The total forces on the cylinder are computed based on the mean curves of C_M and C_D and compared with measured total inline forces. This method permits the verification of the reliability of the measuring system. The lift coefficients and the lift force frequencies are computed from the transverse forces on the instrumented section.

7.2.1 Morison Equations

The Morison equation was introduced in Chapter 2 (Eq. 2.32). However, this empirical form was originally written in terms of the local water particle velocity and acceleration components as:

$$f = C_M \rho \frac{\pi}{4} D^2 \frac{\partial u}{\partial t} + \frac{1}{2} C_D \rho D |u| u \tag{7.1}$$

in which f = force per unit length, and where u and $\partial u/\partial t$ = local water particle velocity and acceleration. This form has since been extended such that the local water particle kinematics (partial derivatives) are replaced by the total water particle kinematics (material derivatives), e.g., the acceleration term $\dot{u}$ replaces $\partial u/\partial t$. The equation then reduces to:

$$f = C_M f_I + C_D f_D \tag{7.2}$$

in which

$$f_I = \rho \frac{\pi}{4} D^2 \dot{u} \tag{7.3}$$

and

$$f_D = \frac{1}{2} \rho D |u| u \tag{7.4}$$

Also, the horizontal velocity components on a vertical cylinder are extended to include normal components on an inclined cylinder

$$\underline{f} = C_M \rho \frac{\pi}{4} D^2 \dot{\underline{w}} + \frac{1}{2} C_D \rho D \left| \underline{w} \right| \underline{w} \tag{7.5}$$

in which $\approx$ denotes a vector and w and $\dot{w}$ are normal components of velocity and acceleration. In this case, the vector components along three orthogonal axis provide the magnitude of these forces.

This formulation does not account for the presence of the transverse or lift forces. For a vertical cylinder, the horizontal transverse force is written similar in form to the drag force as:

$$f_L = \frac{1}{2} C_L \rho D u^2 \tag{7.6}$$

where C_L is the lift coefficient. In this formulation, C_L varies over a wave cycle due to the random nature of the lift force. Therefore, in computing the lift force coefficient from measured data, either the maximum or the rms value of the lift force is considered. The coefficient is referred to as a maximum lift coefficient, C_{Lmax}, or an rms lift coefficient, C_{Lrms}. These are discussed further in Chapter 10. It should be noted that for an inclined cylinder not inline with or normal to the wave direction, the inline and lift forces are mixed together due to the complicated flow problem and can not be easily separated.

There are several factors that require consideration in designing a test to investigate hydrodynamic coefficients. In order to highlight these requirements, a typical test setup is described here which considered a tubular test section at different orientation to regular waves.

7.2.2 Description of a Test Setup

In a test to derive C_M, C_D and C_L, the measurement of forces should be made locally. Typically, the test structure is a cylinder of circular (or square) section and is composed of three sections. The center section of the test structure is an instrumented section of small length. The instrumentation should measure loads in two orthogonal directions. The test cylinder is made of metal or plastic material and of sufficient thickness to ensure rigidity under wave pressures. The cylinder outer surface is sanded smooth to provide smooth cylinder hydrodynamic coefficient values. In order to study the effect of marine growth on structural members, the test cylinder may be roughened with sand particles uniformly coated on its surface. The test section is floated on an internal load cell separated from the lower and upper cylinder sections. The inner part is maintained continuous with the other sections. The gaps between the sections are kept to a minimum, being less than 1.6 mm (1/16 in.). Some tests have been performed by covering the gap with soft elastic materials. However, experience has shown that any attempt to cover the gap with as pliable a material as thin plastic wrap introduces nonlinear behavior during calibration, giving difficulty in arriving at an acceptable scale factor. Since the inner part is full of water, the channeling of flow through these gaps in wave cycles generally has very small effect on the measured forces. Any pressure build-up inside the instrumented section is uniform and does not contribute to the measured forces.

In the laboratory, there are three basic approaches in determining these hydrodynamic coefficients: (a) oscillating the model in a harmonic motion in still water, (b) oscillating the water past the fixed model in a harmonic motion, and (c) generating waves past the fixed model which could be placed vertically or horizontally. There is one other test condition that is seldom employed in a test of this nature because of the complex nature of the flow field, and that is: (d) generating waves past the model that is allowed to oscillate freely in waves, e.g., in the case of an articulated tower. In this case, the added mass and damping coefficients associated with the structure motion are intermingled with the coefficients from the exciting forces on the structure from the waves. At the present time, one of the methods (a) or (c) is used. The methods (a) and (b) are equivalent and will produce the same result.

It is the intent during these test phases to allow the cylinder to experience as wide a range of Keulegan-Carpenter number and Reynolds number as possible. In a typical wave tank test, the KC number based on water particle velocity varied from 0 - 50, and the Re number varied up to 100,000. As discussed in Chapter 2, the difficulty of such tests is the limited range of Re number achievable in a wave tank environment. It is usually two orders of magnitude smaller than the prototype value. Thus, scaling of data from such tests is difficult as the model flow field may be markedly different. These tests are, however, quite valuable in understanding the basic physics of fluid-structure interaction problem in the combined drag-inertia flow regime.

The data on the inline forces are analyzed by a least square technique. The harmonic components of the lift force are obtained by a Fourier series analysis. The details of these analysis methods will be discussed in Chapter 10.

During each wave test run, the following measurements are generally recorded: the inline (x) and transverse (z) loads on the instrumented section, the inline loads at the restraints at the top and bottom of the instrumented cylinder, and the wave profile at the center of the test section. A sample output of the measured data is shown in Fig. 7.1.

The water particle kinematics, namely, the particle velocity and acceleration, may be derived from the wave profile using a suitable (e.g., irregular stream function) theory. The inertia (C_M) and drag coefficient (C_D) are determined for each wave cycle in a test run from the measured loads and the water particle kinematics. In the estimation of C_M and C_D for a given Keulegan-Carpenter number, a mean curve is drawn through the scatter of the variable C_M or C_D when plotted versus KC number. The scatter in the data may be quantified in terms of a coefficient of variation (COV) value at various KC regions (over a small Re range).

Let the collection of estimates for C_M (or C_D) at a given KC number (in reality, a small range about a mean value is chosen) be y_i (i = 1,2,...N) and the chosen mean value of the estimate be $\bar{y}$. Then, the random error of the estimate is defined by the standard deviation of the estimate about the mean value as:

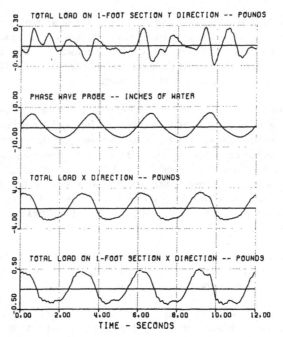

FIGURE 7.1
TYPICAL DATA FOR A FIXED VERTICAL CYLINDER IN WAVES

$$\text{random error} = \sigma_y = \frac{1}{N}\left[\sum_{i=1}^{N}(y_i - \bar{y})^2\right]^{\frac{1}{2}} \qquad (7.7)$$

The coefficient of variation is defined as the normalized random error

$$COV = \frac{\sigma_y}{\bar{y}} \qquad (7.8)$$

Figure 7.2 demonstrates the dependence of mean C_M and C_D values on the KC number for a vertical cylinder [Cotter and Chakrabarti (1984)]. The COV values at increments of KC = 5 are shown on these plots. The general trends of COV values show that the COVs are low for C_M and high for C_D at the low KC values. The COVs are low for C_M at the highest KC values in this example simply because the number of data points in this area is low. In general, scatter in C_D is high at low KC and scatter in

C_M is high at high KC values. This is not deemed to be a significant problem, however, since loading becomes inertia dominated at low KC, and drag dominated at the high KC regions. Thus, large variations of C_D or C_M in these respective regions will have little effect on the total load on the cylinder. Dean (1976) has presented a detailed discussion of the influence of loading regime on the uncertainties in C_D and C_M. The larger scatter in these coefficients could be attributed, at least partially, to a lack of discrimination in the selection of a suitable weighting function for each particular coefficient in these regions. The scatter also seems to be high for both C_M and C_D in the KC range of 10-15, where the values of these coefficients have been found to show the most variation in earlier tests. Variation of C_M and C_D with the Re values has been clearly demonstrated by Sarpkaya (1976) in his tests.

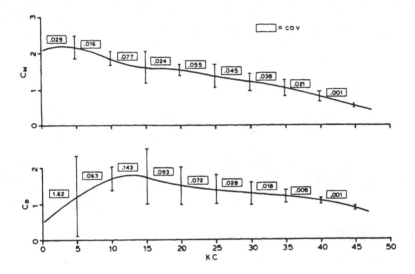

FIGURE 7.2
C_M and C_D VERSUS KC FOR FIXED VERTICAL CYLINDER

These values of hydrodynamic coefficients help compute the wave loading of practical space frame structures. Sometimes, the complete structure is model tested [Allender and Petrauskas (1987)] to verify design or to investigate certain specific area, such as, member interaction effect.

7.2.3 Pressure Profile Around a Cylinder

The total wave force on an offshore structure is developed from the local wave forces on the individual members making up the structure. The local forces described in the preceding section are empirical in nature. Wave forces on a member result from the distribution of pressures on the surface of the member caused by a passing wave. A description of the pressure distribution around the member can provide valuable information in the description of the local force. Such a description may be obtained through model tests.

The cylinder is subjected to regular waves of varying height and period. The distributions of the measured pressures around the cylinder are determined in a combined inertia and drag dominated field. The measured pressure profiles from these tests may help develop a computational tool for these forces eliminating the need for the empirical form of the Morison equation.

In one test, pressures were measured at ten points around the perimeter of a smooth cylinder at an elevation of 0.3 m (1 ft) below the still water level. The vertical cylinder (Fig. 7.3) was fabricated from a clear plexiglas tube divided into three sections: an upper tube section; a lower tube section; and a pressure transducer block mounted between them. The pressure transducer block was machined from a solid piece of clear plexiglas. Ten ports were drilled radially into the block at equally spaced intervals around its circumference. A pressure transducer (Fig. 6.12) was connected to each port to measure the dynamic pressure at that point due to wave action (Fig. 7.3). Commercially affordable pressure transducers are generally not waterproof. These transducers may be individually waterproofed by using commercially available potting compound. Alternatively, they may be encased in a water-tight section (Fig. 7.4).

7.2.4 Multiple Cylinder Tests

Many offshore applications include groups of small cylinders. Riser configurations consist of small diameter tubes in a circular array. Many offshore structures have multiple tubes in close proximity to each other. In particular, the jacket type of drilling platforms that exist in the Gulf of Mexico and in other parts of the world have cylindrical components which experience a considerable interference effect from the neighboring components. Therefore, it is important to know the effect of the neighboring cylinders and the spacing between these cylinders on the forces imposed on a cylinder in the array. Because of the small cross-section of these cylinders, drag and lift forces as well as the inertial forces on these cylinders are equally important.

Testing [Chakrabarti (1982)] with an array of tubes in waves (Fig. 7.5) is similar to the single cylinder tests. In this case, one or more cylinders in the array or group may be

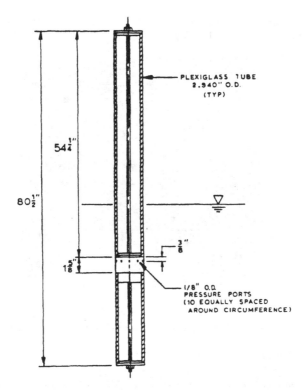

FIGURE 7.3
TEST SETUP FOR PRESSURE PROFILE ON A VERTICAL CYLINDER

instrumented locally for XY loads. Moreover, allowance is given in the setup to provide different spacing among cylinders so that the effect of spacing variations on C_M, C_D, and C_L may be studied. Dependence of coefficients on parameters other than spacing is similar to a single cylinder test.

7.3   SEABED PIPELINE TESTING

Offshore pipelines are used in transporting oil or gas from production facilities to on-shore or offshore facilities for processing and distribution. The pipelines are placed on the seafloor and are anchored or placed in trenches. The tendency of a pipeline to self-bury dictates the trenching requirements for laying the pipe. If the pipe does tend to bury itself in a particular soil or under particular loading scenarios, trenching becomes unnecessary. Typical methods of testing the stability of these pipelines in a wave tank are described.

FIGURE 7.4
CYLINDER BLOCK SHOWING PRESSURE TRANSDUCERS

7.3.1   Theoretical Background

In common practice, pipelines are designed so that they do not move laterally under hydrodynamic loading. In order to prevent motion, the designer specifies a concrete coating or other form of ballast that will prevent sliding by employing a Coulomb friction model:

$$R_s = \mu \left(W_n - F_v \right) \tag{7.9}$$

in which R_s = soil resistance force; μ = friction coefficient; W_n = net submerged weight of the pipe; F_v = vertical component of the hydrodynamic force.

In order to ensure pipeline stability, R_s must be greater than F_H, the horizontal component of the hydrodynamic force. This requires an adequate description of F_v, F_H and μ.

From the test data, the friction coefficient for sliding and ultimate sliding is sought. The coefficient of sliding, μ, is defined as the ratio of horizontal load to vertical load which will cause small displacements of the pipe. In this load regime, the pipe moves a short distance back and forth and builds up a berm at either end of the traverse. The coefficient of ultimate sliding, μ_u, is determined when the pipe begins to displace upward, rides over the sand berm in front of the pipe and begins to slide freely.

FIGURE 7.5
TRANSVERSE THREE CYLINDER ARRAY IN WAVES

The friction coefficient depends highly on local conditions. Lateral sliding tests have shown that the Coulomb friction model is generally valid for smooth pipes sliding on a sand bed and having a value of μ of 0.6 to 0.7. However, when the pipe is artificially roughened, the value of μ may increase by 20% to 30%. On soft clay, the value of μ may range from 0.2 to 0.8 and the value is dependent on the pipe diameter and the net weight, W_n. These tests indicated that pipes do not settle into sand but do settle into clay which changes the sliding behavior of the pipe. Thus, the data indicated that the Coulomb friction model was adequate for specifying the minimum pipe weight on sand but not on soft clay.

Another scenario is possible in which the pipe begins to oscillate harmonically on top of the soil foundation. In this case, the motion can be described by the following equation:

$$F_H = C_M M \ddot{x} + C \dot{x} + K x \qquad (7.10)$$

where M = mass of the pipe; $\ddot{x}$ = horizontal acceleration of the pipe; C = system damping; $\dot{x}$ = horizontal velocity of the pipe; K = system stiffness; x = horizontal displacement of the pipe.

The mass of the pipe is determined during calibration. The mass coefficient is chosen from test based upon the Keulegan Carpenter number determined from the test configuration. The horizontal force, F_H, and the displacement, velocity and acceleration are determined from the test. For each test, the pipe loading, pipe displacement and acceleration are known from direct measurement.   The pipe velocity may be determined by integrating the pipe acceleration measured by an accelerometer mounted on the pipe. A check between the measured acceleration and displacement can be made through a double integration of the acceleration signal, which may establish confidence in the derived velocity signal.   The inertia of the pipe is computed by multiplying the acceleration by the mass of the pipe.

A net load is calculated at each sample time:

$$F_{NET} = F_H - C_M M \ddot{x} \tag{7.11}$$

Assuming harmonic motions, the system stiffness and damping can be determined from the in-phase and out-of-phase components of the Fourier representation of F_{NET}.

Relationships are sought for the friction coefficients, μ and μ_u, for the system damping and stiffness, C and K, respectively, and for trench or burial depth with non-dimensional parameters related to soil properties, the pipe and the load.   These parameters are defined [Karal (1977)] as:

- Parameters related to clay

$$\kappa = \frac{\gamma a^2}{W_n} \tag{7.12}$$

$$\lambda = \frac{S_u a}{W_n} \tag{7.13}$$

where κ and λ are dimensionless parameters related to unit weight and undrained shear strength of the soil respectively, γ = specific weight of submerged soil; a = pipe radius; and S_u = undrained shear strength of soil.

- Parameters related to sand

ϕ = angle of friction

$$\kappa = \frac{\gamma a^2}{W_n} \qquad\qquad (7.14)$$

$$\lambda = \frac{c_u a}{W_n} \qquad\qquad (7.15)$$

where c_u = cohesion of soil.

- Parameters related to the pipe

$$\delta^* = \frac{\delta}{D} \qquad\qquad (7.16)$$

$$\varepsilon = \frac{d_p}{D} \qquad\qquad (7.17)$$

where δ^* = surface roughness coefficient; δ = mean diameter of roughness particles; D = pipe diameter; ε = pipe submergence ratio; d_p = depth of submergence of pipe.

- Parameters related to loading

$$C_W = \frac{F_H}{F_V} \qquad\qquad (7.18)$$

$$S_W = \frac{2a\sigma_u}{F_H} \qquad\qquad (7.19)$$

where C_W = hydrodynamic force ratio; S_W = force coefficient; σ_u = reduction of hydrostatic pressure on the front of the pipe due to water particle motion.

The importance of these various parameters in affecting the soil resistance forces is determined by plotting the measured forces in nondimensional form versus these parameters. The important parameters are thus identified. The relationships between the forces and these test parameters are established in empirical forms by fitting mean curves through these data points. These empirical coefficients may then be used in simple soil resistance formulations.

7.3.2   Model Testing

A test program to measure the soil resistance forces on a submerged pipeline and the tendency of the pipeline to self-bury or create a depressed trench area in waves or under a mechanically imposed hydrodynamic load is described here.

This test incorporates a pipe, modeled for stiffness, diameter and weight, resting on the soil foundation in a wave environment. Current may also be superimposed on the waves. The pipe model will span the soil box and be supported at its ends by springs modeled to represent the restraint of the adjacent pipe. Strain gauges can be applied to the pipeline to measure the time history of axial and bending strain.

The test matrix is developed to determine the effect of Keulegan Carpenter number, pipe diameter, weight and stiffness, and type of soil foundation. Using this test setup, the tendency of the pipe to create a trench or to self-bury is modified by the sand transport past the pipeline due to the wave action. In addition, if scour occurs under the pipeline, the pipeline will sag under its own weight and will modify the scour behavior. The scour, sagging, trenching or self-burial processes will continue until a steady state condition is reached for a given set of environmental conditions.

7.3.2.1   Pipeline Model

The subsea pipe generally has a large span so that the pipeline model may be treated as a two-dimensional structure in a wave tank test. Because of the proximity of the pipe to the seafloor, modeling of the soil characteristic of the foundation may be important.

The weight of the pipe should be adjusted to model the typical submerged weight of pipes of this size. The length of the pipe model should be sufficient to minimize end effects. The pipe model should be capped at the ends. If scouring at the ends of this longer pipe model is considered a problem, vertical walls may be placed close to the pipe ends to simulate the two dimensional effects desired. This should be weighed against possible effects from soil build-up between the pipe ends and walls.

Several pipe models are chosen so that a large range of KC and Re number may be appropriately and uniformly covered by the test runs. Pipe models may be tested in the roughened condition as well as in the smooth condition. The roughness on the pipe surface is applied using commercially available uniform sand particle coated paper.

A small section of each pipe is instrumented to measure loads in two orthogonal directions, horizontal and vertical. The pipe model is made up in three sections as shown in Fig. 7.6. The two end sections are "dummy" sections and are included to eliminate end effect loads on the instrumented section of the pipe. The sections are rigidly attached to the trench bottom. Several different trench geometries may be chosen for these tests. The pipe is assumed to be supported by the trench surface in all cases. Figure 7.7 shows three possible trench configurations.

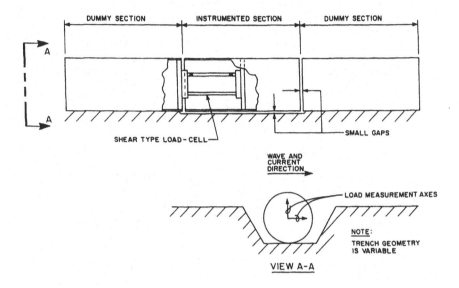

FIGURE 7.6
PIPE MODEL

7.3.2.2 Wave Tests

The water depth is chosen to isolate free surface effects from the test bed and the pipe. One of the test sections may contain a deep box which may be filled with the seabed material, for example sand or soft clay. The box should be flush with the rest of the bottom (see Fig. 7.8). Regular wave tests are recommended with systematic variation of KC and Re. For example, short period waves are generated which have heights that correspond to the same KC number as some of the long period waves. This will allow the correlation of the effect of the orbital shape parameter ($\varepsilon = v_0/u_0$) on the force coefficients. It has been found in certain limited tests of this nature that the coefficients are higher in shallow water ($\varepsilon < 0.7$) in comparison to the corresponding deepwater values ($\varepsilon > 0.9$).

In one series of tests, KC and Re number varied from 3–160 and 0.5 x 10^5–3.6 x 10^5 respectively. Two dimensional effects may be simulated with very long period waves. For example, for a 15 sec, 0.3 m (1 ft) high model wave with a 50.8 mm (2 in.) diameter pipe, a simple calculation based on linear theory gives:

$$KC = \frac{u_0 T}{D} = \frac{2\,ft\,/\,sec \times 15\,sec}{2/12\,ft} = 180 \qquad (7.20)$$

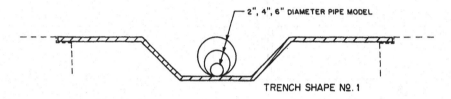

TRENCH SHAPE NO. 1

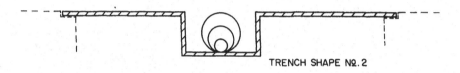

TRENCH SHAPE NO. 2

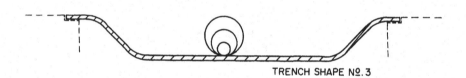

TRENCH SHAPE NO. 3

FIGURE 7.7
SAMPLE TRENCH GEOMETRIES

$$Re = \frac{u_0 D}{\nu} = 2 \times \frac{2}{12} \times 10^5 = 0.33 \times 10^5 \qquad (7.21)$$

For a 152 mm (6 in.) diameter pipe, the Re number will be 1 x 10^5.

7.3.2.3 Force Servo-Control Mechanism

In order to mechanically simulate two-dimensional flow, the pipe model may be oscillated harmonically in water. Two orthogonal oscillating loads are applied

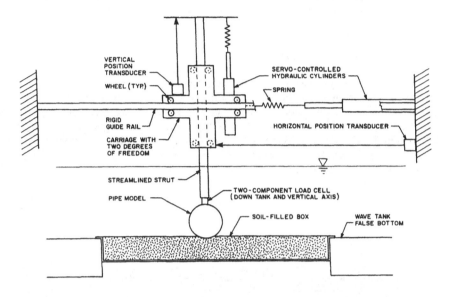

FIGURE 7.8
TYPICAL TEST SETUP OF A PIPE ON SOIL BED

simultaneously to the pipe model while it rests on the soil. The simulated load time histories represent horizontal and vertical loads on the pipe experienced in waves. Both regular and random wave loads may be simulated. The wave loads may be produced mechanically using the system shown in Fig. 7.8. A two axis carriage may be used to support the pipe model so that it can translate in the downtank and vertical direction. The pipe is restrained in the crosstank axis and is fixed against rotation. The simulated wave forces may be applied to the pipe using the loading mechanism shown in Fig. 7.9. The loading mechanism is similar for both axes and consists of a servo-controlled hydraulic cylinder arranged to push or pull the carriage. A spring element with a linear load-displacement relationship connects the cylinder to the carriage. The spring is included in the system to make the forcing system stiffness such that the full displacement of the hydraulic cylinder creates the maximum required force on the model. In this manner, the resolution and stability of the loading system are maximized.

Stable operation of the loading system requires that the pipe does not break free of the soil but is continually restrained to some degree by soil resistance. Small amplitude motions, as from soil compression, will not adversely affect system stability. If the pipe breaks free completely from the soil such that the foundation cannot supply the resistive forces required to balance the driving environmental loads, a servo-control displacement system may be used and the resulting forces may be recorded.

7.4 LARGE FIXED STRUCTURES

Unlike small-membered structures or offshore pipelines, load testing of large fixed structures require a different setup in the tank. For an offshore structure whose submerged member dimensions are large compared to the wave length, the incident waves undergo significant scattering or diffraction in the region of the structure. Diffraction theories are established as valuable tools in the design community and their limitations are known [Chakrabarti (1993)]. Numerous model tests have been performed in the process of designing large fixed offshore structures.

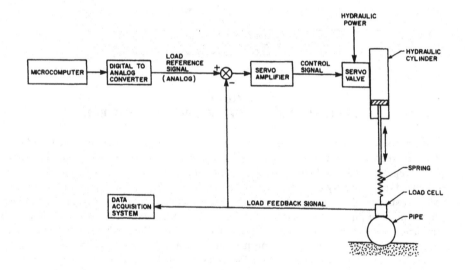

FIGURE 7.9
SERVO-CONTROL LOADING MECHANISM
FOR PIPE MODEL

7.4.1 OTEC Platform

A practical application of diffraction theory was made on a conceptual platform of an Ocean Thermal Energy Conversion (OTEC) system. A model test of a generic OTEC platform was performed in a wave tank (refer to Chapter 3). The generic design of the shelf-mounted OTEC platform consisted of four vertical columns and a large power module of square cross-section. These modules represented the submerged evaporators and condensers and were solid or uniformly perforated. The type (perforated or solid), size and location of the modules were varied during the test. One purpose of the test was to determine the suitability of diffraction theory for computing the global wave forces on the platform. Local dynamic pressures at selected locations on the surface of the modules were also measured.

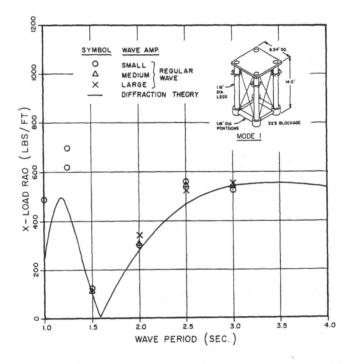

FIGURE 7.10

COMPARISON OF EXPERIMENTAL AND THEORETICAL WAVE LOADS ON MODE 1 OTEC MODEL

Transfer functions (Chapter 10) or Response Amplitude Operators (RAO) were calculated from the measured loads in regular and random waves.   Linear diffraction theory was used to   predict the global wave loads and the   local pressures on the structure.

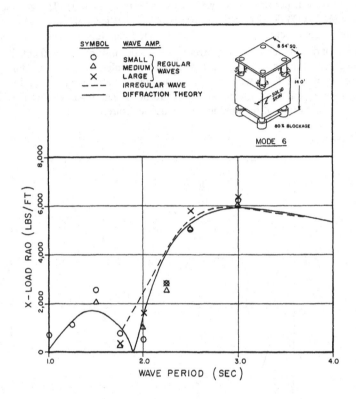

FIGURE 7.11

COMPARISON OF EXPERIMENTAL AND THEORETICAL WAVE LOADS ON MODE 6 OTEC MODEL

Comparisons of the load for two of the structure modes are shown in Figs. 7.10 and 7.11. The two modes chosen are the most transparent (Mode 1) and least transparent (Mode 6) modes. Mode 1 is the basic OTEC platform (Fig. 3.1). Mode 6, on the other hand, is the structure with a large box-type solid module near the surface. The theoretical loads compare very well with the measured loads in both regular and irregular waves. Note that the wave cancellation effect is clear on the four-column (as well as the solid) structure. At a wave period of about 1.60 sec, the column spacing is about half the wave length, such that the wave forces on the columns cancel each other. Thus, a multi-legged structure, e.g., a tension leg platform, may be designed based on a dominant period for a specific offshore location so that the wave forces are small at this period.

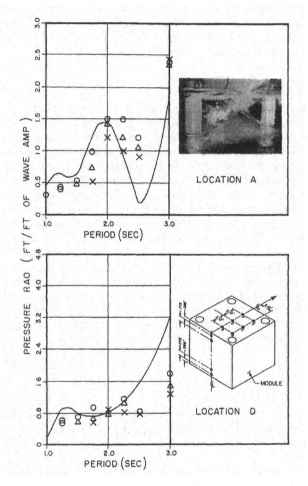

FIGURE 7.12
NORMALIZED DYNAMIC WAVE PRESSURE VERSUS WAVE PERIOD

Figure 7.12 presents a comparison of the experimental and theoretical pressures at two locations on the Mode 6 module. Both pressure transducers are located on the top of the modules. Because of the flat surface on the modules, higher waves were observed to slam on the modules, particularly at higher periods. Considering this large deviation in flow from the potential flow assumption (Fig. 7.12 inset), the correlation is understandably poorer. These pressures, however, make no contribution to the horizontal force on the platform. This example demonstrates that for certain large objects, even though local pressure at certain location may not be represented well by potential theory, the global loads may still be predicted well.

7.4.2 Triangular Floating Barge

Another practical example for the diffraction/radiation theory is a floating structure which examines the motion analysis. The typical floating draft of a large steel gravity platform is quite deep. However, because of the limited water depth normally available at a construction site, this provides a constraint in the design of the deck and the deck equipment. One approach of reducing this draft during towout of the platform is to separate the deck from the platform. Thus, the deck is designed as a floating body and transported by itself to a sheltered deep water site for placement on the platform before subsequent towout to the installation site. One example of this type of structure is the gravity platform for the Maureen Field in the North Sea in which a "hi-deck" was floated separately and was placed as the deck on the installed structure.

A conceptual design of a triangular deck acting as a floatover barge, e.g., a hi-deck design, was tested in a wave tank. This shape is also prevalent in a jack-up drilling unit deck which floats during towout. The shape is rather unique in that the structure is large with a shallow draft and placed near the free surface. Therefore, the applicability of the linear diffraction/radiation theory may be questioned. The purpose of the test was to determine the seakeeping characteristics of the barge.

7.4.2.1 Description of Model

The tribarge is triangular in shape with the corners clipped. It is supported in place as a deck on three corner columns of a platform 9.7 m (32 ft) in diameter. The barge itself is 100 m (328 ft) on each side. The overall height of tribarge is 3 m (10 ft). The prototype barge is shown in Fig. 7.13.

A 1:48 scale model of the triangular floatover barge was built. The model was built out of 19 mm (3/4 in.) marine plywood. Aluminum pipe sections of 203 mm (8 in.) internal diameter were inserted in the three holes for the corner columns. All plywood and pipe sections were glued together and caulked to make them water tight.

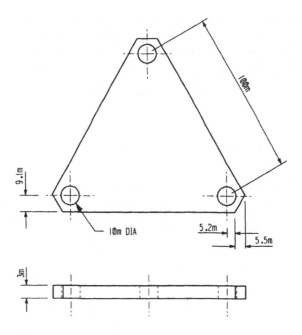

FIGURE 7.13
PROTOTYPE TRIANGULAR BARGE DIMENSIONS

The weight of the barge alone was 1113N (250 lbs). The barge was ballasted to 1736N (390 lbs) to provide a draft of 63.5 mm (2-1/2 in.). The center of gravity and dry moments of inertia of the barge in various planes were scaled by appropriately locating the ballast weights.

7.4.2.2 Test Setup

The barge was tested in two different modes. In the first mode the model was held fixed by two load cells in the vertical direction and one load cell in the horizontal direction. The load cells were ring type strain gauge devices. In-place calibration was performed by hanging weights from the model in the horizontal and vertical direction and recording the loads in the load cells. The wave heights were measured by capacitance type wave probes. Incident wave heights were obtained in the absence of the model in the wave tank, which is a standard practice for this type of tests.

In the second mode of testing the model was set free to heave, pitch and surge. In this case, three ten-turn potentiometers were attached vertically to the edge of holes in the model barge. In addition, a potentiometer was connected horizontally to the model through a pulley. These lines were at a constant tension of 4.45 N (1 lb) each. No springs were present in the lines. The heave, pitch and surge motions of the model were derived from the potentiometer measurements. The natural periods in pitch and roll were determined by pluck test with the floating ballasted model in water. Added mass and damping at the natural periods were determined (Chapter 10) from these data.

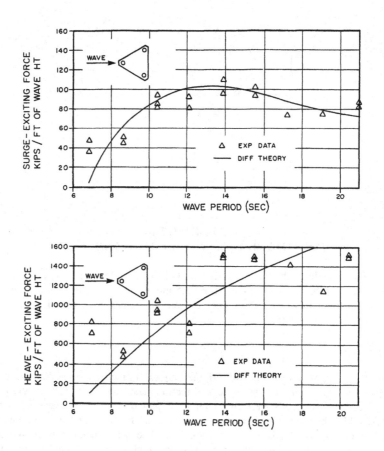

FIGURE 7.14
COMPARISON OF EXCITING FORCES IN SURGE AND HEAVE ON A FIXED TRIBARGE

7.4.2.3 Test Results

The measured forces and motions were compared with the linear diffraction theory results. The motion analysis was performed as a coupled three degrees of freedom system based on the frequency dependent hydrodynamic coefficients derived from the radiation theory.

The surge and heave exciting forces on the barge in the head sea orientation of the model are shown in Fig. 7.14. The data are scaled up to the prototype values using Froude's law. The corresponding theoretical results from the linear diffraction theory are shown in the figures. Note that the scatter in the experimental data is generally small indicating linearity of the responses. The comparison with the theoretical results is good for the surge force on the fixed barge, albeit a scatter in the heave force partly attributable to the flow through the holes in the area of the three platform columns which is not accounted for in the theory.

The motion responses in surge, heave and pitch are compared in Fig.7.15 for the head sea orientation. The normalization of data and comparison in all cases are found to be far better than the exciting force comparisons.

This example demonstrates that the linear diffraction theory may not predict the forces on large flat-bottom, shallow-draft structures well. However, such a structure, when free to move, has a more forgiving nature, and linear theory does a good job in predicting the motion of the structure.

7.4.3 Large-Based Structures

Many offshore structures are designed and constructed on the basis of using the seafloor as a sealed foundation. However, wave force testing of gravity structures [Hansen, et al.(1986)] is typically done by mounting the structure on load cells and positioning the model as close to the model seafloor as possible but still maintaining a uniform gap beneath the structure. It is recognized that this testing method will result in erroneous measurement for a bottom seated gravity structure, resulting in lower vertical forces due to the test setup, and corrective measures in the computation are needed. This technique will be discussed in detail later. In this section, methods of modeling these structures are described.

A cylindrical tank may have use as an offshore storage facility in relatively shallow water. The tank is held in place by gravity and friction which prevent sliding and overturning of the structure from the wave action. It is important to determine these wave loads for such a structure which can be obtained experimentally. Since the model should be positioned on the floor of the wave tank, a special technique is required to model this situation.

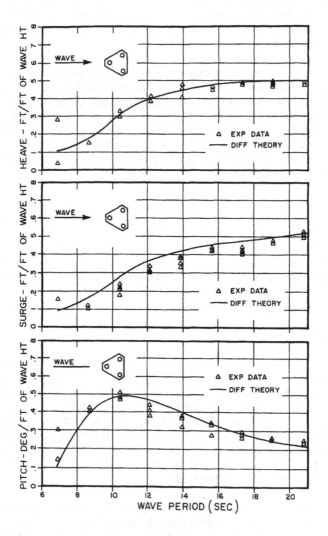

FIGURE 7.15
COMPARISON OF MOTION RESPONSE OF A FLOATING TRIBARGE IN REGULAR WAVES

Such a test [Chakrabarti and Tam (1975)] was performed with a 1:48 scale model of a five hundred thousand barrel cylindrical storage tank. The test was performed in a wave tank installed with false floor (Fig. 4.10). A circle slightly larger than the base of the storage tank model was cut out of the steel section of the false floor. The circular plate was supported by load cells in both the horizontal and vertical directions (Fig.7.16). The ballasted model was placed on this circular plate.

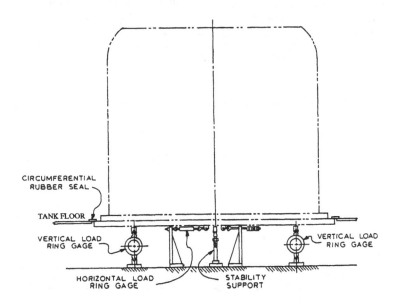

CIRCUMFERENTIAL RUBBER SEAL

TANK FLOOR

VERTICAL LOAD RING GAGE

VERTICAL LOAD RING GAGE

HORIZONTAL LOAD RING GAGE

STABILITY SUPPORT

FIGURE 7.16
LARGE CYLINDRICAL STORAGE TANK MODEL TEST SETUP

The circular plate was attached to two vertical load cells on the inline center line. A third (free) support located on the transverse center line of the model provided stability to the test platform without affecting the load readings. The two vertical load cells, acting as a couple, resisted and recorded the wave overturning moment on the structure. These cells were preloaded by the weight of the test plate, and the model. The horizontal cell was pretensioned with a stack of belleville washers. The spring constants of these washers were carefully selected so as not to attenuate the force reading by the spring-mass system. A soft rubber gasket material was placed at the cutout between the floor and the circular plate to seal the false floor of the wave tank.

The entire model set-up was recalibrated in place. A cable and pulley system was used so that known calibration weights could be used to exert a horizontal force at the top of the model. For each   horizontal loading, a moment   was also determined. Satisfactory reading on the load cells assured proper set up.

Alternatively, another method of testing for the true horizontal and vertical loads on a sealed gravity structure is to use a rolling hydrodynamic seal [Brogren and Chakrabarti (1986)] between the tank floor and the model. The hydrodynamic seal, or gas bag, is made by cutting two thin, soft, plastic sheets in the exact shape of the footprint of the model. The edges of the plastic sheeting may then be welded or glued together and provisions made for the attachment of hoses.

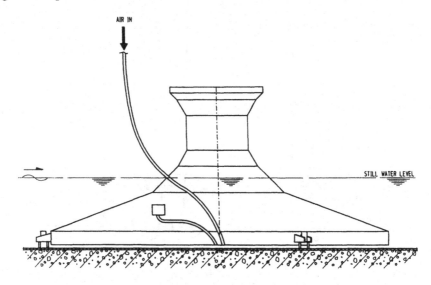

FIGURE 7.17
GRAVITY BASED ARCTIC PRODUCTION PLATFORM
[Brogren and Chakrabarti (1986)]

The seal is placed in the gap beneath the model and the tank floor and inflated with air to a pressure greater than the combination of the static water pressure and the expected dynamic wave pressure. Additional pressure is necessary since water at wave pressure can penetrate between the gas seal bag and the tank floor or between the seal bag and the underside of the model.

In order to verify the bottom pressures underneath the gravity structure and the effect of the seal to the overall loads as well as the bottom pressures, the test on a model of an arctic drilling structure (Fig.7.17) was conducted in two phases. In the first series of test, the gap between the model and the tank floor was left open, and the pressures and loads were recorded. In the second phase, the gap was sealed by a bag and the test was repeated.

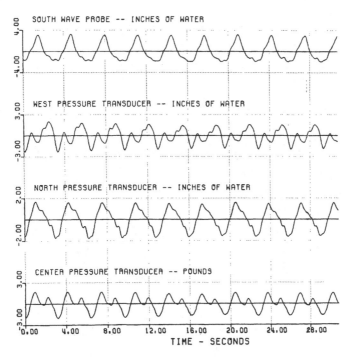

FIGURE 7.18
PRESSURE PROFILES IN REGULAR WAVES UNDER A SMALL GAPPED STRUCTURE

The test results are summarized in Figs 7.18 and 7.19. Figure 7.18 shows the pressures at three points under the model when a gap existed between the model and the tank floor. The gap was on the order of 6.4mm (1/4 in.). The pressures are due to a shallow water regular wave of 3.25 sec period and about 108 mm (4-1/4 in.) high. Note that both the magnitudes and phases of the pressures at different points under the model are different. Thus, it is not easy to correct for the pressures under the model in the measured vertical loads and overturning moments to compute the loads on a corresponding seated model.

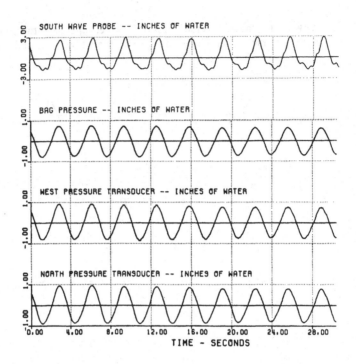

FIGURE 7.19
PRESSURE PROFILES IN REGULAR WAVES UNDER A SEALED
STRUCTURE

Figure 7.19 shows the pressures from the north and west transducers, as well as the bag pressure when the air bag was used to seal the underside. This plot shows that the pressures at different points under the model are nearly equal in amplitude and phase; the small difference being attributed to the elasticity of the bag interfering with the pressure transducers under the model. These results were typical. It is then a straightforward task to correct for the vertical loads and overturning moment by considering the uniform bag pressure. Since the net vertical load on the model and the bag pressure underneath were recorded in amplitude and phase, the corresponding vertical load on the bottom seated model is computed as $F_{vs}(t) = F_{vg}(t) + p_g(t) *$ Area where s and g stand for seated and gapped model, p_g is the gap pressure and Area corresponds to the footprint area of the model. The resulting measured vertical loads would then correspond to the diffraction theory [Chakrabarti (1993)] results.

7.4.4 Open-Bottom Structures

As in the above example, many of the large-volume fixed structures are located near the ocean floor. These could be open-bottom oil storage tanks or flat bottom gravity structures placed on the seafloor. An offshore oil storage tank could act as a sealed structure resting firmly on the bottom under its own weight, or as an open structure (due to scouring, etc.) held in-place by piling. The anchoring requirements are quite different for the two cases and are based on knowledge of how a prototype design will act near the ocean floor. For structures that are slightly open at the bottom, the flow introduces a fluctuating pressure underneath the structure. Due to the presence of this pressure, the vertical forces, in particular, become an order of magnitude smaller over the practical range of wave periods.

Even if an actual gap does not exist nor is created by scouring etc., the pore pressure on a granular foundation under the mat or pads will produce a similar effect. A few analytical approaches to the wave-structure-soil interaction problem have been made [e.g., Monkmeyer (1979)], in addressing this problem. The effect of water waves introduces a pressure field on the seafloor that is predicted by a water wave theory developed on the basis of a flat solid seafloor. This periodic and fluctuating pressure field on a porous sea bed is transmitted to the pore water below the structure. Thus, a gravity structure seating on a porous foundation experiences a pore-pressure field setup underneath the structure's base. Scaling problem with foundation soil will be discussed later in this chapter.

Data from wave tank tests on a slightly open hemisphere [Chakrabarti and Naftzger (1976)] showed that a substantial reduction in the vertical force occurs when the structure is lifted even slightly off the bottom, particularly for long waves relative to the structure size. Moreover, it was shown that as the gap approaches zero, there is a uniform pulsating pressure inside the shell, which appears as a mean of the pressure at the bottom when the body is sealed. The pressure inside a hemispherical shell is approximated quite well by two mean pressures at the bottom of the corresponding sealed hemisphere. One of these mean pressures is the arithmetic mean of the four stagnation point pressures at the bottom, fore and aft and to the sides. The other mean pressure is the first term of the Fourier series expansion of the pressure profile around the bottom periphery.

Thus, if the force and pressure distribution (at the bottom) on a closed circular bottom-seated structure are known (e.g., from linear diffraction theory), it is relatively straightforward to compute the vertical force on an open-bottom structure. Of course, it assumes no interaction with the foundation.

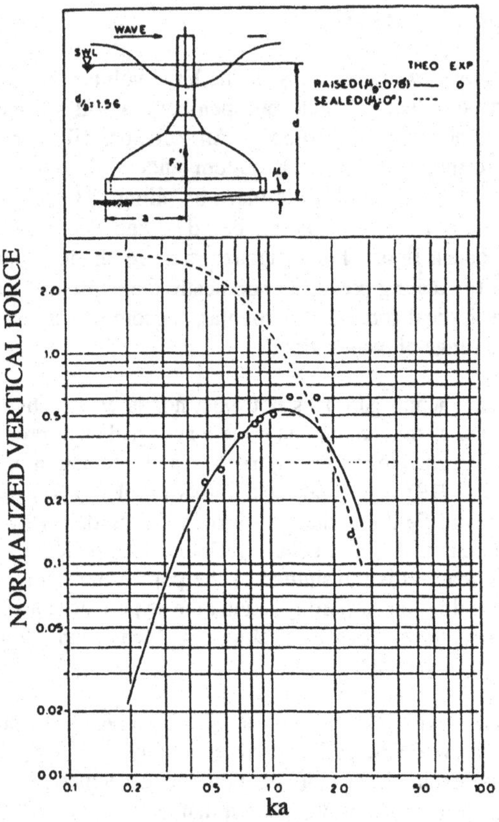

FIGURE 7.20
COMPARISON OF VERTICAL FORCES ON A SEALED VERSUS OPEN-
BOTTOM LARGE UNDERWATER STORAGE TANK
[Chakrabarti and Naftzger (1976)]

The vertical wave forces on a model of an underwater storage tank with a small uniform bottom gap were measured. The results of these tests are compared with the above theory in Fig.7.20. The reduction in the uplift in long-period waves (small ka) due to the opening at the bottom can be seen from the results of the corresponding sealed body. This is investigated further in modeling an offshore production platform.

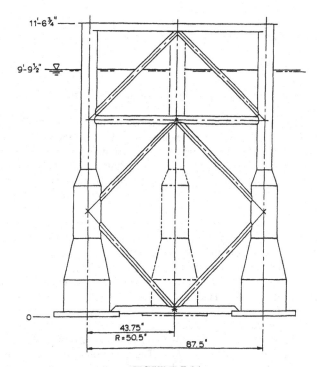

FIGURE 7.21
THREE POSTER GRAVITY DRILLING PLATFORM MODEL

7.4.5 *Gravity Production Platform*

Many submersible gravity offshore structures have been designed and installed in various parts of the world. These structures are used in exploration drilling, production, and storage of crude oil. A typical design of a gravity structure includes a large mat-type base resting on the foundation. The horizontal force due to the waves on the structure is resisted by the ballast in the structure and the friction between the bottom of the base and the foundation. These structures generally consist of a few large members and several small interconnecting members.

Communication between the underside of the structure and outside will introduce pressure through scouring or through porous foundation which will alter the vertical force and overturning moment on the structure due to waves. The effect of the bottom pressure on these quantities was investigated in a test [Chakrabarti (1986)] with a gravity drilling platform model in which both pressures and forces were measured.

7.4.5.1 Model Description

The drilling platform model consists of three columns 191 mm (7.5 in.) in diameter arranged in a 2.22 m (7.3 ft) triangle. The columns increase in diameter successively with depth through conical sections to 356 mm (14 in.) to 572 mm (22.5 in.) and finally to 826 mm (32.5 in.) diameter bottom pads. The pads and vertical legs are ballasted to achieve stability against sliding and overturning moment. The three posts or columns are connected to each other by means of tubular members in various planes. A sketch of the elevation of the three poster drilling platform model is shown in Fig. 7.21.

7.4.5.2 Test Setup

The model was suspended from three vertical load cells connected near the radially inside edge of the bottom pads on each of the three 572 mm (22.5 in.) diameter posts. The points of connection made a regular triangle with its center coinciding with the center of the three legs of the drilling platform. One load cell was used for the horizontal force measurement. The locations of the load cells are shown in Fig. 7.22. The vertical load cells were preloaded to oppose the weight of the model as well as additional weights placed on the platform so that the load cells were maintained in tension during the wave cycle.   The horizontal load cell was pretensioned to about 890 N (200 lbs) with an inline turn buckle. The overturning moment about the bottom center of the platform was calculated from the vertical and horizontal load cell readings using the appropriate moment arms. Note that the model was suspended from the load cells, leaving small gaps between the bottom of the pads and the tank floor which was necessary for accurate load measurement.

The effect of the actual size of the gap on the measured forces is negligible as long as the gap is small. This was verified during testing of a submersible drilling rig model with a large open mat (Fig. 3.7). The load test was performed (Fig. 7.23) at two different gaps of 9.6 mm (3/8 in.) and 3.2mm (1/8 in.) respectively. The difference in the measured forces due to the two gaps was found to be negligible [Chakrabarti (1987a)].

Near the bottom of the three-post platform model several pressure transducers were located to measure the pressure under the pads. The undisturbed bottom pressure was obtained from a pressure transducer placed in front of the model near the floor (about 76 mm or 3 in. off the bottom). Three pressure transducers were located at the bottom center of pads A, B and C (Fig. 7.22).

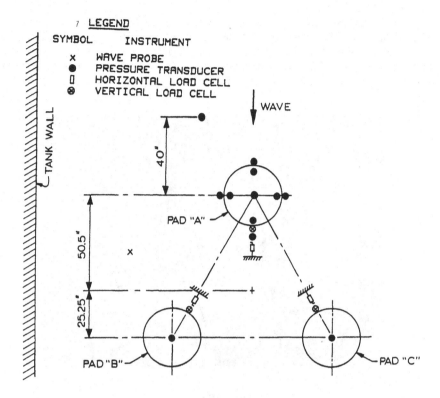

FIGURE 7.22
LAYOUT OF INSTRUMENTS RELATIVE TO MODEL IN TANK

Simulating the case of the model sitting on the floor (sealed case), the total vertical force and overturning moment on the model platform due to waves are obtained from the load cell and pressure transducer readings. The vertical uplift load on the bottom of the pads is calculated from the center pressure and the pad area. Then these time traces of uplift forces are added to the net vertical force traces to obtain the total vertical force on the sealed structure. Similarly, appropriate moment arms from the center of the structure to the centers of the pads are used with these uplift forces and added to the net overturning moment.

The measured pressure amplitudes at the center of the three pads were within five percent of one another. Similarly, the average outside pressure amplitudes from the four pressure transducers near one pad was nearly equal to the corresponding average inside pressure amplitude from the four transducers just inside the pad. These values were slightly lower than the center pressure possibly due to radial flow near the edge of the pads. The comparison of the normalized pressure from the linear incident wave theory and the test data on the undisturbed pressures and center pressures under pad is plotted in Fig. 7.24. The center pressures are found to be close to the undisturbed pressure. This is expected [see Clauss (1987)] for small diameter pads (i.e., small ka).

FIGURE 7.23
SUBMERSIBLE RIG MODEL INSTALLED ON LOAD CELLS

The wave forces on the structure were computed theoretically using a hybrid method. The net horizontal and vertical loads measured on the model are plotted against the corresponding wave periods over the theoretical curves in Figs 7.25 and 7.26. The model had a gap of about 6.3 mm (0.25 in.) between its pads and the foundation. The load cell readings reflect the uplift force under the pads resulting from this gap. The total vertical load on the structure simulating the condition of model seated on the tank floor is calculated from the load cell and pressure transducer records, and the normalized values are plotted in Fig. 7.26.

The test results clearly demonstrate that for a gravity structure exhibiting a gap at the bottom of its pads there is an uplift force which tends to reduce the net vertical load on the structure considerably. The pressure build-up under the relatively small pads (i.e. small ka) is similar to the undisturbed bottom pressure given by the linear wave

theory.   While the horizontal force is not affected by this small gap, the net moment
on the structure is altered due to this uplift force.   Bottom founded structures on
porous foundation are expected to experience similar effects.

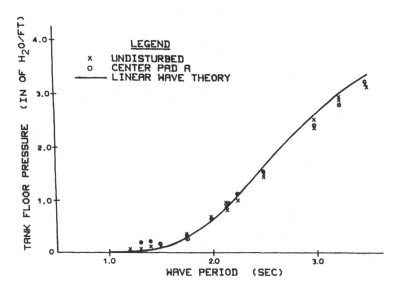

FIGURE 7.24
CORRELATION OF BOTTOM PRESSURES

7.4.6   *Second-Order Loads on a Cylinder*

Measurement of the second-order loads may be important for certain types of
offshore structures.   One of these is the tension leg platform. The TLP experiences
large tendon loads originating from what is commonly known as springing. The second
order tendon responses arise from the second-order wave loads on the platform.   In
order to   examine the second-order wave exciting loads on a TLP, a fixed caisson
representing a typical TLP column may be tested by mounting it on load cells in a
wave tank. Since the TLP is a floating structure in deep water, the caisson requires
mounting from an overhead structure (rather than on the tank floor).

Such a   test was conducted on a 1.22m (4 ft) diameter vertical caisson. The
cylinder was 2.9m (9.5 ft) long and weighed 10346 N (2325 lbs).   The model was
constructed of 6.4 mm (1/4 in.) thick steel plate. It was stiffened circumferentially with
ring stiffeners and with gusset plates at the top end. The purpose was to provide a
model which is stiff enough to consider rigid under wave loading. The external welds

were ground and the model was painted to a smooth finish to minimize surface roughness.

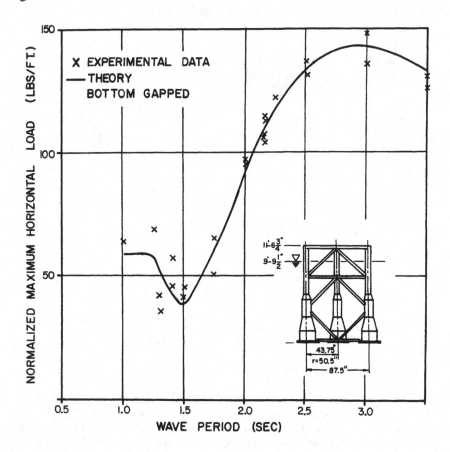

FIGURE 7.25
HORIZONTAL FORCES ON A BOTTOM GAPPED GRAVITY PLATFORM
MODEL

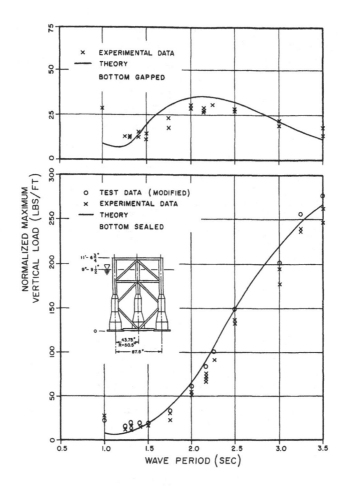

FIGURE 7.26
VERTICAL FORCES ON A SEALED AND BOTTOM GAPPED GRAVITY
PLATFORM MODEL

Three   mounting saddles were   used to attach   the cylinder to   three two-component (XY) load cells. The load cells were pinned to the saddle plates with bolts. The three vertical load cells were bolted to the load cell mounting chairs on T support shown in Fig. 7.27.   The entire T support was attached to the underside of bridge which spans the width of the wave tank.

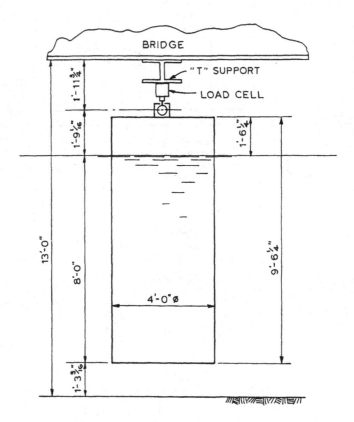

FIGURE 7.27
ELEVATION OF TEST CYLINDER IN TANK

As will be discussed shortly, the test setup introduced difficulties in the measurement of the forces, particularly the second-order component. Therefore, the same test was repeated in two different setups to verify the accuracy of the measurement. For the phase I portion of the test, water was pumped into the cylinder to balance the buoyant force and thus provide a wider measurement range for the load cells. To prevent the ballast water from sloshing, a cover plate was bolted inside the cylinder at the interior water level.

In the phase II tests the cylinder was emptied. This produced a large net vertical load on the cylinder due to buoyancy and the vertical load cells necessarily had to be removed. This was accomplished by blocking the deflection beams of the vertical cells to prevent any damage to the cells. Thus, the setup remained essentially the same and only the horizontal load was measured in phase II.

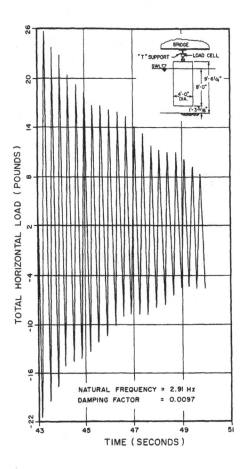

FIGURE 7.28
NATURAL FREQUENCY VIBRATION OF CYLINDER MODEL IN SETUP

After the calibration of the individual load cells was completed on a calibration stand, static in-place calibration was performed with the cylinder in its test position. A steel cable was wrapped around the cylinder and over a pulley. Calibration loads were applied in two ranges: ± 4450 N (1000 lb) with 890 N (200 lb) increments and 890 - 1023.5 N (200-230 lb) with 22.3 N (5 lb) increments.

The higher range was calibrated to insure that the load cells were linear over the whole test range of first and second-order forces. The lower range calibration was performed to insure that the 3% error noticed in the 8900 N (2000 lb) range was also 3% at the 133.5 N (30 lb) range expected for the second-order component. This was shown to be true in both phases of the test.

Since the load cells act as springs in relation to the cylinder mass, the cylinder/load cell system is actually a dynamic system. Therefore, in a test setup, the dynamic characteristics of the cylinder/load cell system must be determined. The transient response of the system was determined in order to compute the natural period and damping of the cylinder/load cell system. A load was applied to the cylinder in a manner similar to the in-place calibration tests, except a string was used instead of the steel cable. Once loaded, the string was then cut and the cylinder's response was measured.

TABLE 7.1
AMPLIFICATION FACTORS FOR MEASURED RESPONSE
NATURAL FREQUENCY = 1.75 Hz
DAMPING FACTOR = 0.0164

Wave Period (sec)	Amplification Factor
3.00	1.169
2.75	1.208
2.50	1.263
2.09	1.431
1.94	1.530
1.83	1.637
1.72	1.787
1.61	2.010
1.50	2.375
1.39	3.066
1.29	4.582
1.16	22.416

For phase I, a natural frequency of 1.75 Hz and a damping factor of 1.64 percent were obtained. In phase II, these values provided a higher natural frequency of 2.91 Hz but a lower damping factor of 0.97 percent. A typical transient response from phase II setup is shown in Fig. 7.28.

The dynamic nature of wave loading causes an amplification in the deflection of the load cell, leading to an increase in the measured load. The first-order loads are generally at frequencies far removed from the natural frequency of the setup so that the amplification effect is negligible. However, the second-order frequencies are high enough to excite the natural frequency of the cylinder/load cell system, and the load measured by the load cells must be divided by the amplification factor (see Section 10.5) of the cylinder/load cell system. For this purpose the damping factor found at the natural frequency is used throughout the entire frequency range. Values of amplification factors vs. wave periods for the phase I findings are given in Table 7.1.

Figure 7.29 shows a time domain plot of the horizontal load due to a regular wave, and its first and second-order components. The lower two traces were generated by applying a band pass filter to the measured data (top trace) in the appropriate frequency domains around the fundamental and second harmonic frequencies. Note the clear presence of the second harmonic load which is about 15% of the first-order load. The steady load appears with the first harmonic trace due to the use of a low pass filter.

The first-order horizontal load compared well with the linear diffraction theory. The deviation of this data when compared to the linear diffraction theory was analyzed. In this error analysis three statistical quantities were calculated: linear error, rms error and bias. The quantities rms error and linear error are a measure of the distance of the data from theory, while bias is a measure of whether or not the data is evenly scattered about the theory, i.e., if it is principally above or below the theory. The equations used are listed below.

$$\text{Linear error} = \frac{1}{N} \sum \frac{|Y_D - Y_T|}{Y_T} \tag{7.22}$$

$$\text{rms error} = \left[\frac{1}{N} \sum \frac{[Y_D - Y_T]^2}{Y_T^2} \right]^{\frac{1}{2}} \tag{7.23}$$

$$\text{Bias} = \frac{1}{N} \sum \frac{Y_D - Y_T}{Y_T} \tag{7.24}$$

in which N = number of test runs in a set, Y_D & Y_T = experimental data and corresponding theoretical estimate, respectively, for the test runs, and the summation is

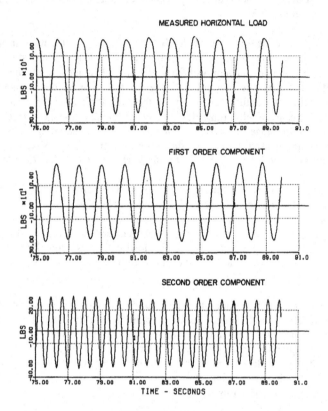

FIGURE 7.29
FIRST AND SECOND-ORDER MEASURED HORIZONTAL WAVE LOADS

understood to run from 1 to N. The results of this error analysis for the first-order horizontal loads are listed below.

	Linear error	rms error	Bias
Phase I	0.096	0.149	0.033
Phase II	0.090	0.107	-0.002

The above table shows that the data is uniformly scattered about the linear diffraction theory, with a linear error less than 10% and an rms error less than 15%. It also shows that the phase II results are slightly better since its natural frequency is further removed from the first-order response.

The results [Chakrabarti (1987b)] from the steady component of the second-order loads are normalized with respect to the square of the wave amplitude and plotted versus ka in Fig.7.30. The top figure corresponds to the phase I test in which the cylinder was ballasted with water. The bottom figure corresponds to the empty cylinder in Phase II tests.

The theoretical curve from Chakrabarti (1990) is superimposed. The theory follows the mean trend of data reasonably well. However, the scatter in the data is quite large and shows no definite trend with wave amplitudes. The scatter is attributed primarily to measurement error. Since the steady loads were typically about 3-5% of the magnitude of the first-order loads, the error in the measurement of this load in the presence of the first-order load (in spite of frequency difference) is expected to be high. Thus, the experimental determination of second-order loads on a large cylinder is a difficult task even in a carefully controlled setup. The major contributing factor to this problem arises from the presence of large first-order loads.

It is somewhat easier to measure the steady and oscillating drift forces which are responses due to second-order loads on a moored floating cylinder (see Chapter 9). One technique for accurate measurement of the second-order load is to suppress the first-order load. Such attempts have been made by researchers. For example, [Eatock Taylor and Hung (1985)], the dynamic motions of an articulated column which is mainly linear, were suppressed by fitting a vibration absorber to the system. The column was allowed to tilt due to the mean drift force which is second-order. The mean drift force was successfully computed from this displacement.

7.4.7 Scaling of Fixed Elastic Structures

Hydroelasticity deals with the problems of fluid flow past a submerged structure in which the fluid dynamic forces mutually depend on the inertial and elastic forces on the structure. It is well known that for long slender structures, the stiffness of the structure plays an important role in the response of the structure in waves.

Consider an offshore structure fixed at the base that is allowed to be displaced in waves due to the elastic support structure. The support structure consists of tubular elastic members which hold a rigid deck above the water surface. If Δx is the maximum displacement of the deck, then a functional relationship for Δx is given by [Dawson (1976)]:

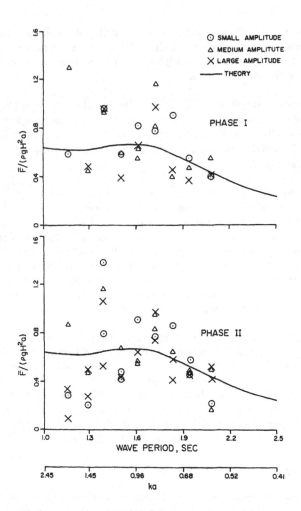

FIGURE 7.30
NORMALIZED SECOND-ORDER WAVE LOADS ON A LARGE FIXED
VERTICAL CYLINDER

$$\Delta x = \phi \left(M_d, M, \rho, g, H, \omega, E, \ell \right) \qquad\qquad (7.25)$$

where M and M_d are the respective masses of the support structure and the deck, E is the elastic modulus of the support structure and ℓ is a characteristic length. Dimensional analysis provides a relationship among the following nondimensional quantities:

$$\frac{\Delta x}{\ell} = \phi \left[\frac{H}{\ell}, \frac{\omega^2 \ell}{g}, \frac{\rho \omega^2 H \ell^3}{E \ell^2}, \frac{M \omega^2}{E \ell}, \frac{M_d \omega^2}{E \ell} \right] \qquad (7.26)$$

Thus, the nondimensional displacement will be invariant with change of scale if every quantity on the right hand side remains invariant. The first term will remain unchanged if the wave height varies as ℓ. The second term is the ratio of wave acceleration to gravitational acceleration and will remain fixed if ω varies inversely as $\sqrt{\ell}$. The third term is the ratio of wave forces to elastic forces. This ratio is proportional to ℓ/E and hence cannot accommodate an arbitrary scale change without changing the model material, i.e. E is proportional to ℓ. The last two terms are the ratio of inertial forces to elastic forces for the structure and deck respectively. These conditions are met if the masses vary as $E\ell^2$ varies. If E varies as ℓ varies then the masses will vary proportional to ℓ^3. In general, however, these conditions are not met.

Some simplification can be made if, for elastic deformation, the structure displacement is assumed to be linearly proportional to the total wave force. Since the maximum wave force f is proportional to $\rho\omega^2 H \ell^3$, Eq.7.26 may be divided by $f/E\ell^2$ and this term may be eliminated from the right hand side. Additionally, if the offshore structure experiences vibration near the fundamental mode, the structure mass may be incorporated into the deck mass to produce an effective mass, M_e [Dawson (1976]. Then

$$\frac{E\ell(\Delta x)}{f} = \phi \left(\frac{H}{\ell}, \frac{\omega^2 \ell}{g}, \frac{M_e \omega^2}{E \ell} \right) \qquad (7.27)$$

The left-hand side will remain fixed if $\ell_r \sim \lambda$, $H_r \sim \lambda$, $\omega_r \sim \sqrt{\lambda}$ and $(M_e)_r \sim E_r \lambda^2$, where the subscript r stands for ratio between prototype and model. This relationship does not follow Froude scaling because of the presence of E_r. Further assuming $E_r \sim \lambda$, the Froude scaling law is satisfied. Since f varies with ℓ^3, then Δx will vary with ℓ^2/E or ℓ (for $E_r \sim \lambda$). The characteristic strain will vary as ℓ/E (or 1 for $E_r \sim \lambda$) and the maximum stress, σ will vary as $E(\ell/E)$ or linearly with λ. It is sometimes assumed that

$$M_e = M_d + \frac{1}{5} M \qquad (7.28)$$

for uniform mass and quadratic relation for deflection. Noting that $M = \rho_s \ell^3$, the deck mass must satisfy

$$\frac{M_d}{M_{dp}} = \lambda^2 \frac{E}{E_p}\left(1 + \frac{1}{5}\frac{M_p}{M_{dp}}\right) - \lambda^3 \frac{1}{5}\frac{\rho_s}{\rho_{sp}}\frac{M_p}{M_{dp}} \qquad (7.29)$$

where the subscript p stands for the prototype value ($\lambda=1$).

The relationship between the various parameters for different scale factors is shown in Fig. 7.31 for $E/E_p = \rho_s / \rho_{sp} = 1$ and $M_p/M_{dp} = 0.5$.

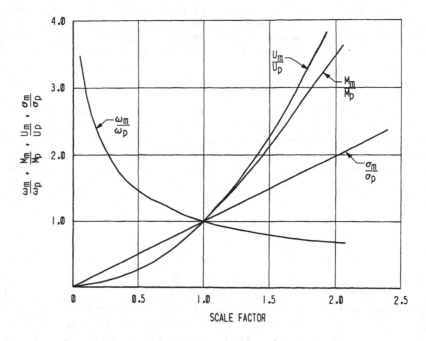

FIGURE 7.31
SCALING RELATIONS FOR IDENTICAL MATERIALS IN MODEL AND PROTOTYPE [Dawson (1976)]

In order to illustrate the above scaling laws, a simple example of a uniform circular cylinder of length ℓ is chosen as a support structure. The loading on the cylinder is assumed to consist of inertial forces as determined by linear theory for a water depth, d, a wave of frequency ω and wave number k. Neglecting the damping term, the governing equation for the deflection x at (y,t) is written as

$$EI\frac{\partial^4 x}{\partial y^4} + m\frac{\partial^2 x}{\partial t^2} = f_0\cosh ky \cos \omega t \qquad (7.30)$$

where EI is the flexural rigidity of the cylinder and m is its mass per unit length. The amplitude of the maximum wave force f is given by:

$$f_0 = \frac{kf}{\sinh kd} \qquad -d < y < 0 \qquad (7.31)$$

The boundary conditions are assumed as follows:

- x = x' = 0 at the bottom, where x' is the spatial derivative with respect to the vertical (y),

- the displacement and its three spatial derivatives are continuous at the still water level, and

- x" = 0, EIx''' = $M_d\ddot{x}$ at the cylinder top.

The steady-state solution for displacement is written as

$$x = S(y) \cos \omega t \qquad (7.32)$$

where $S(y) = A_1 e^{by} + B_1 e^{-by} + C_1\cos by + D_1\sin by + G\cosh ky$ for $-d<y<0$ (7.33)

and $\quad S(y) = A_2 e^{by} + B_2 e^{-by} + C_2 \cos by + D_2 \sin by$ for y>0 (7.34)

The quantities b and G are obtained from the following:

$$b^4 = \frac{M_d\omega^2}{EI} \qquad (7.35)$$

and

$$G = \frac{EI}{\ell}\frac{k}{k^4 - b^4}\frac{f}{\sinh kd} \qquad (7.36)$$

When the eight boundary conditions are applied, the eight unknowns A_i, B_i, C_i & D_i(i = 1,2) may be determined.

The dimensionless deck deflection $EI\,x/f$ is uniquely determined from the following dimensionless quantities:

$$I/\ell^4, M/M_d, M_d\omega^2\ell^3/EI, d/\ell \ and\ k\ell \tag{7.36}$$

where $M(=m\ell)$ is the total structure mass. In order to establish the error in the scaling laws for various scale factors, the following values of the last four nondimensional quantities are assumed for the prototype: $M/M_d = 0.5$, $M_d\omega^2\ell^3/EI = 1.92$, $d/\ell = 0.80$, and $k\ell = 4\pi$. Note that the last two quantities are invariant with a change of scale. The wave frequency ω is assumed to be 0.8 of the massless structure resonant frequency. Since ω varies as $1/\sqrt{\ell}$, from Eq. 7.29

$$\frac{M}{M_d} = \lambda / (2.2 - 0.2\lambda) \tag{7.37}$$

$$M_d\omega^2\ell^3/EI = 1.92(1.1 - 0.1\lambda) \tag{7.38}$$

The error in the scaling laws for various scale factors when compared to the analytical results is shown in Table 7.2. For a $\lambda = 0.1$, the error in the scaling law is only about 4 percent.

TABLE 7.2
COMPARISON OF ANALYTIC (x_a) AND SCALING-LAW (x_s) RESULTS UNDER CHANGE OF SCALE [Dawson (1976)]

λ	$\dfrac{M}{M_d}$	$\dfrac{M_d\omega^2\ell^3}{EI}$	$\dfrac{IE\ell x_a}{\ell^4 F}$	$\dfrac{IE\ell x_s}{\ell^4 F}$	Error (%)
1.0	0.500	1.920	0.699	0.699	0
0.8	0.392	1.959	0.693	0.699	0.91
0.6	0.288	1.998	0.687	0.699	1.80
0.4	0.189	2.037	0.681	0.699	2.69
0.2	0.093	2.071	0.675	0.699	3.55
0.1	0.046	2.092	0.673	0.699	3.96

7.4.8   Elastic Storage Tank Model

In some instances, stiffness may be an important factor for a large diameter structure. One example may be found in the study of the standing waves in a storage tank. Liquid sloshing inside a partially filled storage tank is a well-known phenomenon

[Abramson (1963)]. For an offshore storage tank, the liquid inside the tank is excited by the motion of the tank wall from an external force from waves or earthquakes. In addition to the inertial force, the elastic force is considered to be important. In this case, the deflection of the tank wall takes place due to the dynamic pressure distributions around its outside; so that in modeling this phenomenon, the stiffness of the walls should be modeled as well. From geometric similitude, the ratio of the shell deflection should linearly follow the scale factor. For an elastic structure, the Cauchy number must be satisfied and the modulus of elasticity for the model should scale directly as λ (Chapter 2).

To evaluate the significance of the error due to any out-of-scale elastic modulus, it is necessary to look at the contribution of the elastic modulus, E, to the different modes of shell deflection. The total deflections at a point on the model shell is the sum of the deflections due to axial stiffness, bending stiffness and buckling stiffness. All are dependent upon E. Deflection due to buckling is not considered to be a significant contributor to total shell deflection over the working range of a well-designed shell.

Both axial and bending stiffness can not be modeled at the same time if the elastic modulus does not scale as λ. Bending to give the shell an oval shape appears to be the most likely source of radial shell deflection and also of sloshing wave action due to deflection. The flexural stiffness in bending is given by EI, where I is the sectional moment of inertia. Stiffness scales as

$$\left(EI\right)_p = \lambda^5 \left(EI\right)_m \tag{7.39}$$

For a given scale factor and a given Young's modulus for a model material, the above relationship may be satisfied by choosing an appropriate value of the model wall thickness. However, this model will not satisfy the scaling of the axial stiffness, EA, at the same time.

Such scaling of an elastic model is illustrated by a large offshore oil storage tank model. The model was tested for sloshing waves when subjected to incident external waves in a wave tank environment. In order to determine the effect of elasticity of the wall of the tank on the amplitude of the standing waves inside, a Froude model was built which, in addition, followed the Cauchy similitude. The problems associated with Cauchy-Froude similitude in such modeling is discussed through this example.

7.4.8.1. Elastic Model Design

A range of scales from 1:30 to 1:48 was examined for the elastic model. A scale of 1:48 was selected, because it provided the smallest model dimension with minimum tank blockage for which reasonable wave heights could be generated over the desired range of wave periods.

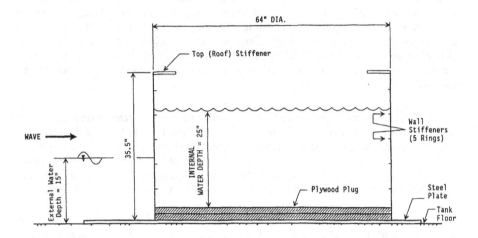

FIGURE 7.32
SETUP OF ELASTIC CYLINDER MODEL IN THE WAVE TANK

For a prototype steel structure and a scale factor of $\lambda = 48$, the modulus of elasticity, E, of the model should be 625,000 psi. This value of E is higher than most plastics, but is lower than common metals. Fiber reinforced plastic could have been selected and a formula developed to yield a product with the desired stiffness. However, it was desirable to build the model from commercially stocked materials. This resulted in a selection of a material whose elastic moduli distorted the stiffness scaling somewhat.

After studying a tabulation of common plastic materials, their tensile strengths, moduli of elasticity and available sizes and thickness ranges, Lexan unfilled polycarbonate thermoplastic was selected. This material exhibited the most isotropic properties with the narrowest range of published values. For this material, the modulus of elasticity in tension is 3.5×10^5 psi, and in bending the flexural modulus is 3.4×10^5 psi. With this material and proper modeling of the bending stiffness, the scaling ratio between axial and bending is 1.49:1. In other words, the model shell is 1-1/2 times stiffer in axial tension than desired when the bending modulus is properly scaled. The smaller stiffeners present in the prototype were not modeled in the test. However, their stiffness was distributed over the shell in the computation of the shell thickness for the model. The larger stiffeners were modeled (Fig. 7.32) as annular rings of rectangular sections.

Figure 7.33 shows a plot of the comparison of the simulated deflection of the radial shell

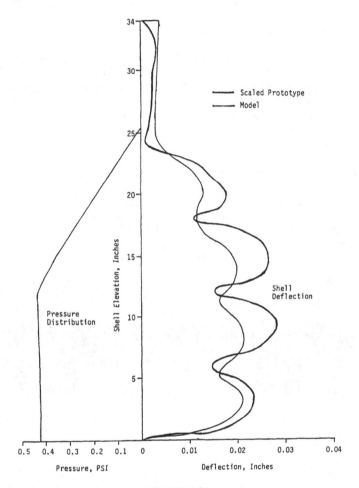

FIGURE 7.33
PRESSURE DISTRIBUTION AND SHELL DEFLECTION AT 0 DEG. FOR A
REGULAR WAVE OF 2.02 SEC PERIOD AND 11 IN. HEIGHT

at 0 deg azimuth between the model and the scaled down prototype when subjected to a regular wave. The shapes of the deflected curves are similar, but the model shell simulation does not show the circumferential growth that the prototype simulation does. Much of this error can be attributed to the distorted axial stiffness of the model design.

7.4.8.2. Elastic Model Tests

The elastic model was tested (Fig. 7.34) in regular waves. However, no sloshing waves of significant height were generated during any of the test runs on the elastic shell mounted on a rigid foundation. Neither changing the shell stiffness by removing the roof stiffener nor lowering the contained water elevation resulted in significantly increased wave heights. In no test run was the standing wave amplitude greater than 6.3 mm (0.25 in.) and generally, it was less than 3.1 mm (0.12 in.) for incident wave amplitude up to 102 mm (4 in.).

In order to study the effect of various movements of the model on the standing waves, the elastic model tests were carried out on different foundations. The model was first placed on a fixed base which restricted its motion except for the walls (Case 1). In the second series of tests, the model was placed on horizontal and vertical springs (Case 2). These springs were subsequently replaced with softer springs in both directions (Case 3). In the final phase, the horizontal springs were removed leaving only the vertical springs (Case 4). In this case, the horizontal movement of the model was unrestrained.

FIGURE 7.34
ELASTIC MODEL UNDER TEST IN WAVE TANK

Larger standing waves were observed with the tank model supported on foundation springs than with the tank mounted on a rigid foundation. As the foundation stiffness decreased from Case 1 through Case 4, the standing wave heights

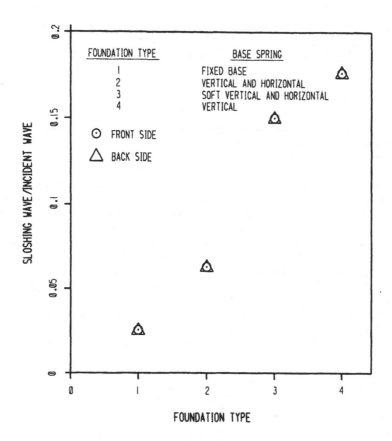

FIGURE 7.35
EXAMPLE STANDING WAVES IN AN ELASTIC MODEL ON VARIOUS
FOUNDATION TYPES

were found to increase. This is illustrated in Fig. 7.35 for a regular wave of 1.40 sec, which was the fundamental natural sloshing period of the cylindrical tank. It is clear that the higher wave amplitudes observed as the foundation was softened are due to the rigid body movement of the tank at its foundation to a much greater extent than they are due to the flexing of the tank shell.

This example shows the difficulty of modeling large elastic structures having both axial and bending stiffness. While the example shows that the elasticity is not an important cause of sloshing compared to the foundation movement, it is not suggested that such scaling is not necessary. In many cases it should be explored carefully during planning a model test. Long models are scaled for bending in the mid-section which is an important design criterion for many structures, such as tankers, riser, SPMs etc.

7.5   SCOUR AROUND STRUCTURES

The scour at the foundation of a gravity-based structure or at a buried pipeline is an important consideration for its design. The general aspects of the scour and associated scaling problem are covered briefly in this section.

The purpose of testing scour in a wave tank is to determine the depth and spread of scour about the periphery of a structure under simulated wave conditions and then projecting them to the actual prototype conditions. This allows the designer to examine the long-term stability of the structure on its foundation. Bottom-founded gravity structures are vulnerable to the scour action at their bases. Model tests are usually carried out to determine the size, type and extent of scour. A bed of model foundation material is prepared at the bottom of the wave tank flush with its floor.   The model is placed on top of this bed material. The model bed must be thick enough to allow ultimate scour depth. The segmented false bottom in the wave tank is very suitable for this type of testing.   Such a setup is shown in Fig. 7.36. Sometimes, preventive measures are taken to avoid excessive scour detected in a model test, and these preventive solutions may be tested to verify their effect on the scour problem.

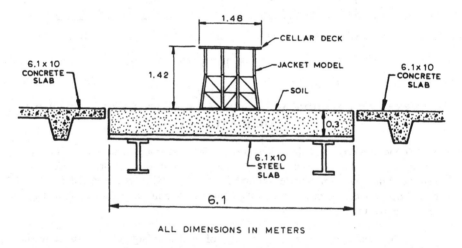

ALL DIMENSIONS IN METERS

FIGURE 7.36
SCOUR TEST SETUP

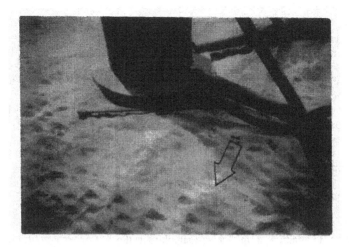

FIGURE 7.37
**SCOUR PATTERN AT THE CIRCULAR BASE OF GRAVITY PRODUCTION
PLATFORM**

Effects of scour are dependent on the bed material, namely, its type (e.g., noncohesive sand, etc.) and size (i.e., mean grain diameter). Therefore, the bed material should be properly chosen in model scale. Scaling of bed material is the single most important factor in the study of scour. It is also the most difficult task.

In planning such a scour test, several materials are investigated in the search for a way to model a prototype noncohesive material. Plastic material is often found to satisfy both the sediment number criterion for noncohesive material (e.g., sand) and the Froude scaling of the other prototype parameters. Whole grain fine silica sand is also selected to represent prototype material.

A photograph of scour pattern at one of the circular bases of a three-post gravity platform is shown in Fig. 7.37. The scour is the result of regular waves of a given frequency (after about 1,000 wave cycles). Considerable sand transport is noticed around the model base.

7.5.1 Factors Influencing Scour

There are three primary factors that govern the scouring process; the prevalent sea condition, the structure geometry and the sediment type. Whenever a structure is placed at the bottom in the ocean, the ambient flow patterns are changed, resulting in localized areas of high fluid velocity. The fluid velocities are what determine the "disturbing" forces acting to dislodge the sediment particles close to the structure. The actual disturbing force is the viscous shearing force (or stress) imparted from the fluid to the sediment. This shear stress is proportional to the square of the fluid velocity; thus, if the velocity (due to current and waves) in a certain area is doubled due to the presence of the structure, the viscous shearing stress in that area is four times the ambient shear stress. Although excessive scour can occur in high shear stress areas, there are forces determined by the sediment that resist the "disturbing" forces. The sediment particle weight and the intergranular reactions are the main restoring forces of a noncohesive soil, whereas a cohesive sediment has an additional force from cohesion. It is widely accepted that a cohesive soil exhibits more resistance to scour than a noncohesive soil [Herbich, et al. (1984)].

7.5.2 Scour Model Tests

At present, the most reliable approach to predicting scour around an offshore structure is to conduct a model test simulating the real ocean conditions. Although this method is the best available approach, scale effects introduce a major problem when trying to predict prototype scour depths from the model test. A scale model can reproduce the prototype flow patterns around the structure, but the effect of the flow patterns on the real ocean sediment can not often be reproduced due to sediment scaling problems.

Another problem inherent in scour model testing is its functional dependence on the fluid flow properties. Two of the dimensionless numbers that scour depends on are the Froude number and the Reynolds number based on the structure size and particle velocities. This means that in order for the model to correctly simulate the real ocean condition, these two numbers must be kept constant from the prototype to the model. The problem arises from the fact that the real ocean Reynolds number can not be reproduced in the laboratory.

In order to illustrate the effect of these two nondimensional quantities on scaling, let us consider a simple example. Assuming a scale factor of 1:64, the following relationship results for the prototype and model fluid velocities respectively:

$$U_m = \frac{1}{8} U_p \qquad (7.40)$$

$$U_m = 64 U_p \qquad (7.41)$$

Equation 7.40 is the result of the Froude scaling (i.e., holding the Froude number constant from model to prototype), and Eq. 7.41 is the result of Reynolds scaling (i.e., holding the Reynolds number constant from model to prototype). If an ocean current of 0.6 m/s (2 ft/sec) exists at the prototype site, then Froude scaling would indicate a model speed of 0.08 m/s (0.25 ft/sec); however, Reynolds scaling would indicate a model speed of 39 m/s (128 ft/sec). A laboratory can produce conditions

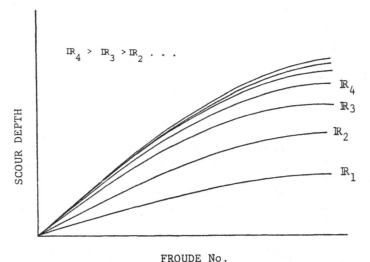

FROUDE No.

FIGURE 7.38
EXPECTED SCOUR-FROUDE NUMBER-REYNOLDS NUMBER
RELATIONSHIP

following the Froude scale, but it obviously can not produce conditions satisfying the Reynolds scale. In many instaces, this does not pose to be a problem in a scour model.

Many researchers feel that the Froude criterion is the more important of the two scaling laws when considering scour effects. It is believed that the relationship between the Froude number, the Reynolds number and the resulting scour is of the form shown in Fig. 7.38. This figure is a graph of scour versus Froude number for constant values of Reynolds number. According to the relationship, as the value of the Reynolds number increases, its influence on the scour-Froude number relationship decreases until finally a limiting Reynolds number is attained.

At a Reynolds number above the limiting Reynolds number, the scour is independent of the Reynolds number, and it depends only on the Froude number. It must be emphasized that this is the expected relationship. Insufficient data exists to confirm this, but strong arguments can be presented to reinforce the notion that this relationship does exist [Carstens (1966)]. Assuming this to be the case, the problem is reduced to finding the limiting Reynolds number. Once this number has been defined, all model tests should be conducted at a Reynolds number above the limiting Reynolds number. Since the prototype is at a Reynolds number far above the limiting value, it is only Froude dependent; therefore, the model must also be only Froude dependent. If the model is tested at a Reynolds number below the limiting value, "Reynolds effects" will distort the scour-Froude relationship and the results will be unreliable.

If the "Reynolds effects" problem can be solved, one is still left with an improperly scaled sediment. This aspect may be more favorably dealt with by defining the Froude number based upon the sediment characteristics.

7.5.3   *Scaling of Soil-Structure Interaction*

Consider a structure resting on a foundation soil and subjected to waves and current. In scaling such a system, the environment, the structure, and the soil need to be scaled down to the model size. In hydrodynamic testing, the environment and the structure are generally scaled following Froude's law. During these tests, the Froude numbers between the model and the prototype based on the structure dimension are kept the same. For the soil, however, the similarity of Froude number based on the soil particle size is not adequate to assure dynamic similarity.

A Froude-scaled model can reproduce the prototype flow patterns around a relatively large structure, but the effect of the flow patterns on the real ocean sediment cannot be properly reproduced due to sediment scaling problems. Often, a noncohesive sediment, when scaled down, produces a sediment size that can only be fulfilled by a cohesive sediment. However, a cohesive soil possesses properties vastly different from a noncohesive soil. For this and other reasons to be stated shortly, distorted sediment scaling is the accepted form of scaling in a model test simulating scour. This section outlines the difficulties associated with the scaling of soil-structure interaction, the associated scaling laws and an efficient way of choosing the soil model to assure adequate dynamic similarity between the model and the prototype.

7.5.3.1 *Noncohesive Soil*

Often - desired requirements for modeling scour materials are as follows:

• Modeling of Free Fall Velocity

The free fall velocity of a particle of sediment is given by

$$u_F = \frac{\rho(s-1)d_{50}^{\ 2}}{18\nu} \tag{7.42}$$

where d_{50} = mean particle diameter, s = specific gravity of the sediment particle and ν = kinematic viscosity of water. This equation is valid for fine grained ($d_{50} < 0.13$mm), lightweight sediment [Hallermeier (1981)]. It is sometimes necessary that the free fall velocity of the particles between the model and the prototype be scaled following Froude's law.

$$u_{F_m} = \sqrt{\lambda}\, u_{F_p} \tag{7.43}$$

- Shear Velocity > Particle Fall Velocity

The shear velocity is defined as

$$u^* = \sqrt{\tau / \rho} \tag{7.44}$$

where τ = bed shear stress and ρ = fluid density. In current, the shear stress is given by

$$\tau = \rho g \left(\frac{U}{C}\right)^2 \tag{7.45}$$

where C = Chezy coefficient. The Chezy coefficient is estimated from

$$C = \sqrt{8g / f_f} \tag{7.46}$$

where f_f is the friction factor as a function of Reynolds number, Re. The values of f_f are given as functions of IR (i.e. Re) and R/d_{50} in Fig. 7.39. In the figure, R is the hydraulic radius defined as the ratio of wetted cross-sectional area to the wetted perimeter of the channel (in which current flow takes place).

- Rough Turbulent Flow Around Structure

In order to model both the boundary layer and the wave motion simultaneously and correctly, the boundary layer motion in the model must also be a fully developed rough turbulent flow. Under the assumption that the prototype and model flow conditions are turbulent, the dependence on the Reynolds number may be waived.

It should be assured that the scale is large enough for the Reynolds number to produce a rough turbulent flow in the model scale. This requires that the model Reynolds number, based on the model dimension and flow velocity, be at least 30,000.

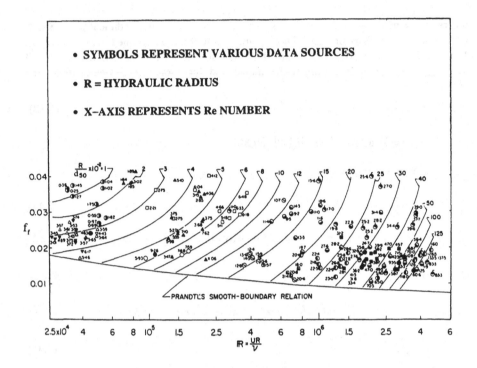

FIGURE 7.39
FRICTION FACTOR PREDICTOR FOR FLAT-BED FLOWS IN ALLUVIAL
CHANNELS [Lovera and Kennedy (1969)]

- Modeling of Grain Size Reynolds Number

It is desirable to minimize the distortion in the grain size Reynolds number. This allows the model sand particle to behave similar to the prototype sediment material. In order to achieve this similarity, the model grain size Reynolds number should be at least one-fourth of the prototype grain size Reynolds number. The grain size Reynolds number for this purpose is defined as

$$N_R = \frac{u * d_{50}}{\nu} \tag{7.47}$$

Moreover, the model bed material should be such that the material is not too small in size to be trapped by the surface tension of the model.

• Modeling of Sediment Number

The sediment number is defined as the ratio of the flow velocity to the turbulent settling velocity.

$$N_s = u_0 / u_T \qquad (7.48)$$

The turbulent settling velocity, u_T, is computed from

$$u_T = \sqrt{g d_{50}(s-1)} \qquad (7.49)$$

In theory [Carstens (1966)], the sediment number represents the ratio of the disturbing forces (e.g., the viscous shear force) of the flow to the restoring forces of the sediment. The surface of the bed material is the interface between the liquid phase and the solid-particle phase. At the interface, exchange of sand grains occurs when the lift and drag forces acting on the surface grains are sufficient against the restoring force to roll them over or to lift them into the overlying water oscillation. The restoring forces in this context are the sediment particle weight and the intergranular reactions for a noncohesive soil. For a cohesive soil, the force of cohesion is an additional restoring force.

In modeling scour, it is desirable to keep the sediment number equal between the model and the prototype. Since the velocities are generally Froude scaled, this requires that the diameter, d_{50}, and the specific gravity, s, of the model material are chosen such that the product $(s-1)d_{50}$ (see Eq. 7.49) satisfies Eq. 7.48. This may be accomplished by choosing a lighter material for the model so that a particle larger than the Froude scale material may be used. This, in turn, may avoid choosing a particle whose properties may fall into the cohesive zone. An example is provided in a latter section to illustrate this area.

• Modeling of Entrainment Function

The entrainment function is defined as the ratio of the square of the shear velocity to the square of the turbulent settling velocity

$$\xi = \frac{u^{*2}}{u_T^{\;2}} \qquad (7.50)$$

A chart showing the general and local bed load transport area for the range of values of the entrainment function vs. grain size Reynolds number is given by Fig. 7.40.

In modeling the particle size, the values of the entrainment function should remain in this range of bed load transport.

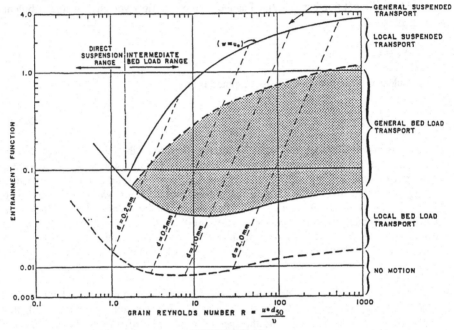

FIGURE 7.40
SEDIMENT TRANSPORT REGIMES [Martec, Canada]

- Modeling of Pore Pressure

When water waves propagate over a porous seabed, the wave pressures induced on the sea bottom are transmitted directly to the pore-water below the soil surface or mudline. Thus, the underside of a structure firmly placed on the mudline will experience an uplift force from this pore pressure field. The pore pressure follows Darcy's law as long as flow through the porous medium is laminar which is dependent on Reynolds number. In this case, similitude is difficult. When the sediment transport occurs in the rough turbulent range, the viscosity scale effects are not too influential.

7.5.3.2 Model Calculations

The various modeling criteria discussed above are summarized in Table 7.3. It should be obvious that all of these criteria may not be satisfied simultaneously in a single test setup with a model bed material. However, they should be examined in choosing a model bed material and attempts should be made to maintain the order of magnitude of the quantities similar so that similar results may be expected in the model

TABLE 7.3
SUMMARY OF NONCOHESIVE SOIL MODELING CRITERIA

ITEM NO.	DESCRIPTION
1	Model Material Should be Noncohesive
2	$u_{F_m} = u_{F_p}$
3	$u* > u_F$
4	$Re > 30,000$
5	$N_{R_m} \sim N_{R_p}$ (order of magnitude)
6	$N_{s_m} = N_{s_p}$

test. While a direct scale-up is often not possible, the model test will provide the general pattern of scour, erosion, etc., and provide an order of magnitude expected in the prototype when scaled up.

In this section, computations are carried out to investigate how certain prototype materials may be scaled in a model test. A sample of sand material $d_{50} = 0.1$ mm is selected here. A scale factor for the model is taken as $\lambda = 36$. The water particle velocity at the model base is 0.073 m/s (0.24 ft/sec) (for a prototype value of 0.44 m/s or 1.43 ft/sec). The model current velocity is 0.085 m/s (0.28 ft/sec) (for a prototype value of 0.51 m/s or 1.68 ft/sec). These are based on Froude's law of similitude.

The model materials are chosen from sand and crushed walnut shells. The results are summarized in Tables 7.4 and 7.5. The smallest whole-grain sand particle has a mean diameter of 0.01 mm while the smallest crushed walnut was found to be about 0.2 mm. These materials are chosen for the comparison in the tables.

Table 7.4 summarizes the properties of model material of sand having a grain size of 0.01 mm used to model a prototype sand material of mean diameter of 0.1 mm. It is assumed that fresh water will be used in the test. The table shows that the particle fall velocities and the grain size Reynolds number do not scale. The scaling of the particle fall velocity may not be very important as long as the shear velocity of the model material is larger than the model particle fall velocity. This latter criterion is satisfied for the selected model material. Similarly, the modeling of the entrainment function and sediment number is generally met. Assuming that Re > 30,000, the viscosity scale effect is small and the pore pressure will be adequately duplicated in the model. One might expect that the model test will produce results characteristic of the prototype behavior.

TABLE 7.4

SUMMARY OF MATERIAL PROPERTIES FOR PROTOTYPE AND MODEL SAND FOR A SCALE OF 1:36

Quantity	Prototype	Model	Remarks
Fluid Kinematic Viscosity, v (ft^2/sec)	1.59 E-3	1.22 E-3	Sea vs. Fresh Water
Fluid Density, ρ (slugs/ft^3)	1.98	1.94	Sea vs. Fresh Water
Material Size, d_{50} (mm)	0.1	0.01	Smallest Whole Grain Sand
Specific Gravity, s	2.65	2.65	
Particle Fall Velocity, u_F (ft/sec)	0.0191	0.00025	Does Not Scale
Shear Velocity, $u*$ (ft/sec)	0.0458	0.01	$u* > u_F$
Turbulent Settling Velocity, u_T, (ft/sec)	0.1305	0.0413	Same Order of Magnitude
Grain Size Reynolds Number, N_R	0.923	0.026	Does Not Scale
Entrainment Function, ξ	0.123	0.058	Roughly Equal
Sediment Number, N_s	10.96	5.82	Roughly Equal

The same prototype material of $d_{50} = 0.1$ mm is also modeled by crushed walnut shells which are lighter and have a specific gravity of 1.35. The results are summarized in Table 7.5. This modeling is better since it scales all items quite well except the entrainment function.

TABLE 7.5
SUMMARY OF MATERIAL PROPERTIES FOR PROTOTYPE AND MODEL WALNUT SHELL FOR A SCALE OF 1:36

Quantity	Prototype	Model	Remarks
Fluid Kinematic Viscosity, ν (ft^2/sec)	1.59 E-3	1.22 E-3	Sea vs. Fresh Water
Fluid Density, ρ (slugs/ft^3)	1.98	1.94	Sea vs. Fresh Water
Material Size, d_{50} (mm)	0.1	0.2	Crushed Walnut Shells
Specific Gravity, s	2.65	1.35	
Particle Fall Velocity, u_F (ft/sec)	0.0191	0.021	Scales Well
Shear Velocity, u* (ft/sec)	0.0458	0.01	u* > u_F
Turbulent Settling Velocity, u_T, (ft/sec)	0.1305	0.085	Scales Reasonably Well
Grain Size Reynolds Number, N_R	0.923	0.524	Scales Well
Entrainment Function, ξ	0.123	0.014	Does Not Scale Well
Sediment Number, N_S	10.96	2.82	Same Order of Magnitude

Even though the model sand scales several of the criteria set forth in Table 7.3, the size of the model sand of d_{50} = 0.01mm is much smaller than the lower limit of noncohesive sediment which is generally assumed to be about 0.08mm. Therefore, the model sand may not be suitable to model the noncohesive prototype sediment. On the other hand, the model walnut shell scales the majority of the criteria in Table 7.3 and as such is suitable for modeling the prototype sediment and duplicating prototype scour effects.

The above is a brief and simple explanation on scaling of scour in a model test. If one is interested in the details of sediment transport similitude requirements for movable-bed models, reference is made to the works of Kamphuis (1982) and Yalin (1971). They have explained the various similitude requirements for this type of modeling and the consequence of not fulfilling some of the criteria.

7.5.3.3 Cohesive Soil

In structure-soil interactions, the cohesive soil resistance to the structure is significantly less than the resistance created by cohesionless soils. This conclusion is based on the assumption that the soil is undrained, and the additional loading is balanced by an increase of pore pressure. Structures exposed to wave loading oscillations may develop an almost undrained soil response. Modeling of cohesive soil in a scale model is a complex problem. While tests have been done at small scale to establish cohesive soil properties, structural modeling with cohesive soil is uncommon. There are significant time effects on the lateral soil resistance to structures supported by cohesive soils. The effects are caused by several factors, such as

- Aging of the disturbed soil;
- Pore pressure dissipation from the strained soil; and
- Rate of external loading on the structure (due to waves).

The question of time scale for scour is not answered very easily without close correlation with prototype observations, and few, if any, are available. For this reason small scale testing with cohesive soil is avoided.

7.5.4 Scour Protection

When a structure is considered vulnerable to scour that may result in loss of stability, protective measures are taken to ensure that stability is maintained. Scour protection can be classified as either active or passive. Active scour protection is defined as any kind of protection that reduces the scour potential of the flow near the seabed (i.e., reduces the disturbing forces). Passive scour protection is any method that increases the ability of the foundation to resist scouring elements (i.e., increases the restoring forces).

Many devices are available commercially that may be placed at the ocean floor to reduce the fluid flow near the structure base and lower the potential of scour. Model tests are often conducted to investigate the effect of a scour protecting device. Tests are conducted with and without the device to study the extent of scouring and its prevention.

7.6   WIND TUNNEL TESTS

The superstructure, such as the exposed deck of an offshore structure experiences wind loads. Often, these structures are tested in wind tunnels to determine such loads in model scale. This area of testing has been considered to be outside the scope of this book. The following briefly describes wind tunnel testing.

The effect of wind on the exposed offshore platform is an important design consideration. When the wind flows over the platform, its flow pattern changes, introducing a force on the platform superstructure. The adverse wind condition also affects several operational efficiencies of the platform, such as the drilling operation, transport operation, etc. A realistic assessment of these problems is made [Littlebury (1981)] by model studies in a boundary layer wind tunnel. The conventional aerodynamic wind tunnel generating laminar flow is generally unsuitable for offshore platform tests.

Geometric similarity of airflow in the boundary layer over the sea is obtained in terms of several nondimensional parameters, such as wind speed profile, intensity of turbulence and length scale ratio of turbulence. A typical scale for a wind tunnel test of an offshore platform is between 1:100 and 1:200. To generate a stable boundary layer within this range of scale, the boundary layer is started upstream of the flow development section so that a reasonably stable boundary layer may be achieved at the test section. For this purpose a barrier of spires is used shaped in such a way that the required amount of shear is produced. The tunnel floor is roughened to stabilize the turbulent shear flow. For a model scale of 1:100 to 1:200, it is virtually impossible to maintain the Reynolds similitude. However, for bluff bodies representing the offshore platform, the flow effect is independent of the Reynolds number so that the distortion in Reynolds scaling may not be important. The tunnel wind speed is often about 100 m/s. An electronic balance (dynamometer) is normally fitted in the tunnel to measure the forces and moments acting on the model structure.

7.7   REFERENCES

1.   Abramson, H.N., "The Dynamic Behavior of Liquids in Moving Containers", Applied Mechanics Reviews, Vol. 16, No.7, 1963, p. 501.

2.   Allender, J.H., and Petrauskas, C., "Measured and Predicted Wave Plus Current Loading on a Laboratory-Scale, Space Frame Structure," Proceeding on Nineteenth Annual Offshore Technology Conference, Houston, Texas, OTC 5371,1987, pp. 143-151.

3.   Brogren, E.E., and Chakrabarti, S.K., "Wave Force Testing of Large Based Structures," Proceedings of Twenty-first American Towing Tank Conference, Washington, D.C., August, 1986.

4.   Carstens, M.R., "Similarity Laws for Localized Scour", Journal of the Hydraulics Division, ASCE, Vol. 92, No. HY3, May, 1966.

5.   Chakrabarti, S.K., "Inline and Transverse Forces on a Tube Array in Tandem with Waves," Applied Ocean Research, Vol. 4, No. 1, 1982, pp. 25-32.

6.   Chakrabarti, S.K., "Recent Advances in High-Frequency Wave Forces on Fixed Structures," Journal of Energy Resources Technology, Trans.ASME, Vol. 107, Sep., 1985, pp. 315-328.

7.   Chakrabarti, S.K., "Wave Forces on Offshore Gravity Platforms," Journal of Waterway, Port, Coastal and Ocean Engineering, Vol.112, No.2, March 1986, pp. 269-283.

8.   Chakrabarti, S.K.,"Wave Forces on an Open-Bottom Submersible Drilling Structure," Applied Ocean Research, Vol.9, No.1, 1987, pp. 1-6.

9.   Chakrabarti, S.K., "Correlation of Steady Second-Order Force on a Fixed Vertical Cylinder," Applied Ocean Research, Vol.9, No. 4, 1987, pp. 234-236.

10.   Chakrabarti, S.K., Nonlinear Methods in Offshore Engineering, Elsevier Publishers, Netherlands, 1990.

11.. Chakrabarti, S.K., Hydrodynamics of Offshore Structures, 2nd Edition, Computational Mechanics Publications, Southampton, U.K., 1993.

12.   Chakrabarti, S.K., and Naftzger, R.A., "Wave Interaction with a Submerged Open-Bottom Structure," Proceedings on Eighth Annual Offshore Technology Conference, Houston, Texas, OTC 2534, 1976, pp. 109-123.

13.   Chakrabarti, S.K. and Tam, W.A., "Interaction of Waves with Large Vertical Cylinder," Journal of Ship Research, Vol.19, March, 1975, pp. 23-33.

14.   Clauss, G.F. "Flat Foundations for Offshore Structures," Marine Technology, Vol. 18, 1987, pp. 23-31.

15. Cotter, D.C., and Chakrabarti, S.K., "Wave Force Test on Vertical and Inclined Cylinders," Journal of Waterway, Port, Coastal and Ocean Engineering, ASCE, Vol. 110, No.1, Feb., 1984, pp. 1-14.

16. Dawson, T.H., "Scaling of Fixed Offshore Structures," Ocean Engineering, Vol. 3, 1976, pp. 421-427.

17. Dean, R.G., "Stream Function Representation of Nonlinear Ocean Waves", Journal of Geophysical Research, Vol. 70, No. 18, 1965, pp.4561-4572.

18. Dean, R.G., "Methodology for Evaluating Suitability of Wave and Force Data for Determining Drag and Inertia Forces," Proceedings on Behavior of Offshore Structures, BOSS ' 76, Vol. 2, Trondheim, Norway, 1976, pp. 40-64.

19. Eatock Taylor, R., and Hung, S.M., "Mean Drift Forces on an Articulated Column Oscillating in a Wave Tank," Applied Ocean Research, Vol.7, No. 2, 1985, pp.66.

20. Hallermeier, R.J., "Terminal Settling Velocity of Commonly Occurring Sand Grains," Sedimentology, Vol. 28, No. 6, 1981, pp. 859-865.

21. Hansen, D.W., Chakrabarti, S.K., and Brogren, E.E., "Special Techniques in Wave Tank Testing of Large Offshore Models," Presented at Marine Data Systems International Symposium, New Orleans, Lousiana., April 1986, pp. 223-331.

22. Herbich, J.B., Schiller, R.E., Watanabe, R.K. and Dunlap, W.A., Seafloor Scour, Marcel Dekkar, Inc., New York, N.Y., 1984.

23. Kamphuis, J.W., "Coastal Mobile Bed Modelling from a 1982 Perspective," C.E. Research Report No. 76, Queen's University, Kingston, Ontario, 1982.

24. Karal, K., "Lateral Stability of Submarine Pipelines," Proceedings on Ninth Annual Offshore Technology Conference, Houston, Texas, OTC 2967,1977. pp. 71-78.

25. Littlebury, K.H., "Wind Tunnel Model Testing Techniques for Offshore Gas/Oil Production Platforms", Proceedings on Thirteenth Annual Offshore Technology Conference, Houston, Texas, OTC 4125, 1981, pp. 99-103.

26. Lovera, F. and Kennedy, J.F., "Friction-Factors for Flat-Bed Floors in Sand Channels," Proceedings ASCE, Vol. 95, 1969, pp. 1227-1234.

27.   Monkmeyer, P.L. "Wave-Induced Seepage Forces on Embedded Structures", Proceeding of Civil Engineering in the Oceans IV, ASCE, San Francisco, Calif., 1979.

28.   Sarpkaya, T., "In-Line and Transverse Forces on Cylinders in Oscillatory Flow at High Reynolds Numbers," Proceedings on Eighth Annual Offshore Technology Conference, Houston, Texas, OTC 2533, 1976, pp. 95-108.

29.   Sarpkaya, T. and Isaacson. M., <u>Mechanics of Wave Forces on Offshore Structures</u>, Van Nostrand Rheinhold, New York, New York, 1981.

30.   Vasquez, J.H. and Williams, A.N., "Hydrodynamic Loads on a Three-Dimensional Body in a Narrow Tank," Proceedings of Offshore Mechanics and Arctic Engineering Conference, ASME, Calgary, Canada, 1992, pp. 369-376.

31.   Yalin, M.S., <u>Theory of Hydraulic Models</u>, MacMillan Press, London, England, 1971.

32.   Yeung, R.W. and Sphaier, S.H., "Wave Interference Effects on a Truncated Cylinder in a Channel", Journal of Engineering Mathematics, Vol. 23, 1989.

CHAPTER 8

MODELING OF OFFSHORE OPERATIONS

8.1   TYPES OF OFFSHORE OPERATIONS

Offshore structures are built on shore or in sheltered coastal waters and transported to site by many different means. After the structures reach their destination, they are submerged and installed at site at the prescribed location. Thus, a completed offshore structure goes through three distinct stages before it is ready for operation at the site. These stages are transportation, launching and submergence. This chapter deals with these three operations for various offshore structure configurations.

There are two common types of transportation operations for offshore structures. If an offshore structure is small in size or if the structure does not have a large buoyancy module, it is carried on a barge. One such structure is a jacket type drilling and production structure. These structures are built horizontally on launch pads near the water. They are pulled onto a barge which may be temporarily ballasted to be at proper level to accommodate the structure. The barge is then towed to the installation site before launching the structure at its final location. The launching of the jacket from the barge, i.e., sliding it down into water from the rails on the deck on which it sits and the critically controlled prescribed submergence onto the bottom in a vertical position are very tricky operations. They almost invariably require model testing before the structure is actually taken out from the launching pad. Even though this procedure for smaller jackets has become a routine procedure, for larger structures, model testing is a must, considering the investment and risk involved.

The second type of transportation involves large volume structures which can be towed by themselves either vertically or horizontally at a shallow draft. Structures that are large near their bases, such as gravity production and storage structures, are generally built vertically in a graving dock. Before towout to the final destination of the structure, the graving dock is flooded and opened to the sea. Tug boats are used to tow the structure into the sea. Sometimes when the structure reaches deep water, it is ballasted down further for added stability during the towing operation. Some of the large concrete structures are actually built on water in deep water fjords. In this case, the structure is continually submerged as it is built up vertically. The buoyancy sections are pressure tested before final towing. The structure is deballasted to a pre-determined towing draft before towout. After final destination is reached these structures receive ballast water in a prescribed manner in their ballast chambers and vertically settle down to the sea floor.

There are other large volume structures that are built horizontally in a graving dock. An example of this type of structure is a buoyant articulated tower.   These structures are towed in a near horizontal position to their final destination by tug boats. If the tower carries a gravity base connected by a universal joint, the base is anchored to the tower by temporary tie rods (to prevent damage to the universal joint). At the site, the tie rods are released, and the tower is swung down into position by flooding the ballast tank. The tower assumes different angles during swing down. These operations of towing and submergence for both vertical and horizontal modes are routinely model tested.

The towing, launching and submergence model testing for different types of offshore structures are described in the following sections. Some of the modeling problems and scaling techniques of evaluating model data are discussed.

8.2   TOWING OF A BARGE

In this section, the technique of towing a barge model and the scaling up of the data for evaluating the prototype behavior is described. Tests of tanker models are similar.

A Froude model of a barge is towed in loaded as well as various ballast drafts at constant towing speeds. This test determines experimentally the towing resistance of the hull of the model. Accordingly, the load encountered by the towing mechanism is recorded. This load is scaled up to the prototype value. The resistance characteristics of the prototype barge is presented as a function of its speed.

8.2.1 Scaling Technique

A ship in uniform horizontal motion, with its vertical plane of symmetry parallel to its velocity, experiences a hydrodynamic force (resistance) in a direction opposite to its velocity and lying in the plane of symmetry. A small lift force in the vertical direction is also experienced by the ship, which, however, is usually ignored.   This resistance principally arises from two major categories: (i) one of viscous origin which essentially is due to the shearing forces acting tangetially to the hull surface, and (ii) one of pressure origin which is due to a component resulting from all pressure forces acting normally to the hull surface. Though both these categories interact with each other, they are usually considered separately. The hull being at the free surface, the variation of pressure forces over the hull generates a system of surface waves on water and the pressure distribution itself is altered due to these waves. The wave system which the ship carries with it represents a drain of energy necessary to maintain it and therefore this is called wave-making resistance. It turns out that wave-making resistance is the predominant part of the second category.

In scaling the model barge tow test results to the prototype, we assume that the total resistance R_t can be decomposed into

$$R_t = R_f + R_r \qquad (8.1)$$

where R_r is frictional (viscous) resistance and R_r is residuary resistance whose dominant part is wave-making resistance. Usually, the resistances are expressed in terms of coefficients which are normalized values of the resistances with respect to $0.5\rho A U^2$. Therefore, from Eq. 8.1 the corresponding coefficients are related by

$$C_t = C_f + C_r \qquad (8.2)$$

It may be noted that only Froude's similarity is used and Reynolds number dependent part of drag is calculated using a friction formula. This is because if both Fr and Re are to be kept constant for both prototype and the model, for ship problems one needs a fluid whose kinematic viscosity is about $1/\lambda^{3/2}$ of that of water . For $\lambda = 25$ this represents a factor of 1/125th, and no such fluid is known. The frictional resistance coefficients, C_f are functions of Re, while the residual resistance coefficient C_r is principally a function of Fr.

The geometrically similar model is towed at speeds corresponding to the prototype obeying Froude's similarity. The residuary resistance coefficient, C_r, being a function of the Froude number will have the same value between the model and the prototype:

$$C_{rp} = C_{rm} \qquad (8.3)$$

The friction coefficient C_f for both the model and the prototype is calculated using either ITTC formula [Muckle and Taylor (1987)]:

$$C_f = \frac{0.075}{\left(\log_{10} Re - 2\right)^2} \qquad (8.4)$$

or the Schoenherr line. For the prototype, an allowance of about 0.0004 is usually added to the above formula to improve model-prototype correlation. The following steps then give the scaling procedure:

- measure R_{tm} at U_m & compute C_{tm} by dividing by $\frac{1}{2}\rho_m A_m U_m^2$

- compute C_{fm} for $Re_m = U_m \ell_m / \nu_m$ using Eq. 8.4

- $C_{rm} = C_{tm} - C_{fm} = C_{rp}$

- compute C_{fp} for $Re_p = U_p \ell_p / \nu_p$ using Eq. 8.4 & augment it by +0.0004.

- $C_{tp} = C_{fp} + C_{rp}$

- Compute R_{tp} (i.e., multiply by $\frac{1}{2}\rho_p A_p U_p^2$) which is the total resistance of the prototype.

    The plot of Eq.8.4 is shown in Fig.8.1. It should be noted that ℓ used in the formula is usually taken to be the length at waterline, which therefore can vary with the draft. The density and the viscosity difference of sea water and the water medium where towing tests are done are accounted for in the above.

8.2.2 Barge Test Procedure

    A typical towing test set-up is shown in Fig. 8.2. The barge model is towed by the overhead bridge using a hollow rectangular towing staff. The lower end of the staff is attached near the middle of the model at the SWL by a pivot connection so that the barge model is free to pitch. The upper end is free to move up and down through roller guides to allow the barge to heave. By proper choice of the rollers and guides, the friction in the system is minimized. Thus, the staff allows two degrees of freedom.

    The towing load is measured by a load cell which increases the measured load by a factor of 6 through the placement of the load cell and moment arms (Fig.8.2). The load cell is calibrated in place in the dry by imposing loads at the bottom of the staff with known forces.

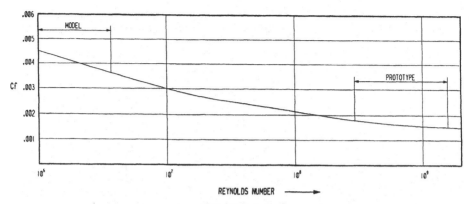

FIGURE 8.1
SKIN FRICTION DRAG COEFFICIENT

The scaling procedure described in the previous section is illustrated with the results of a towing test ($\lambda = 48$) performed in a towing tank. The speed ranged from 2.1 to 8.4 m/s (4.1 to 16.4 knots) in the prototype scale (0.31-1.22 m/s or 1-4 ft/sec model scale). These tests were repeated by towing the free model with a bridle arrangement so that the model had all six degrees of freedom. A sketch for this test set-up is shown in Fig. 8.3. The bridle is taken over a pulley to a ring load cell with a spring located in between the load cell and the point of attachment to reduce shock loads.

After the tests with the undisturbed flow were completed, a 19 mm (3/4 in.) strip of silver duct tape was attached just behind the bow thruster and another one about 0.6 m (2 ft) further behind in order to assist in the flow separation without introducing significant additional resistance. This is one of the commonly accepted methods of artificially stimulating turbulence in the flow in this type of tests. This method tends to equalize the distortion in the Reynolds number scaling discussed above. Studs are also used on the model surface for this purpose.

8.2.3 Data Analysis And Results

The length, ℓ at the waterline is used as the characteristic length in evaluating the Froude and Reynolds numbers. The mass density, and the corresponding kinematic viscosity of fresh water are corrected to provide the sea water values. Table 8.1 presents the towing speed, resistance coefficients and resistances for the model and prototype respectively at a given draft. The table was prepared following the scaling procedure outlined earlier. At a towing speed of 0.84 m/s (2.75 ft/sec) corresponding to 5.8 m/s (11.28 knots) for the prototype or less, the Froude number is 0.15 or less, and accordingly in these cases the wave-making resistance should be small. For Froude numbers higher than 0.15, the total load will consist of viscous drag and pressure drag..

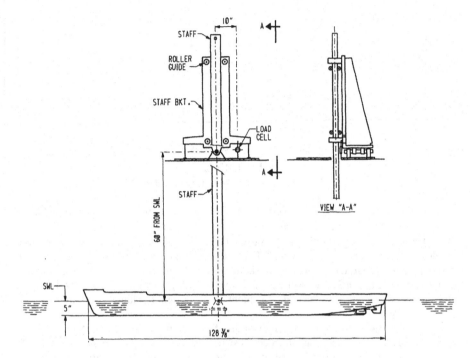

FIGURE 8.2
TOWING OF A MODEL WITH TOWING STAFF

Plots of friction drag loads by the Schoenherr line as well as the measured loads versus model towing speeds (and Froude number) are shown in Fig. 8.4 for the loaded barge. Note that the experimental data are closer to the Schoenherr curve for Froude number less than 0.15 and drag is primarily frictional. Small bow waves were observed to form during these tests. The towing test data using both towing staff and a bridle arrangement are shown. The difference between the measured data and the calculated friction drag is the residual resistance (e.g., wave-making drag). This residual resistance is multiplied by λ^3 where λ = 48 to scale up to the prototype value. Then the total towing load for the prototype is obtained by adding the residual

TABLE 8.1
MODEL AND PROTOTYPE RESISTANCE ON BARGE AT LOADED DRAFT

Wetted Area, A_m = 24.2 sq.ft

Length, ℓ_m = 10.7 ft

Mass Density of Water, $\rho_m = 1.94$; $\rho_p = 1.98$ lbs-sec^2/ft^4

Kinematic Viscosity, $\nu = 1.174 \times 10^{-5}$ ft^2/sec (@62°F)

TOWING SPEED		MODEL RESISTANCE				PROTOTYPE RESISTANCE			
U_p	U_m	R_{t_m}	C_{t_m}	C_{f_m}	$C_{r_m} = C_{r_p}$	C_{f_p}	$C_{f_p} + 0.4$	C_{t_p}	R_{t_p}
ft/sec	ft/sec	lbs	x10^{-3}	x10^{-3}	x10^{-3}	x10^{-3}	x10^{-3}	x10^{-3}	lbs
7.3	1.05	0.14	5.40	4.73	0.67	1.77	2.17	2.84	8,283
9.0	1.30	0.20	5.03	4.51	0.52	1.72	2.12	2.64	11,803
10.4	1.50	0.26	4.91	4.38	0.53	1.69	2.09	2.62	15,642
12.5	1.80	0.37	4.86	4.22	0.64	1.65	2.05	2.69	23,201
14.2	2.05	0.48	4.86	4.11	0.75	1.63	2.03	2.78	30,942
15.9	2.30	0.58	4.66	4.02	0.64	1.60	2.00	2.64	36,841
17.7	2.55	0.73	4.78	3.93	0.85	1.58	1.98	2.83	48,940
19.3	2.78	0.85	4.68	3.86	0.82	1.56	1.96	2.78	57,160
21.1	3.05	0.98	4.48	3.80	0.68	1.55	1.95	2.63	64,633
22.9	3.30	1.20	4.70	3.74	0.96	1.53	1.93	2.85	82,499
24.9	3.60	1.45	4.76	3.68	1.08	1.51	1.91	2.99	102,330
26.7	3.85	1.64	4.71	3.63	1.08	1.50	1.90	2.98	117,266
28.4	4.10	2.00	5.06	3.59	1.47	1.49	1.89	3.36	149,592

1 FT = 0.3M ; 1 LB = 4.45 N ; °C = 5 (°F - 32)/9

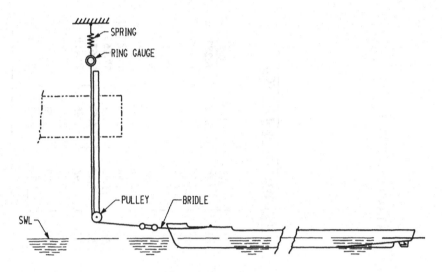

FIGURE 8.3
TOWING OF A MODEL WITH A TOWING BRIDLE

resistance to the prototype friction drag (which uses Schoenherr friction coefficient plus 0.0004 for roughness allowance). Test runs with the bridle arrangement which allowed all six degrees of freedom for the barge model showed similar results without appreciable difference. The resistance with the turbulence simulator was generally found to be about 5-10 percent higher.

8.3   SUBMERSIBLE DRILLING RIG TOWING TESTS

Unlike a self-propelled barge or ship, large offshore structures for the most part require external power for transportation to the offshore site. Gravity submersible drilling rigs with large mats (Fig.3.7) are designed to be towed by tugs to the work site and be submerged to rest on the bottom by controlled flooding of ballast chambers in the mat and caissons. The following describes a test designed to study such a rig's towing characteristics under different environmental conditions. The test was done at a scale of 1:48. The model was constructed with the mat, caissons, and superstructure all reasonably representing the prototype dynamically as well as statically.

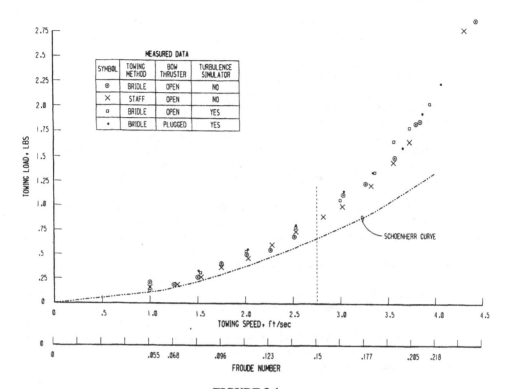

FIGURE 8.4
FRICTION DRAG FORCE ON MODEL AT LOADED DRAFT

8.3.1 Test Description

To test the rig's towing characteristics, the model is pulled along the tank with an overhead bridge. In this respect it should be noted that the large bridge moving on rails fails to model the characteristics of a tug boat. Thus, the interaction effect between the two floating bodies is not accounted for and the effect of the motion of the tug on the towing load is not considered in these tests. Nonetheless, such tests to represent towing by tugs are quite common, and determine the powering requirements.

The model is ballasted to the desired towing draft by filling various ballast chambers. The nonlinear stiffness characteristics of the towing line is modeled. The towing line is usually attached to a bridle which, in turn, is attached to the model. In the case of the aforementioned rig model, the towing draft was set at 64 mm (2.5 in.)

which is 3m (10 ft) in the prototype. A nylon line of diameter 3 mm (1/8 in.) was used with the bridle arrangement.

The model is towed in still water, in waves and, sometimes, with wind as well. The survivability of the rig is also tested with the model held stationary in waves by the tow line. The test is designed to give an indication of the maximum sea state the rig can endure while in transit. The significant height of the sea state is increased until the rig's stability appears to be marginal. In some instances, the rig actually capsizes with increased wave height during testing. The critical towing line load is monitored during these tests. The damage stability of the rig is also examined during this type of test. To simulate damage, one critical tank in the model is flooded. The rig's behavior in waves in this damaged condition is examined.

In towing tests, the speed is increased in increments up to a maximum design speed for the rig. The towing tests are repeated in waves. The waves are chosen to represent the expected sea states enroute during the period of towing. The rig model under discussion was designed to be towed at (prototype) speeds between 0.51 m/s and 2.56 m/s (1 and 5 kt) and a wave of significant height of 1.2 m (4 ft).

Wind loads are generated using fans. To determine the wind loads actually seen by the model, the fans or blowers are aimed at the model while it is stationary in still water. The mean wind load is taken to be equal to the mean load recorded by the tow line load cell. The above model was towed at a speed of 2 m/s (4 kt) in a 30.7 m/s (60 kt) wind with a head sea and at 1 and 2 m/s (2 and 4 kt) in a 10.2 m/s (20 kt) wind with a 30° heading.

8.3.2 Test Results

While the results of these tests are specific for the particular structure tested, they are described here to illustrate the type of results expected from these tests. The results are useful in determining the power requirement of the tug boat as well as the stability of the structure in the environment encountered during towing to the site. The adequacy of the towing line is also determined.

The results of towing test in still water are presented in Fig 8.5, which is a plot of the average tow line load versus the towing speed for the five runs that were made. The results shown have been scaled up from the model test using Froude scaling, and therefore represent prototype values. The scaling distortion due to viscous effects has not been included in this scaling.

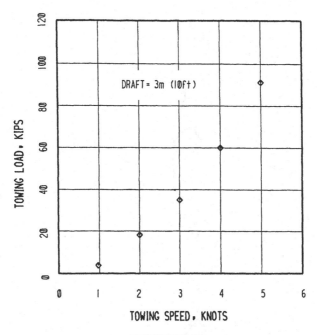

FIGURE 8.5
TOWING TESTS OF A RIG MODEL IN STILL WATER

8.4 TOWING OF A BUOYANT TOWER MODEL

A large buoyant tower is transported from its construction site in floating condition with the help of a tug boat. Depending on the available draft and route of tow, the tower may be towed horizontally or vertically. The towing characteristics of a buoyant tower model is studied in a wave tank test. In one such test, the buoyant tower consisted of a gravity base connected to the tower by a universal joint. The tower itself had three sections separated by bulkheads: a large buoyant section at top, a middle section that is usually flooded during operation, and a lower ballast section. The ballast, usually concrete, is placed in an annular space in the ballast section before towing. The ballast section is flooded at site for swingdown and submergence. The tower has a fluid swivel at the top which allows the tanker to weathervane about the tower.

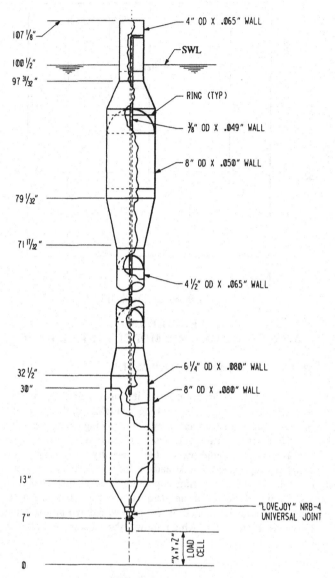

FIGURE 8.6
MODEL OF A BUOYANT ARTICULATED TOWER

8.4.1 Model Particulars

Models of this type are generally made of standard aluminum tubes and sheets. The outer dimensions are scaled following the chosen scale factor. The overall weight and location of the C.G. are adjusted to conform to the scaled down prototype values. The bulkheads are located at the scaled locations. The thickness of the materials used at various elevations and stiffening rings inside the tower are not usually scaled. The yoke and swivel connection is placed on the top of the model. The boat bumper is placed on the top shaft near the still water level (SWL). A drawing of the current model is shown in Fig. 8.6. The yoke (not shown) made a 48° angle to the tower centerline during the tow out operation.

Ballast weights in the form of lead shots are placed at selected locations to match the scaled-down prototype weight, center of gravity in air (and the natural period in air as well) for the condition tested. It is advisable to avoid placing ballasts on the outside of the submerged portion of the model. In the present case ballast weights in the form of lead shots inside 19 mm dia x 0.3 m long (3/4 in. x 1 ft) PVC tubes were placed in the annular space in the ballast tank.

A base was constructed from the same material. The base was attached to the bottom of the tower by means of a commercially available universal joint. For all the tests performed here, the ballast in the base was adjusted so that the base remained level under water and had a small net downward load.

8.4.2 Towing Tests

The model test simulated a towing scenario where two identical tugs towed the structure in parallel by means of two identical towing lines. The stiffness of the towing lines was simulated in the model, with strands of rubber bands of unequal lengths which modeled their nonlinear behavior. Each scaled line had a total length of 427 m (1400 ft) including a 64 mm (2-1/2 in.) dia 305 m (1000 ft) long steel wire in series with two 279 mm (11 in.) circumference, 40 m (131 ft) long nylon ropes. The theoretical load-elongation curve for one such line and the actual calibration curve of the model towing line are shown in Fig. 8.7.

In order to study the optimum tow orientation of the tower, several horizontal and vertical tow positions are tested. In the horizontal position, the tow line may be attached to the base end as well as the yoke end. In the vertical position, the tow line may be attached at different points on the tower. The overhead bridge is normally used as the carriage for towing. If a tug boat is employed for towing the structure, it allows the study of the snap loads on the line due to the relative motions of the boat and the tower.

The vertically floating tower was towed using a tug boat model to determine if the relative motion of the boat and the tower produced an increase in the towline load. The tug boat was remotely controlled. The simulated towline described before was connected to a bridle located 102 mm (4 in.) (4.9 m or 16 ft prototype) above the bottom of the buoyancy chamber of the vertically floating tower. The draft of the tower was 2210 mm or 87 in. (106 m or 348 ft prototype). The yoke was turned towards the towing direction so that the tower made an angle of 7.7 degrees in the equilibrium position with the line slack. A sketch of the test setup is shown in Fig. 8.8.

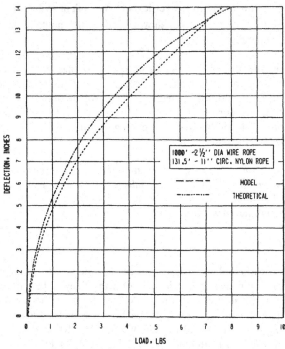

FIGURE 8.7
CALIBRATION CURVE FOR TOWING LINE

Before starting the test, the speeds of the tug boat pulling the tower and the bridge were synchronized. As expected, the maxima appeared as sharp peaks indicating an impact type loading due to the variation in the motions between the tug boat and the tower. The magnitudes and duration of these snap loads are difficult to scale, however. The design of the tow line should accommodate these snap loads. In the design, it is advantageous to choose a towing speed such that the line does not become slack from the oscillating wave loads during towing in expected sea states.

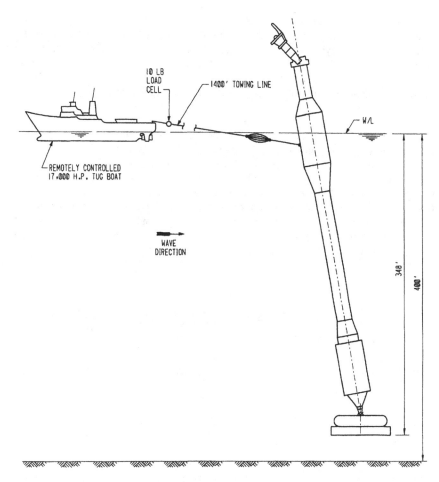

FIGURE 8.8
TOWING TEST SETUP FOR A VERTICALLY FLOATING SPM

8.4.3 Bending Moment Tests

During the horizontal tow in waves, the tower experiences a bending stress (due to hogging and sagging along its length) in the central shaft that may be severe from the design point of view. A model test may be designed to determine the bending moment in the central shaft of the model. Two methods may be employed to determine this moment in a model test. The first one is to build a model that scales the stiffness of the prototype in bending. Desired locations on the tower may then be strain gauged directly. This is expensive and difficult to achieve (See Chapters 7&9). A second method is to segment the model at the desired point of measurement and a load cell is used to connect the two, which allows the measurement of the moment. One such test is described below.

The model is segmented at the points where the bending moment is desired, and the sections are pinned and mounted on bearings (Fig. 8.9) so that they are free to move in the vertical plane at these points. Load cells are attached off-center between the separated sections to record the loads due to bending in the vertical plane. Thus, the recorded load times the distance of the load cell from the center of the shaft produces the bending moment. Two points on the tower were chosen - one 953 mm or 37.5 in. (45.7 m or 150 ft prototype) from the bottom U-joint pin and the other 1270 mm or 50 in. (61 m or 200 ft prototype) from the pin. The center of the load cells was located at 84 mm (3-5/16 in.) from the center of the shaft.

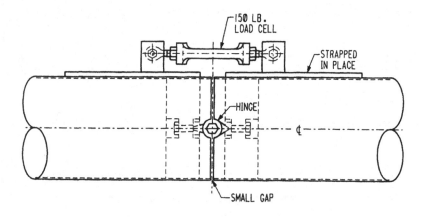

FIGURE 8.9

ARRANGEMENT OF LOAD CELL FOR THE BENDING MOMENT TEST

It is critical that the two shafts at the cut do not touch during the test runs. The effect of the friction in the bearings and hinges on the load cells is usually not known. However, in order to avoid any effect from this uncertainty, the load cells should be re-calibrated in place for appropriate scale factors. The in-place calibrations for the tower showed a small amount of hysteresis and the best slopes (in the least square sense) were considered to be the calibration constants. Figure 8.10 shows the floating tower being tested in regular waves with the bending moment cells.

8.4.4 Scaling To Prototype

In this type of tests, the model is built and the tests are performed on the assumption that the Froude law is applicable. It is known that in an inertia predominant system, the data between the model and prototype may be scaled following Froude's law. This assumes that the forces due to waves are mostly inertial, and any nonlinear drag effect is negligible. In the present case, this latter assumption is not necessarily true. Since the drag coefficients in the prototype (in the turbulent region) are generally smaller than those in the model, Froude scaling is expected to predict higher loads, thus giving conservative results. This is, however, not necessarily true for the motion and other responses. In this type of tests, if the model test results are scaled up to the prototype values irrespective of the inertia or drag predominance, care should be exercised in interpreting the prototype results.

FIGURE 8.10
SHAFT BENDING MOMENT MEASURED ON SPM IN WAVES

An example of how the towing loads should be scaled up if the drag coefficient (C_D) is known as a functions of Reynolds number (Re) is shown here. In a flow normal to the axis of a circular cylinder, the drag coefficient as a function of Reynolds number is known (Fig. 2.2). Note that this graph is not strictly applicable for the buoyant vertical tower since the tower was not of uniform diameter nor was it strictly vertical during towing. Therefore, the values obtained from the use of this graph are for illustrative purposes only.

The tubes making up the tower model and their frontal areas are tabulated below.

Tube Dia	Frontal Area	Reynolds Numbers (x10⁻⁴) at Speeds		
(in.)	(in.^{2})	0.49 ft/sec	0.73 ft/sec	0.93 ft/sec
8.0	184	2.6	3.0	5.2
6.3	46	2.0	3.0	4.1
4.5	170	1.5	2.2	2.9
5.0	14	1.6	2.4	3.2

The drag coefficient in this range of the Reynolds number is about 1.2. For the base, approximate drag coefficients are obtained from Hoerner (1965) as follows:

Shape	Frontal Area (in.^{2})	C_D
pyramid	14.9	1.0
base edge	49.5	0.5
cylinders	39.3	0.5

These values are assumed not to vary with scale.

In case of the prototype, the corresponding drag coefficients for the tower are shown in the following table.

Tube Dia. (ft)	Frontal Area (ft^2)	Speed: 2 kts		Speed: 3 kts		Speed: 4 kts	
		Re x10⁻⁶	C_D	Re x10⁻⁶	C_D	Re x10⁻⁶	C_D
32	2944	8.7	0.43	13.0	0.41	17.0	0.41
25	736	6.8	0.42	10.0	0.43	14.0	0.41
18	2720	5.0	0.42	7.3	0.42	9.6	0.43
20	224	5.3	0.42	8.0	0.43	10.0	0.43

During towing, the tower was not vertical. The frontal area was adjusted due to the angle of inclination of the tower. The towing load was calculated from

$$F = \frac{1}{2}\rho U^2 C_D A \cos\theta \qquad (8.5)$$

where U = towing speed, A = frontal area, and θ = angle into the direction of tow from vertical.

The above formula was used to compute towing loads in the model scale. These values are compared with the actual measured model loads in the following table:

U	θ	Force in Pounds (model)			
(ft/sec)	(deg)	Tower	Base	Total	Experimental
0.47	19.7	0.74	0.09	0.83	0.86
0.71	29.6	1.54	0.20	1.74	1.56
0.96	37.6	2.57	0.37	2.94	2.41

The calculated values compare favorably with the experimental data despite the approximation involved in the evaluation of C_D. A similar table is then developed for the prototype:

U	θ	Force in Kips (prototype)			
(kts)	(deg)	Tower	Base	Total	Model Projected
1.93	19.7	30.3	10.0	40	95
2.91	29.6	68.1	22.1	90	173
3.94	37.6	105.1	40.9	146	267

The last column in the preceding table represents the scaled up model data using Froude scale and disregarding the Reynolds number effect. As noted earlier, these values are much higher than the calculated prototype values using appropriate drag coefficients. Thus, the model data should be scaled up not by Froude scaling but by proper care of the C_D values.

The tests showed that a horizontal tow should be used to avoid the potential risk of high tow line loads associated with the vertical tow. The tow line loads can be reduced by using a softer spring in the tow line. This permits limited tower motion and minimizes peak loads on the tow line.

8.5 LAUNCHING OF OFFSHORE STRUCTURES

Most offshore structures are built on land close to water. Use of a graving dock is a popular facility for building and launching large-volumed structures offshore before towing to the site for installation. The Khazzan oil storage structures described earlier were built this way. Some of the large concrete structures in Norway are built in deep water in fjords by ballasting the structure as the construction progressed. Then the structures are deballasted to a shallow draft before towing out of the fjords.

8.5.1 A Unique Launch

Submersible drilling rigs with large bases were built by CBI at its Pascagoula Construction Yard in Mississippi. In the yard the launching dock had a much greater elevation than the existing water level adjacent to the dock. Therefore, a unique, patented launching procedure was developed. Open-bottom buoyancy cans (up to 10.7m or 35 ft in diameter) were designed to support the structure on a cushion of compressed air during launch and load out in the sheltered bay off the Gulf of Mexico. Before the final towing of the rig under its own buoyancy in deep water, the system of buoyancy cans was disconnected from the rig by de-airing and thereby submerging the cans in place.

Since this launching procedure had not been used in the past, a test was conducted to investigate the launching sequence and the stability of the rig on the flotation cans. A simple model of the rig was built at a scale of 1:48. Nine buoyancy can models for supporting the rig were also built. The procedure outlined in the launching sequence was followed in the test in 15 discrete steps, and the pressures in each buoyancy can were monitored and adjusted as necessary. Air supply reservoirs were connected to each flotation can to accurately model the soft volume needed in the test with respect to the prototype can volume. (Note that soft volume refers to the air trapped in an open-bottom enclosure as opposed to hard or buoyant volume in a closed container or vessel). The seakeeping characteristics of the rig were studied while supported by all nine cans in wind and waves. Finally, the cans were ballasted beneath the structure, allowing the rig model to submerge to its design towing draft.

8.5.1.1 Modeling of Soft-Volume Cans

Under normal testing conditions, the atmospheric pressure in the laboratory is close to the atmospheric pressure prevailing at the site. The hydrostatic head in a soft volume depends on the pressure difference between the inside and outside air pressure and, as such, poses no scaling problem. However, this head is directly related to the compressibility of air inside the soft volume which depends on the absolute pressure. Assuming that air follows Boyle's gas law,

$$PV = \text{constant} \tag{8.6}$$

where P = absolute pressure (atmospheric + hydrostatic) of air in the soft volume, and V = trapped air volume. This law must be met in the model as well as the prototype. If the geometric similarity is maintained between the model and prototype, then the volume, V can be modeled properly. However, the pressure, P, includes the atmospheric pressure which does not scale. The pressure in the model environment is higher than the scaled down value since the scale factor is greater than one. In fact, the distortion is greater as the scale factor becomes larger. There are two ways to circumvent this problem. In the model, the soft volume members may be placed under vacuum (of appropriately scaled negative pressure or below atmospheric pressure) which is difficult to accomplish. The other method is to distort the soft volume so that the gas law is satisfied in the model scale. This is achieved by connecting additional volume to the soft volume members. It is clear that as the pressure changes, the amount of this extra volume also changes. Thus, in a test, e.g., during deballasting, a variable reservoir volume is needed for each can.

After the submersible rig has been launched from the dock, it is supported by freely floating soft volume cans. These support cans are modeled as shown in Fig. 8.11. Let us consider a prototype can of unit cross sectional area and small thickness initially floating in Position 1 with its freeboard equal to Hop. The internal pressure, ·P1p, corresponds to a differential head, H1p, between the water level inside and outside. If Po represents atmospheric pressure, then

$$P1p = Po + H1p \qquad\qquad (8.7)$$

and the air volume in the can is V_{1p},

$$V1p = Hop + H1p \qquad\qquad (8.8)$$

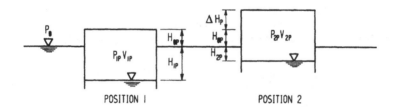

FIGURE 8.11
MODELING AN OPEN-BOTTOM SUPPORT CAN

Note that the volume has been normalized by the cross-sectional area of the cans.

If the floating can is displaced by an amount, ΔHp, to Position 2, its internal pressure will now be P2p.

$$P2p = Po + H2p \tag{8.9}$$

where H2p is the differential head between the water level inside and outside of the can. The new volume is V2p.

$$V2p = Hop + \Delta Hp + H2p \tag{8.10}$$

Assuming the products PV to be constant between Positions 1 and 2,

$$(Po + H1p)(Hop + H1p) = (Po + H2p)(Hop + \Delta Hp + H2p) \tag{8.11}$$

Solving this equation for the differential pressure head at Position 2, we get

$$H_{2p} = \frac{1}{2}\left(\left(P_o + H_{op} + \Delta H_p\right)^2 + 4\left\{P_o\left(H_{1p} - \Delta H_p\right) + H_{1p}\left(H_{op} + H_{1p}\right)\right\}\right)^{1/2}$$

$$-\frac{1}{2}\left(P_o + H_{op} + \Delta H_p\right) \tag{8.12}$$

When the can is modeled with a scale factor of λ, the gas law gives us

$$\left(P_o + \frac{H_{1p}}{\lambda}\right)\left(\frac{H_{op}}{\lambda} + \frac{H_{1p}}{\lambda} + V_r\right) = \left(P_o + \frac{H_{2p}}{\lambda}\right)\left(\frac{H_{op}}{\lambda} + \frac{\Delta H_p}{\lambda} + \frac{H_{2p}}{\lambda} + V_r\right) \tag{8.13}$$

where V_r is the volume of air that must be added to the can volume to compensate for the fact that P_o will not be scaled in the model. Solving for this added volume, we get

$$V_r = \frac{P_o(\lambda - 1)\left(H_{1p} - H_{2p} - \Delta H_p\right)}{\lambda\left(H_{2p} - H_{1p}\right)} \tag{8.14}$$

Since the above relation assumes the density of the supporting liquid to be equal for the prototype and model and we are representing a prototype that floats in seawater by a model floating in fresh water, the equation for V_r must be modified to account for this difference.

In the following, the subscript sw refers to seawater and the subscript fw refers to fresh water. If we consider the same prototype can floating in both seawater and fresh water with an identical mass of air in the can for both cases, then

$$Ppsw\,Vpsw = Ppfw\,Vpfw \qquad\qquad (8.15)$$

For the can to be floating in equilibrium in both cases, the upward force on the can must be the same whether the can is floating in seawater or fresh water. If the upward forces are equal then the following must be true assuming that the buoyancy force from the submerged can wall is nearly the same for both cases:

$$Ppsw = Ppfw \qquad\qquad (8.16)$$

and

$$Vpsw = Vpfw \qquad\qquad (8.17)$$

The can pressures are taken to be

$$Ppsw = P_0 + H1psw\,\gamma sw \qquad\qquad (8.18)$$

and

$$Ppfw = P_0 + H1pfw\,\gamma fw \qquad\qquad (8.19)$$

where γ is the water density. Solving for the fresh water differential head we get

$$H_{1pfw} = \frac{\gamma_{sw}}{\gamma_{fw}} H_{1psw} \qquad\qquad (8.20)$$

which shows the fresh water differential head to be equal to its corresponding sea water head multiplied by the ratio of seawater density to fresh water density. If we multiply the differential pressure head terms in Eq. 8.14 by the ratio $\gamma sw/\gamma fw$, we get

$$V_r = \frac{P_o(\lambda-1)\left[\dfrac{\gamma_{sw}}{\gamma_{fw}}\left(H_{1psw} - H_{2psw}\right) - \Delta H_p\right]}{\lambda\dfrac{\gamma_{sw}}{\gamma_{fw}}\left(H_{2psw} - H_{1psw}\right)} \qquad\qquad (8.21)$$

which gives the reservoir volume that should be added to the model can floating in fresh water in terms of the differential head of the prototype can floating in saltwater. The reservoir volume, V_r, allows us to correctly model the compressibility of the soft air volume.

Modeling the prototype seawater with fresh water also affects the model's freeboard. If we again consider the same prototype can floating in both seawater and fresh water, then

$$Vpsw = Hopsw + H1psw \qquad (8.22)$$

and

$$Vpfw = Hopfw + H1pfw \qquad (8.23)$$

Using Eq. 8.20 for H1pfw and the equality in Eq. 8.17, and solving for Hopfw gives us

$$H_{opfw} = H_{opsw} + H_{1psw}\left(1 - \frac{\gamma_{sw}}{\gamma_{fw}}\right) \qquad (8.24)$$

The fresh water freeboard of the model, Homfw, will, therefore, be

$$H_{omfw} = \frac{H_{opfw}}{\lambda} \qquad (8.25)$$

or

$$H_{omfw} = \frac{H_{opsw} + H_{1psw}\left(1 - \frac{\gamma_{sw}}{\gamma_{fw}}\right)}{\lambda} \qquad (8.26)$$

Since the model should float at

$$H_{omfw} = \frac{H_{opsw}}{\lambda} \qquad (8.27)$$

to correspond to the prototype, the model must be raised in the water by an amount ΔHomfw where

$$\Delta H_{omfw} = \left(\frac{H_{opsw}}{\lambda}\right) - \left(\frac{H_{opsw} + H_{1psw}\left(1 - \frac{\gamma_{sw}}{\gamma_{fw}}\right)}{\lambda}\right) \qquad (8.28)$$

or

$$\Delta H_{omfw} = -\frac{H_{1psw}\left(1-\dfrac{\gamma_{sw}}{\gamma_{fw}}\right)}{\lambda} \qquad (8.29)$$

To raise the model to the correct elevation we can subtract the amount ΔH_{omfw} from the reservoir volume. This will in effect move the volume ΔH_{omfw} from the reservoir into the can, causing it to float at the correct freeboard. The compressibility of air is still correctly modeled because the can-reservoir system volume remains constant. Our equation for V_r is now

$$V_r = \frac{P_0(\lambda-1)\left[\dfrac{\gamma_{sw}}{\gamma_{fw}}\left(H_{1psw}-H_{2psw}\right)-\Delta H_p\right]}{\lambda\left(\dfrac{\gamma_{sw}}{\gamma_{fw}}\right)\left(H_{2psw}-H_{1psw}\right)} + \frac{H_{1psw}\left(1-\dfrac{\gamma_{sw}}{\gamma_{fw}}\right)}{\lambda} \qquad (8.30)$$

which gives us the reservoir volume that will allow us to correctly model the compressibility of air and cause the model to float with the correct freeboard. The volume calculated by Eq. 8.30 will have dimensions of cubic units per square unit of model cross sectional area. An example of the extra volume needed for proper modeling is given as follows.

Let us assume a scale factor of 1:48 for the model. Also, consider a prototype freeboard for a can to be 2.6 m (8.5 ft) and the prototype pressure to be 1.78 m (5.85 ft) of seawater. Then, the calculation based on the above formulas requires an extra reservoir volume in the model (with fresh water) of 0.24 m^3 (8.55 cu. ft). This is a substantially large volume compared to the model volume. This volume should be provided in the form of closed air reservoir on land which is connected to the open-bottom can by flexible pipes.

8.5.1.2 Submersible Rig Model

The section of a submersible drilling rig model near and under the water was modeled using a scale factor of 1:48. The geometric similarity was maintained only where it was required for this test. The legs and super-structure of the rig were not modeled. The top and bottom of the mat structure were fabricated from single pieces of plexiglas and secured to the side walls with adhesive and screws. A three legged adjustable platform was constructed. Weights were bolted to the platform and properly positioned to place the center of gravity of the completed structure at the point shown in Fig. 8.12.

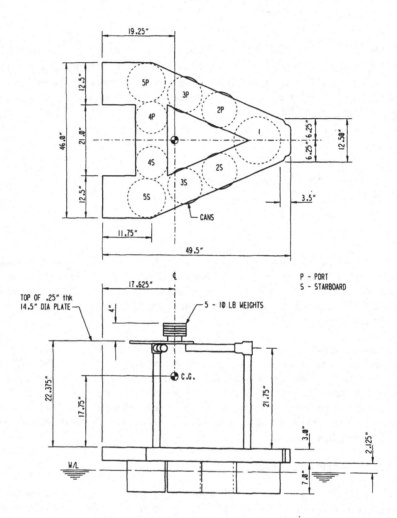

FIGURE 8.12
COMPLETE PLATFORM MODEL WITH BALLASTS

Since the top and bottom of the mat were transparent, positioning of the launch cans was accomplished by marking circles of various colors in the appropriate places

TABLE 8.2
LOCATION OF ROD AND RIG ON LAUNCH PLATFORM

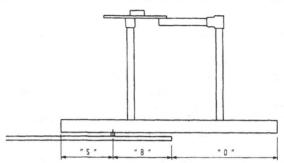

STEP	" 0 " OVERHANG (IN)	" S " CG FROM STERN (IN)	" B " CG FROM BULKHEAD (IN)
1	14.75	19.25	15.50
2	14.75	16.83	17.93
3	19.25	16.00	14.25
4	23.75	14.48	11.28
5	23.75	13.70	12.05
6	28.25	12.53	8.73
7	33.00	10.93	5.58
8	33.00	8.93	7.58
9	37.50	7.40	4.60
10	42.00	3.75	3.75
11	42.00	3.73	3.78
12	43.75	2.94	2.81
13	45.50	2.00	2.00
14	45.5	2.18	1.83
15	FLOATING ON LAUNCH CANS		

and matching and centralizing larger colored circles placed on the top surface of the launch cans. Each launch can was handled individually and was connected to its reservoir placed outside the tank via tygon tubings.

Each launch can pressure was measured by a 203 mm (8 in.) well type, Dwyer Manometer. The launch cans were scaled geometrically. The compensating scaled reservoir volumes were provided with valved bulkhead fitting at the bottom of the container. Water was then added to the container via this fitting to provide a liquid piston to vary the volume of the reservoir for various stages of the launching procedure. The water level of the wave tank was adjusted to 54 mm or 2-1/8 in. below the top of the leveled launching platform. This corresponded to a prototype dock height of 2.6 m (8.5 ft) above the water level.

8.5.1.3 Launching of Mat on Cans

During the launching, the mat was supported at the bow by the floating soft volume cans and at the stern by a 13 mm (0.5 in.) diameter aluminum rod. The aluminum support rod at the stern was positioned to support the model at the centroid of the launch beam reaction. The launch procedure was simulated in fifteen discrete steps (Table 8.2). The structure's overhang past the bulkhead face, the launch beam reaction centroid location, and the pressure in each can were pre-calculated.

For each launch step, the rig and the support rod were positioned at the desired locations and the mat was held level by hand as the cans were pressurized to the specified values. The mat was then released and the angle from horizontal was noted. The pressures in the cans were then adjusted as necessary for the cans to support the model parallel to the water and the new pressures were noted. On the final launch step the model was allowed to float free of the bulkhead, supported entirely by the nine soft volume cans. The submersible model was successfully launched several times following the procedure outlined through computation. The pressures in each can during the launching steps were noted. Observing the can pressures during the launch serves basically to check the statics calculations made earlier. Throughout the launching, the model appeared to be quite stable.

8.5.1.4 Seakeeping of Rig on Cans

The seakeeping of the rig when supported by the cans was tested by floating the model in waves. The model was oriented so that the waves approached it from several different directions. The rig exhibited good seakeeping characteristics when exposed to wind and waves. The model experienced little motion in roll or pitch.

8.5.1.5 Deballasting of Cans

Lowering the rig from its position on top of the cans to the point where it floats on its own requires that air be removed from the cans. The model was lowered into the water by venting air from several different combinations of cans. During each deballasting test, the rig began to level itself as soon as the mat entered the water and was supported by its own buoyancy.

To observe the behavior of the rig in a situation where air is lost from only one can while the rig is supported on the cans, the rig was floated level using the can pressures determined from the launching tests and then unbalanced by venting air from one can at a time. A stability problem was noted when the #5s can was lost. After the rig tilted approximately 4-6 degrees, the #2p and #3p cans were vented to bring the rig back to a nearly level position.

A few tests were performed initially without the auxiliary volumes. The system appeared very stiff and a small change in pressure in the cans affected the behavior of the rig on cans visibly. After comparing the behavior of the model during the stability test with and without the auxiliary volumes containing air, it was apparent that the extra air volumes did influence the model behavior to a considerable extent.

The operational tests performed on the launching of the submersible rig model was generally qualitative in nature. However, they were valuable in learning the limits of the operation and provided the operations crew a first-hand experience in the rigs behavior during the various stages of the launching procedure.

8.6   JACKET STRUCTURE INSTALLATION

The installation of a steel jacket platform requires a three-step procedure sequentially [Bhattacharyya (1984)]. These are load-out, launching and upending. The loadout is carried out on launch barge. In this case, the jacket is fastened to the launch barge forming a coupled jacket-barge system. In launching, the jacket slowly slides off the barge. Until this moment they act as a unit. During upending of the jacket, the dynamics of the jacket in swingdown is an important consideration. For this phase, a derrick barge is often used to install small jackets.

8.6.1 Scaling of Jacket Installation Parameters

During load-out, the parameters that are of importance and need consideration in model testing are the ballasting, tension in the derrick winch, fender reactions, mooring line forces and rocker arm reactions [see Graff (1981) p. 113]. In launching, the surge of the barge is important. In addition, the rocker arm reactions, the trajectory entry velocity, and sinking of the jacket, trim and draft of the barge are also of interest. The forces during this operation include viscous, inertial and gravitational forces and static and dynamic pressure forces. Both Reynolds and Froude similitudes are not possible. The viscosity is considered to have a modest influence on the results of such a test using a Froude model. All other parameters, e.g. displacement, velocity, friction factor, trim angle, hook elevation must follow Froude's law. Since the launch velocity is low, the flow is laminar, in which regime the Reynolds force cannot be ignored. However, jacket geometry is prone to induce turbulence. Additionally, turbulence flow may be artificially induced as is done in a ship model test.

In upending, the crane hook load, hook elevation, jacket ballast and orientation, righting moment and seabed bearing load are important. In this case, the forces include buoyancy and jacket inertia. In launching, the modeling of the jacket trajectory is of fundamental importance including the attitude of the jacket along the trajectory.

Geometric similitude is assumed in all cases. In load-out, the winch, fender and mooring line forces must be related to the jacket weight and the friction forces. Thus, a characteristic equation can be written as

$$\frac{F_m}{W} = f\left(\mu, \frac{F_w}{W}, \frac{F_f}{W}\right) \tag{8.31}$$

where F_m, F_w, F_f = mooring line, winch and fender forces, respectively, W = jacket weight and μ = coefficient of friction between the jacket and the barge launch rails. For the barge ballast and rocker arm hinge reaction, the characteristic equations are

$$\frac{F_b}{W} = f\left(\frac{x}{\ell}\right) \tag{8.32}$$

and

$$\frac{F_r}{W} = f\left(\frac{x}{\ell}\right) \tag{8.33}$$

where F_b, F_r = barge ballast and rocker arm reaction and x = jacket travel.

While the model must be geometrically similar to the prototype for load-out and upending tests, certain less important functional fittings, e.g. bollard eyes, structural members, rocker arm hinges, deck openings, winch assembly, etc. may be designed only from the point of view of practical convenience. Similarly, in load-out, only the weight similarity of the jacket is necessary, while, in upending, the buoyancy distribution of the jacket is also necessary. If the model material and fluid density are different from the prototype, proper adjustment in the member dimensions (e.g. thickness of tubular members) is needed to model weight and buoyancy simultaneously. For ballasting the model, a different liquid density may have to be chosen for filling the tubular members such that the modeling of the tubular cross-section is satisfied. In order to maintain the weight, C.G. and moment of inertia of the jacket at each level, additional members in the form of flat plates, rods, lump weights, etc. may have to be added and some smaller members omitted.

8.6.2 Launching Test Procedure

One of the most critical steps in the operation of an offshore steel jacket is the installation operation [Graff (1981)]. The installation operation consists of load-out, launching and upending in sequence. Due to the high risk involved, a physical simulation with a scale model is often recommended to verify the intended procedure. Model testing of the three step procedure is considered a necessary and powerful tool before the full scale structure is loaded out.

A model test on launching helps address the following areas:

- amount of launching force required to initiate and then maintain the motion of the structure, taking the draft and trim conditions of the barge into account.

- structural strength of the rocker arm and barge hull necessary during launching.

- difference in the trajectory between a single-hinged and a double- hinged rocker arm.

- required buoyancy and its distribution over the structure, including any added buoyancy.

- values of hydrodynamic force coefficients (C_D,C_M) needed to analyze the structure trajectory.

Launching structures from a barge at site is a common procedure offshore in the installation of a structure on the ocean bottom. The procedure involves pulling or pushing the structure longitudinally on skid rails using jacks and/or winches. In load-out and launching operations, the jacket and launch barge work as a system. In model testing, the jacket and the barge must be properly modeled.

Jacket models are usually made from plastic or aluminum alloy. Standard tube sizes are chosen in the jacket model. Therefore, this criterion generally governs the actual model scale after preliminary scale selection based on the facility. This may lead to some peculiar scale factor. The members which cannot be modeled even by the smallest tube size available are often omitted or modeled incorrectly. The similarity in buoyancy is achieved closely, but not necessarily exactly. Certain additional members such as flat plates, and solid rods are included judiciously so that the level and face configurations remain more or less unaltered. Additional lumped masses are attached to the jacket model to satisfy the overall requirements of weight, inertia and C.G. In order to satisfy the Cauchy scaling law, the model material should have a lower density than the prototype. When this is desired, perspex is often found to be the suitable material for both the jacket and the barge.

For the load-out operation [Bhattacharyya, et al (1985)] a quay of suitable height is built on one side of the wave tank. The winch drive may be simulated by a variable speed electric motor fitted at forward of the barge model. The fender reaction and winch pull are monitored by load cells during the load-out. The rocker arm hinge reaction is also recorded (Fig. 8.13). The load-out operation is performed in discrete steps. For a given jacket, the tide range within which it can be loaded out can be found from the simulation results. This, in turn, gives the ballast rates needed for any given tide range to accomplish the load-out.

During the launching operation, the history of the jacket trajectory is of importance. The entry velocity of the jacket is its velocity at the instant of launch and is of great significance [Bhattacharyya, et al. (1985)]. The estimation of the prototype launch velocity, however, is not easy because it depends on the coefficient of friction between the surface of the barge launch beams and the jacket launch trusses, variable trim angle as well as the travel. These parameters are not well defined quantitatively. Typical dynamic friction coefficient may be taken as 0.02 and an initial trim angle as 3 degrees.

Since the full scale friction is often an unknown, it is customary to verify that the jacket launching is possible over a range of friction coefficients. This requires controlled variation of friction between the sliding surfaces. In one method [Rowe and Clifford (1981)] the launch slideways are replaced with free running rollers which may be rotated about a vertical axis thus pointing all the way from along to across the running axis of the roller. When the rollers line up, the slideway friction is negligible. When they are turned 90° on the barge, full material friction is encountered. Friction may be varied between these two values. A system of side rollers is used to guide the jacket. In an alternate method, Bhattacharyya, et al. (1985) employed small rollers fitted at suitable intervals over the length of the launch beam. To achieve a predetermined entry velocity, a pulley was fitted onto the deck in a reverse rigging arrangement. The jacket was pushed by an L-shaped hook so that it was disconnected the instant the rocker arm tilted. The speed of the electric motor was calibrated for specific entry velocity. A limit switch was fixed onto the side of one of the rocker arms so that it recorded the pulses caused by the contact with a few curved protrusions fitted at several points to the side of the launch truss along its length. Entry velocity during launch could be obtained from the relative positions of these pulse records.

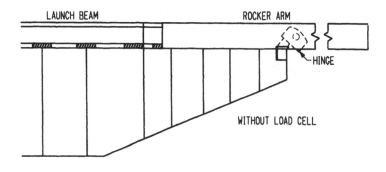

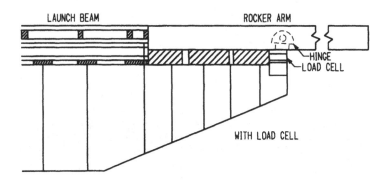

FIGURE 8.13
ROCKER ARM ARRANGEMENTS IN MODEL
[Bhattacharyya, et al. (1985)]

Upending is done in the following steps: the main legs of the jacket are ballasted; the derrick-barge crane is then used to lift the jacket until it is nearly vertical. At this point it is further lowered to the seafloor by controlled flooding of the jacket legs and auxiliary buoyancy tanks (if present). The procedure is followed in stages in a model test. Since the force system represented by the ballasts and crane hook load are similar, the prediction to the prototype values from the model test results is straightforward.

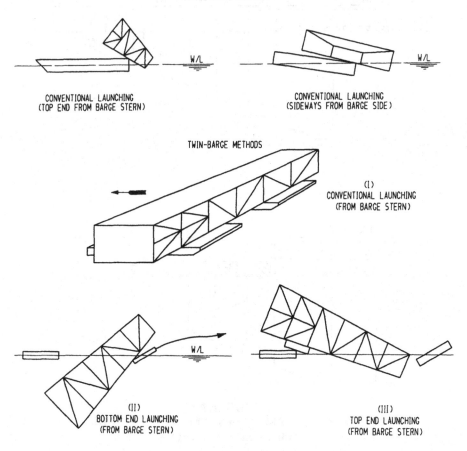

FIGURE 8.14
VARIOUS LAUNCH TEST SCHEMES
[Sekita, et al. (1980)]

In the model, caps are fitted to the top ends of the main legs and flexible hoses are connected to them for ballasting. Air vents are provided to the legs for air escape. Alternately, flood valves in the model may be operated remotely, although this is difficult to control. Slings are attached and connected to a single lifting hook representing the crane. The hook line is fitted with a load cell for the measurement of the hook load. The ballasting is done in incremental steps, the steps being smaller near large expected moments. Draft marks may be placed on four corners of the jacket to determine trim, heel and draft during upending.

8.6.3 Side Launching of Structures

As the structures become large and heavy, longer and stronger launching barges with large winch capacity is required. Side launching is an alternative procedure that may reduce these excess requirements imposed on jacks or winches.

Launching of structures from barges may be accomplished by many different methods. Model tests to compare five of these methods were carried out by Sekita, et al. (1980) on a template type jacket at a scale of 1:60. The conventional method consists of top end launching from the barge stern. Variations of this method include side launching from a single barge by heeling the barge over, side launch from twin barges astern, bottom and top end launching respectively from barge side. These are illustrated in Fig 8.14.

The measurement system is schematically illustrated in Fig. 8.15. The jacket is moved on the barge by a direct-torque motor whose driving power and speed are controlled electrically by voltage and mechanically by gears. Both single-hinged and double-hinged rocker arms and skid rails with and without non-friction coating may be used in the tests.

The side launching from a single barge seemed quite favorable from the model test results. In this case, the heeling characteristics of a barge may be advantageously harnessed to deploy the structure, thus reducing the required capacity of the launching equipment. Also, the vertical motion of the structure was small minimizing the risk of ramming the structure into the seabed. The twin-barge launching of the structure seemed to have special advantages as well, requiring smaller barges, more control of the structure and lower requirement of launch equipment capacity.

8.7 STAGED SUBMERGENCE OF A DRILLING RIG

The upending or submergence test of an offshore structure model follows one of the following methods: dynamic and staged. For certain structures, staged upending provides the necessary information to verify the success of the planned procedure. It is generally easier to accomplish. In the case of the jacket structure model, a staged upending is generally preferred. For certain other structures, dynamic upending is preferred to investigate any instability during this procedure. The rate of flooding of the structure model is an important consideration in these types of tests. In either case, the model is built as a dynamic model in which the weight, C.G., moment of inertia and GM are modeled accurately.

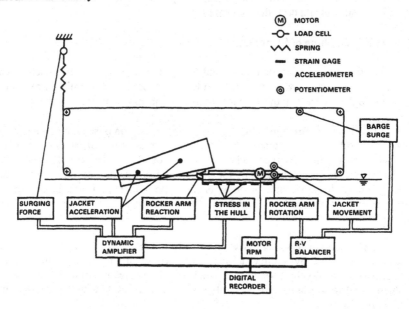

FIGURE 8.15
MEASUREMENT SYSTEM DURING JACKET LAUNCHING
[Sekita, et al. (1980)]

The technique for the introduction of ballast depends on whether the ballasting procedure models dynamic or staged upending scenarios. In the staged upending, the rate of ballast addition is not as important as the seakeeping characteristics at the successive ballast conditions. In this type of testing program, the model is held in-place and lump weights, such as, lead sheets are attached to the model where the ballast needs to be introduced. The model is then released, and the motions in that simulated environment are monitored. A practical example of a planned submergence of a drilling structure is illustrated here.

The submersible drilling rig described earlier was tested for submergence in shallow water. In this case a complete model with legs and superstructure was required. Moreover, the dynamic characteristics of the rig was also modeled. A staged submergence test was carried out by ballasting the rig in controlled steps (Fig. 8.16). The model was submerged in the tank following a controlled ballasting procedure. The structure had 28 ballast tanks in the mat and required an 11 step ballasting procedure to bring it to the bottom from its loaded draft. During the testing, the lower variables (e.g., drill material, and fuel oil) were modeled by completely filling the forward pair of drill water tanks and the forward pair of fuel oil tanks. This was preferred rather than distributing the lower variable among all the drill water and fuel tanks in order to minimize the free surface effect of the liquid variables.

FIGURE 8.16
BALLAST TANKS OF SUBMERSIBLE FILLED IN PRESCRIBED SEQUENCE

In modeling the prototype weight, the model weight was not directly scaled from the prototype value but was corrected to account for the fact that the model floats in fresh water rather than seawater. This was accomplished by multiplying the scaled prototype weight by the ratio of fresh water density to seawater density. This allowed the prototype draft to be correctly modeled.

8.7.1 Test Results

The model was submerged successfully following the prescribed submergence procedure. The model did, however, exhibit some behavior characteristics worth noting. During submergence, the side to side list of the model was difficult to control. Unlike the prototype control of filling rates, there was no way to control the filling rate of the model ballast tanks and the amount of ballast in each tank was estimated by visual observation. Therefore, most of the list may have been due to an imbalance of ballast in the model. Also contributing to the list was the ballast free surface effect. The effect on the model's list was greatest at the draft where the top of the mat was just below the water surface. At this draft, the model would not float level but would list to one side or the other. With the ballast evenly distributed side to side, the model could be made to list to either port or starboard by tilting the model by hand. The fact that the model would list to either side but not remain level in the water at this draft illustrates the effect of the ballast shifting from one side of the tank to the other. Although the ballast tanks contain baffles that reduce sloshing in the prototype, the baffles have holes in them to allow the tanks to be filled using one water inlet and to be vented at one outlet. These holes allow water to flow through the baffles fast enough to produce instability due to free surface in the ballast compartments as the model lists slowly to one side or the other. When the model was tested with the full upper variable load in place, the maximum list observed was approximately 2 degrees to the side. The bow to stern trim of the model appeared not to be affected much by the ballast free surface, but depended mostly on the fore-aft balance.

8.8   DYNAMIC SUBMERGENCE OF A SUBSEA STORAGE TANK

The rate of ballast addition is important in a dynamic upending test. If this type of test is required, two valves are installed in the ballast chamber during model construction. One valve is placed at the bottom of the chamber for ballast water entry. The second valve is installed at the top of the chamber for air venting. The water level in the chamber is monitored with a wave probe. Alternatively, the drop in water level in an outside storage tank containing ballast water is monitored. The valve setting is modified until the desired flow rate is achieved.

Three "Khazzan Dubai" subsea oil storage tanks were submerged and anchored to the floor of the Arabian Gulf. Chamberlin (1970) described the design, construction, and installation of these structures between the late sixties and early seventies. The submergence of this structure required a unique technique and the method was verified through scaled model testing before the actual operation. The structure is composed of an open inverted funnel-shaped outside surface, a center bottle and a ringwall.

Khazzan Dubai tanks were constructed on shore in a graving dock complete with a large topside platform. The resulting higher center of gravity increases the importance of an accurate prediction of structure's stability during submergence. Figure 8.17 shows a schematic of the tank as it arrived at the site floating on a pressurized cushion of air [Burns and Holtze (1972)]. The center of gravity of the structure at this orientation is substantially above the center of buoyancy. However, the structure remains vertical because its outer ringwall which is ballasted with concrete raises the metacenter sufficiently to provide stability.

As air is vented from the roof (Fig. 8.18a), the structure's draft increases until the ringwall is totally submerged. The metacenter at this point is suddenly lowered, and the structure slowly begins to tilt, raising the wall partially out of water to regain stability. As more air is vented, the tilt becomes large enough to allow air to escape from under the wall as shown in Fig 8.18b. The sudden release of air reduces the pressure supporting the structure, and the unbalanced weight of the structure causes a dynamic descent. This process continues with an increasing draft and angle of tilt until the structure is supported by the central pressure vessel, referred to as the "bottle". Angular moments carry the structure beyond the point of static equilibrium to the most severe draft and heeling angle, as seen in Fig. 8.18c. The structure starts to oscillate about the point of static equilibrium point until the motion is damped out.

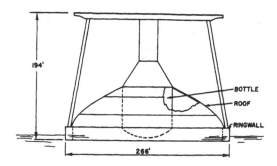

FIGURE 8.17
500,000 BBL SUBMERGED STORAGE STRUCTURE

When the structure comes to rest, as in Fig. 8.18d, water is pumped into the bottle to reduce the effective buoyancy until the center of gravity is below the center of buoyancy (Fig. 8.18e). At this time the structure returns to the vertical position and continued water pumping causes a controlled descent to the seafloor.

8.8.1 Model Testing

A 1:48 scale model of the storage tank was constructed using Froude similitude. The model materials were chosen so that each component would have approximately the correct density and center of gravity. The result was a model with accurate weight, center of gravity, and dimensional properties. The dimensional accuracy is required to insure that buoyancy forces are correct. The proper weight distribution yields a correct mass moment of inertia, and avoids the necessity of using large weights to adjust the center of gravity.

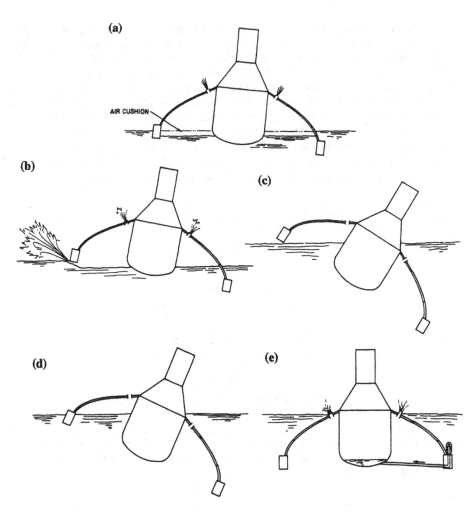

FIGURE 8.18
SUBMERGENCE SEQUENCE OF A STORAGE TANK

The ringwall was formed from a rectangular aluminum bar, and holes were drilled to simulate pile sleeves. An extra set of holes was drilled and filled with foam to adjust the overall density to that of the prototype ringwall. A thin molded fiber glass roof was used to give the proper thickness. This roof was not sufficiently strong to support the ringwall; therefore, wire ties were added as shown in Fig. 8.19.

The platform and supports were made from PVC plastic, and the remainder of the structure from aluminum.

After construction, the ballasted model was carefully weighed and leveled in the water. Removable weights in the shaft allowed for modeling both Khazzan Dubai 2 and 3, which have slightly different centers of gravity.

The model test was performed in a wave tank (Fig. 8.20). Motion picture film was used to record the tests, and frame-by-frame analysis of the film yielded the time history plot in Fig. 8.21. The behavior of the prototype subsequently found during submergence of Khazzan 2 & 3 in the Persian Gulf confirmed the observed mode of submergence of the storage tank model.

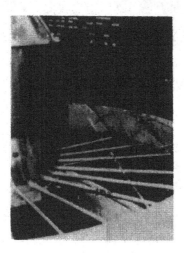

FIGURE 8.19
INTERNALS OF STORAGE TANK MODEL

8.9 OFFSHORE PIPE LAYING OPERATIONS

In offshore pipe laying operations, complete similarity cannot be achieved in a model scale. Distortion is necessary in proper scaling of the pipeline. The modeling should include not only the steady-state condition but the dynamic response of the pipeline under prevailing environmental condition as well. In modeling offshore pipe

FIGURE 8.20
FINISHED STORAGE TANK MODEL BEING TESTED

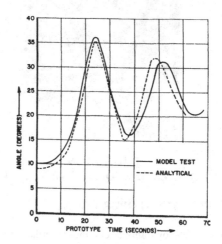

FIGURE 8.21
SCALED-UP MODEL TEST RESULT ON SUBMERGENCE ANGLE VERSUS
TIME [Burns and Holtze (1972)]

laying, there are several scaling laws to be considered and complete geometric similarity
is not   possible   by   fulfilling these   laws. Therefore, compromise is made in this

selection. Some of the scaling laws are relaxed while others are satisfied by changing the model pipe geometry.

8.9.1. Pipeline Similarity Laws

In a pipeline operation there are four major forces that should be considered in the scaling laws. They are the gravity forces, inertia forces, elastic forces and frictional forces. The ratio of these forces between the prototype and model may be conventionally expressed in terms of scale factors. Assuming two separate scale factors, one the linear dimensions, λ and one for time, τ, the ratio between the inertia forces on the prototype and the model [Clauss and Kruppa (1974)] is:

$$F_{Ir} = \frac{\rho_p}{\rho_m} \frac{\lambda^4}{\tau^2} \tag{8.32}$$

where ρ is the mass density of the surrounding fluid, F_I is the inertia force and r denotes ratio between the prototype and model values. Similarly, the ratios for the gravity, elastic and frictional forces are

$$F_{Gr} = \frac{\gamma_p}{\gamma_m} \lambda^3 \tag{8.34}$$

$$F_{er} = \frac{E_p}{E_m} \lambda^2 \tag{8.35}$$

and

$$F_{fr} = \frac{v_p \rho_p}{v_m \rho_m} \frac{\lambda^2}{\tau} \tag{8.36}$$

respectively, where γ and E are the specific gravity and Young's modulus of pipe material. For perfect similarity, the ratio of these forces to the inertia force must be unity. Thus, for the frictional (viscous) forces,

$$\frac{F_{fr}}{F_{Ir}} = 1 \tag{8.37}$$

which gives

$$\tau = \frac{\lambda^2}{v_p / v_m} \tag{8.38}$$

This is equivalent to stating that the Reynolds number for both the model and prototype is the same. As illustrated earlier, in hydrodynamic model testing the viscous forces cannot be scaled adequately and are generally corrected for in scaling up the model test results.

Considering the similarity of gravity and elastic forces, we have

$$\tau = \left[\frac{(\rho_p / \rho_m)}{(\gamma_p / \gamma_m)} \lambda \right]^{1/2} = \left[\frac{(\rho_p / \rho_m)}{(E_p / E_m)} \right]^{1/2} \lambda \tag{8.39}$$

The first equality corresponds to the Froude number while the second one refers to the Cauchy number. These identities can, therefore, be fulfilled simultaneously, if the model and prototype materials are not identical. In fact, the material must fulfill the following requirement from Eq.8.39:

$$\lambda = \frac{(E_p / E_m)}{(\gamma_p / \gamma_m)} \tag{8.40}$$

This limits the choice of model scales. One can show [Clauss and Kruppa (1974)] that there is no suitable material which permits testing at model scales $10 \le \lambda \le 100$.

8.9.2 Partial Geometric Similarity

Full scale pipes are generally coated with a thick concrete jacket for added weight and stability. The coating, however, does not contribute to its section modulus for bending stress calculations. The elastic deflection on a pipe element is obtained from

$$\frac{d^2 M_B}{dx^2} + \frac{W}{\cos\theta} - F_H \frac{d^2 y}{dx^2} = 0 \tag{8.41}$$

where M_B and F_H are the moment and horizontal force on the element and W is the pipe weight per unit length. Nondimensionalizing with respect to the water depth d (so that $\bar{x} = x / d$ and $\bar{y} = y / d$).

$$\frac{d^2 M_B}{d\bar{x}^2} + W d^2 \left[1 + \left(\frac{d\bar{y}}{d\bar{x}} \right)^2 \right]^{1/2} - F_H d \frac{d^2 \bar{y}}{d\bar{x}^2} = 0 \tag{8.42}$$

For large elastic deflections, the beam theory suggests that

$$M_B = \frac{EI}{d} \frac{d^2\bar{y}/d\bar{x}^2}{\left[1+\left(\frac{d\bar{y}}{d\bar{x}}\right)^2\right]^{3/2}} \tag{8.43}$$

where I is the cross sectional moment of inertia. If this expression is substituted in Eq.8.42 and the equation is normalized by dividing throughout by EI/d, then the nondimensional coefficients of the second and third terms in Eq. 8.42 become

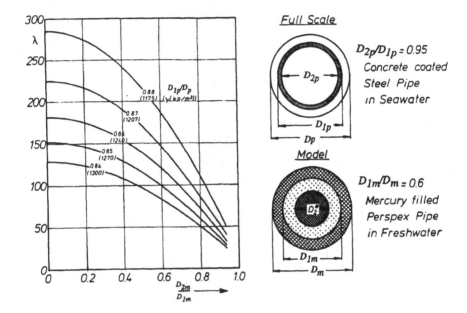

FIGURE 8.22
SCALING PARAMETERS FOR MODEL PIPE DIAMETERS
[Clauss & Kruppa (1974)]

$$C_1 = \frac{Wd^3}{EI} \quad \text{and} \quad C_2 = \frac{F_H d^2}{EI}$$

(8.44)

This requires that three quantities Wd, F_H and EI/d^2 have the same ratios between the model and prototype. These quantities are the gravity, external and elastic forces, respectively.

The outer diameter of the model and prototype pipes are scaled according to λ (= D_p/D_m). Assuming a model cross-section similar to Fig. 8.22 [Clauss and Kruppa (1974)] for the corresponding prototype pipe cross-section, the hydrodynamic inertia forces will scale properly. In this case, Eq.8.32 becomes

$$F_{Ir} = \bar{\rho}_e \frac{\lambda^4}{\tau^2}$$

(8.45)

where

$$\bar{\rho}_e = \frac{\rho_p}{\rho_m} \frac{\bar{D}_{1p}{}^2 - \bar{D}_{2p}{}^2 + \frac{\rho_c}{\rho_p}\left(1 - \bar{D}_{1p}{}^2\right) + \frac{\rho_{fp}}{\rho_p}}{1 - \bar{D}_{1m}{}^2 + \frac{\rho_{fim}}{\rho_m}\left(\bar{D}_{1m}{}^2 - \bar{D}_{2m}{}^2\right) + \bar{D}_{2m}{}^2 + \frac{\rho_{fm}}{\rho_m}}$$

(8.46)

and $\bar{D}1 = D1/D$, $\bar{D}2 = D2/D$, ρ_p = mass density of steel, ρ_m = mass density of model pipe and insert rod, ρ_c = mass density of concrete coating, ρ_f = mass density of surrounding fluid, ρ_{fi} = mass density of liquid filling. This equation ensures similar mass per unit length of pipeline including an allowance for the hydrodynamic mass.

The force ratio for the gravity force is similarly obtained

$$F_{Gr} = \bar{\gamma}_e \lambda^3$$

(8.47)

where

$$\bar{\gamma}_e = \frac{\gamma_p}{\gamma_m} \frac{\bar{D}_{1p}^2 - \bar{D}_{2p}^2 + \frac{\rho_c}{\rho_p}\left(1 - \bar{D}_{1p}^2\right)\frac{\rho_{fp}}{\rho_p}}{1 - \bar{D}_{1m}^2 + \frac{\rho_{fim}}{\rho_m}\left(\bar{D}_{1m}^2 - \bar{D}_{2m}^2\right) + \bar{D}_{2m}^2 - \frac{\rho_{fm}}{\rho_m}} \tag{8.48}$$

Similarly, the ratio for the elastic forces is derived as

$$F_{er} = \bar{E}_e \bar{I}_e \lambda^2 \tag{8.49}$$

where

$$\bar{E}_e = \frac{E_p}{E_m} \tag{8.50}$$

and

$$\bar{I}_e = \frac{\bar{D}_{1p}^4 - \bar{D}_{2p}^4}{1 - \bar{D}_{1m}^4 + \bar{D}_{2m}^4} \tag{8.51}$$

in which the effect of the concrete jacket has been neglected as not contributing to the prototype section modulus. Since all force ratios have to be identical, one obtains

$$\tau = \left(\frac{\bar{\rho}_e}{\bar{\gamma}_e}\lambda\right)^{1/2} = \left(\frac{\bar{\rho}_e}{\bar{E}_e \bar{I}_e}\right)^{1/2} \lambda \tag{8.52}$$

and

$$\lambda = \frac{\bar{E}_e}{\bar{\gamma}_e} \bar{I}_e \tag{8.53}$$

Obviously, for complete similarity,

$$\bar{\rho}_e = \bar{\gamma}_e = \bar{E}_e = \bar{I}_e = 1 \tag{8.54}$$

The model scale and time scale are affected by the material properties as well as the model pipe cross sectional geometry. For the horizontal tension forces

$$F_{Hr} = \overline{\gamma}_e \lambda^3 \qquad (8.55)$$

Other forces will be hydrodynamic in nature and may depend on Reynolds number, for example the transverse current force.

An example of the scale factor for a model pipe is given here. The prototype pipe is considered to be of steel construction coated with concrete such that D2p/D1p = 0.95. The specific gravity of the empty prototype pipe in air is given by

$$\gamma_e = \gamma \left\{ \overline{D}_{1p}^2 - \overline{D}_{2p}^2 + \frac{\rho_c}{\rho_p}\left(1 - \overline{D}_{1p}^2\right) \right\} \qquad (8.56)$$

The model pipe is represented by a perspex pipe and an insert rod, the annular space being filled with mercury. In this case, $\overline{D}_{1m} = 0.6$. For a range of $\overline{D}_{1p}$, the possible values of λ as a function of D2m/D1m are shown in Fig. 8.22. Thus, the model scale can be varied by changing the diameter of the insert rod. For a given pipeline specific gravity, γ_e, the corresponding values of time τ and force F_{Ir} are given in Fig. 8.23.

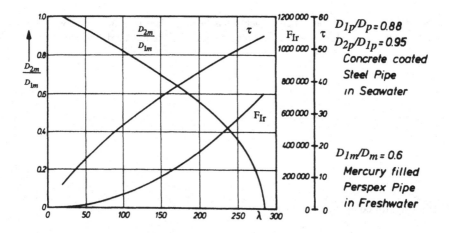

FIGURE 8.23
SCALING PARAMETERS FOR TIME AND FORCE
[Clauss & Kruppa (1974)]

8.10   REFERENCES

1.   Bhattacharyya, S.K., "On the Application of Similitude to Installation Operations of Offshore Steel Jackets," Applied Ocean Research, Vol.6, No.4, 1984, pp.221-226.

2.   Bhattacharyya, S.K., Idichandy, V.G., and Joglekar, N.R., "On Experimental Investigation of Load-Out, Launching and Upending of Offshore Steel Jackets", Applied Ocean Research, Vol.7, No.1, 1985, pp.24-34.

3.   Burns, G.E., and Holtze, G.C., "Dynamic Submergence Analysis of the Khazzan Dubai Subsea Oil Tank," Proceedings on the Fourth Annual Offshore Technology Conference, Houston, Texas, OTC 1667, May, 1972, pp. II-467-478

4.   Chamberlin, R.S., "Khazzan Dubai 1: Design, Construction and Installation," Proceedings on the Second Annual Offshore Technology Conference, Houston, Texas, OTC 1192, May 1970, pp-I-440-454.

5.   Clauss, G., and Kruppa, C., "Model Testing Techniques in Offshore Pipelining," Proceedings on the Sixth Annual Offshore Technology Conference, Houston, Texas, OTC 1937, May, 1974, pp. 47-54.

6.   Graff, W.J., Introduction to Offshore Structures, Design, Fabrication, Installation, Gulf Publishing Co., Houston, TX., 1981.

7.   Hoerner, S.F., Fluid Dynamic Drag, Published by the Author, Midland Park, New Jersey, 1965.

8.   Muckle, W. and Taylor, D.A., Muckle's Naval Architecture, 2nd Edition, Butterworths & Co. Ltd., London, England, 1987.

9.   Rowe, S.J. and Clifford, W.R.H, "Model Testing of Launching and Installation of Steel Production Platforms," Offshore Structures: The Use of Physical Models in their Design, Construction Press, Lancaster, 1981.

10. Sekita, K., Sakai, M., and Kimura, T., "Model Tests on Various Launching Methods for Large Offshore Structures," Proceedings on the Twelfth Annual Offshore Technology Conference, Houston, Texas, OTC 3836, May, 1980, pp. 369-378.

CHAPTER 9

SEAKEEPING TESTS

9.1 FLOATING STRUCTURES

Seakeeping characteristics of a floating structure that is in motion under its own power or moored to either the seafloor or to another structure by some mechanical means determine its ability to survive the environment. The motion and component loads of the floating system are generally computed analytically and verified with model tests. While the seakeeping of rigid structures are commonly model tested, flexible structures are modeled as well. In general, the structure is allowed to undergo six independent degrees of motion - three transitional and three rotational. These motions are schematically shown in Fig. 9.1. Often the structure is restrained to have fewer degrees of freedom due to the type of mechanical connection used to fasten it to the sea-floor or to another vessel or structure.

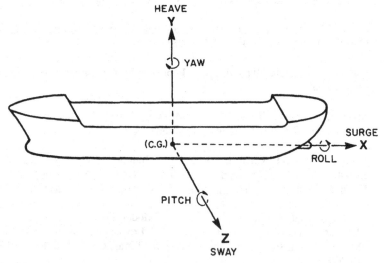

FIGURE 9.1
DEFINITION OF SIX DEGREES OF MOTION OF A FLOATING BODY

Examples of such structures are tankers, barges, buoys, compliant towers, floating production systems and tension leg platforms. Compliant towers are typically designed for deep water applications and are allowed to flex in the waves. In designing compliant towers, the structural stiffness in sway is kept low such that its first natural period is outside the range of large wave energy that may excite the structure. Floating Production Systems (FPS) consist of a semi-submersible platform or a Mobile Offshore Drilling Unit (MODU) which is moored in deep water with mooring lines. The Tension Leg Platform (TLP) is a semi-submersible floater that is held on location by a vertical mooring system called tendons. Drilling and production facilities are supported on the floater. The TLP has gained widespread acceptance as the best choice for the development of deepwater oil fields.

In order to design any of the above structures, the motions of the structure should be known in addition to the wave forces acting on it. The motion response of these structures is routinely obtained through model testing. Modeling is quite straightforward if the structure size is large such that the inertial force predominates. The scaling generally follows Froude's law even though special measures are sometimes taken to correct for the Reynolds scaling effects. For flexible members, Cauchy scaling is additionally superimposed. Model tests of a few of these structures and some of the problems associated with them are discussed in this chapter.

9.2   METHOD OF TESTING FLOATING STRUCTURES

A floating platform model is generally placed near the middle of the test basin and anchored to the basin floor (or its sides or another structure model) by means of anchorline models. Of general interest are the motions of the platform and the loads experienced by the mooring lines when subjected to the model waves, wind and currents. The motion and associated loads occur at several frequencies from very low to very high, depending on the sea state, type of the platform and the mode of anchoring. The modeling of the natural period of motions of the platform and the corresponding damping is extremely important in the test set-up. The system is often nonlinear due to nonlinear mooring line characteristics. One should take special care in modeling these lines within their range of interest.

The offset of a moored floating structure and the mooring line loads are just two of several aspects determined by model tests. These quantities are highly dependent on the damping of the system. Therefore, modeling the damping experienced by a floating structure is an important consideration. One possible source of error in modeling damping is the mechanical friction introduced in the test set-up. As discussed earlier (Chapter 5), a weight hanging over a pulley system has been a common method of introducing steady wind and current loads. The lines used for this system (or just to keep a model in position, for example, from any sway motion) may introduce additional unwanted friction in the system. The mechanical connection used for the motion measurement may also provide additional frictional force in the system. These sources of friction could be significant for the low frequency motion of a moored floating vessel which is very sensitive to the overall damping in the system. It will be shown through computations and experiments that in model testing it is safest and best to avoid mechanical friction whenever possible. If mechanical connections are unavoidable due to other constraints, then they should be thoroughly investigated for suitability and error before embarking on a test. The following analysis closely follows the work of Huse (1989).

Let us consider a single degree of freedom system having hydrodynamic as well as friction type damping. The differential equation is written as

$$M\ddot{x} + b_0 \dot{x} / |\dot{x}| + b_1 \dot{x} + b_2 |\dot{x}| \dot{x} + Kx = F_0 \cos \omega t \tag{9.1}$$

where b_0, b_1, and b_2, are the coefficients of zeroth (Coulomb friction type), linear and quadratic damping terms. The linear and quadratic terms appear from the hydrodynamic damping while a mechanical system in the set up is the source of zeroth damping. The energies associated with these forces are respectively:

$$E_0 = 4b_0 x_0 \tag{9.2}$$

$$E_1 = \pi b_1 \omega x_0^2 \tag{9.3}$$

$$E_2 = (8/3)b_2 \omega^2 x_0^3 \tag{9.4}$$

whereas the energy input into the system from the extinction force is

$$E_f = \pi F_0 x_0 \tag{9.5}$$

The quantity x_0 refers to the amplitude of the displacement x. The nonlinear damping term has been linearized in the above expression. By equating the input energy to the dissipated energy, the external force, F_0 becomes:

$$F_0 = \frac{1}{\pi}\left(4b_0 + \pi b_1 \omega x_0 + \frac{8}{3}b_2 \omega^2 x_0^2 \right) \tag{9.6}$$

Now consider two systems in a test setup where one setup introduces mechanical damping while the other does not. Writing the amplitudes of oscillation for the two systems as x_{of} and x_{on} respectively, and applying Eq. 9.6 for both x_{of} and x_{on}, we have

$$x_{of} = \left[-u_1 \pm \left(u_1^2 - 4u_2 u_3 \right)^{\frac{1}{2}} \right] \Big/ (2u_2) \tag{9.7}$$

where

$$u_1 = \pi b_1 \omega \tag{9.8}$$

$$u_2 = \frac{8}{3} b_2 \omega^2 \tag{9.9}$$

$$u_3 = 4b_0 - u_1 x_{on} - u_2 x_{on}^2 \tag{9.10}$$

This expression provides the error in the motion amplitude due to the additional friction term.

Let us consider a numerical example for a tanker model undergoing low frequency oscillation in still water. A typical model of a 100,000 dead weight ton (dwt) tanker at a scale of 1:65 has the following dimensions:

displacement	$0.49m^3$
length between perpendiculars	3.76m
breadth	0.61m
draft	0.26m
wetted surface area	$4.0m^2$

A surge oscillation is carried out under the following condition to determine viscous damping of the hull in still water:

stiffness, K	22.61N/m
surge amplitude without friction, x_{on}	0.1m
linear damping coefficient, b_1	0.0
quadratic damping coefficient, b_2	$38.4 \, Ns^2/m^2$

The undamped natural period in surge is 30 seconds. A correction factor R is defined as the ratio between the surge amplitude measured without and with friction. The values of R for different values of b_0 are computed from Eq.9.7:

b_0(N)	0.001	0.002	0.005	0.01	0.02
R	1.05	1.10	1.34	3.02	infinite

Note that as b_0 becomes large, R becomes infinite due to the linearization in Eq. 9.9. From the above table, even for a small amount of mechanical friction ($b_0 = 0.001$N) the error in the measured surge amplitude is about 5 percent. Pulleys and potentiometers used in a test set up may provide additional friction in the system, significantly affecting the model motions and may be unsatisfactory for this type of tests.

9.3   SINGLE POINT MOORING SYSTEM

Single point mooring terminals provide a safe environment for mooring a vessel and a linkage for the transport of crude oil. These are facilities of small horizontal dimension to which large vessels are moored by means of a bow hawser or other methods allowing the vessel to weather vane about the mooring terminal. There are many types of single point mooring terminals as shown in Fig. 9.2. The Catenary Anchor Leg Mooring (CALM) system consists of a flat cylindrical buoy anchored to the seafloor by catenary chains. The Single Anchor Leg Mooring (SALM) system consists of a cylindrical buoy attached to a heavy base on the sea floor by a single pre-tensioned chain. Others are Exposed Location Single Buoy Mooring (ELSBM), Buoyant Towers, and the SPAR buoy.

Designers of these systems have often made use of model tests to gain reliable data on the behavior of single point mooring terminals [Pinkster and Remery (1975)]. These tests on scale models are carried out in large basins in which the site environmental conditions are carefully reproduced. While two-dimensional tanks are frequently used for long-crested waves and colinear wind and current, the smaller terminals are vulnerable to short-crested seas and non-colinear currents for which three-dimensional basins are desired. Important test conditions for these systems include three separate scenarios: (1) survival condition of the terminal by itself during extreme environment; (2) extreme operating condition under which the tanker can stay moored; and (3) general operating condition for fatigue and "wear and tear" data.

In setting up a model test program, the environmental conditions are duplicated. Waves in terms of significant height and mean period as well as frequency distribution model must be specified. For short-crested seas, the directional distribution should also be known. Ideally, such information should be obtained from measurements taken at or near the proposed site over a reasonable period of time. However, input is often obtained from published data and hindcasting. Current speed, vertical profile and direction with respect to the predominant wave direction, and wind speed and direction must also be specified. A calibration of the environment is usually performed prior to testing by simulating the environment in the tank without the presence of the model.

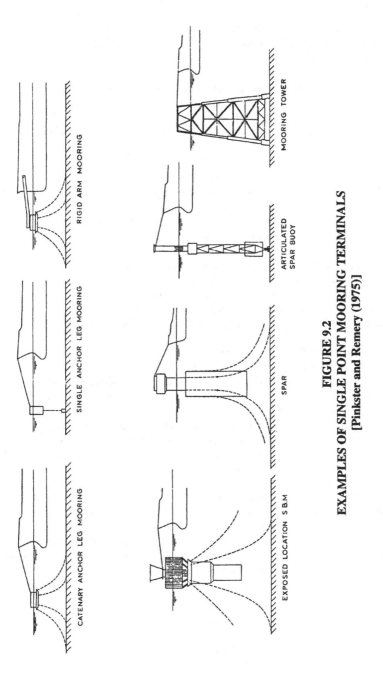

FIGURE 9.2
EXAMPLES OF SINGLE POINT MOORING TERMINALS
[Pinkster and Remery (1975)]

In addition to the environmental conditions, the characteristics of the single point mooring terminal must be known. For example, for a CALM system, the following information is needed:

- buoy: dimension, weight, center of gravity, bow hawser location,

- anchor chains: number, length, weight per unit length, breaking strength, elastic limit and elongation, pre-tension and buoy connection points,

- moored vessel: principal dimensions, displacement, draft, center of gravity, righting moments, radii of gyration, and natural period, and

- bow hawser: length, weight per unit length, hawser connection point and load versus elongation curve.

The model scale for a given basin is chosen based on water depth, wave generating capability and accuracy and magnitude of measurements, as usual. Froude's law is used throughout for modeling an SPM terminal. For testing in wind, waves and current, it is a normal practice to use a scale factor ranging between 50 and 85. This range of scale factor allows accurate measurement of the important variables.

Models of SPM terminals are constructed of different types of materials: wood, metal, synthetic foam, plastic, etc. In practically all cases, components are made as rigid as possible since the tests are aimed at the determination of rigid body behavior and not the elastic effect of construction elements. The center of gravity of the model is determined in air and adjusted, if necessary, by an inclining test (Chapter 3). The mass moment of inertia is checked by a pendulum test. The model anchor chains are made to conform to correct scaled length and weight. The elastic stretch of the model chains is achieved by adding a coil spring at the end of the chain. Thus, the caternary characteristics and elastic properties are simultaneously scaled.

The model of the moored vessel, such as a tanker, is commonly constructed of wood and fiberglass. In most cases, available stock models are used which may determine the model scale for the SPM test. Tankers are fitted with deck and superstructure. The geometry and location of the superstructure is important for the simulation of wind loads. The mass moment of inertia in yaw and pitch are adjusted in air by the pendulum method. The transverse stability is determined by inclining tests in calm water. Similarly, the roll natural period in water is modeled at the same time.

In some SPM model tests, underbuoy hoses or floating hoses are also modeled. These hoses consist of strings of large diameter (305-610 mm or 12-24 in.) flexible hoses. The elements of these strings are 0.9-12 m (3-40 ft) long, bolted together. A hose element consists of a flexible middle section made of steel coils embedded in rubber and rigid end connection ending in a flange. Models of such hoses may be made in a similar manner. They consist of coil springs covered with latex. The extremities of each model element end in a rigid part with a flange, to which the next section may be connected.

The measurements during SPM tests generally consist of (1) forces in anchor chains and bow hawsers, (2) motions of the buoy, (3) axial forces and bending moments in underbuoy and floating hoses, and (4) tanker motions. Before the wave tests commence, the static values of forces and motions are recorded. The SPM system is displaced from the equilibrium in an incremental manner and the overall stiffness of the system is established (static load - deflection curve). Also, the current loads and the combined wind and current loads are recorded.

9.3.1 Articulated Mooring Towers

An articulated loading tower capable of mooring large oil-export tankers operate in moderate water depths (up to about 150 m or 500ft). The towers are maintained upright as a result of their buoyancy sections, and are considered large enough to be treated as rigid body system without appreciable deformation of their members. The motion of the tower and the tanker as well as the nonlinear restoring force characteristics of the hawser are needed for the design of this system. The tower must survive the largest storm (e.g., 100 year) expected during its lifetime. The tower-tanker system must operate at a given sea state based on the location of operation.

Model tests of these articulated towers require the measurement of the motions of the model in two perpendicular planes and the three-component loads at the universal joint. An XYZ load cell can accomplish the necessary load measurement when mounted on the base and connected to the universal joint of the tower (Fig 9.3). The tower motion may be measured either by a direct contact method or can be measured indirectly. In the direct contact method, the displacement at the tower top or the rotation at the base of the tower may be measured. In the latter case, two RVDT's (section 6.4) may be placed at right angles to each other. In this case, the RVDT's must be waterproofed properly for underwater application. Displacement at the top of the tower may be measured in several different ways. Potentiometers (such as 10-turn pots) are used for this purpose (with negators on the opposite side to maintain steady tension in the line). Two potentiometers at right angles to each other will measure the displacement at the top of the tower. The error due to the three-dimensionality of the tower top motion may be minimized by making the horizontal lines to the potentiometers long at the equilibrium position so that the effect of the vertical displacement of the tower top is negligible.

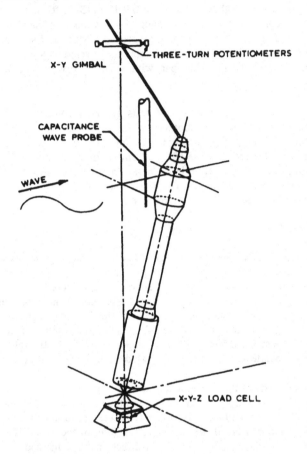

FIGURE 9.3
AN ARTICULATED TOWER TEST SETUP

Another direct contact measurement is a gimballed staff attached to the tower top with a universal joint (Fig 9.3). RVDTs and potentiometers may be employed in the dry for measurements of two angles as well as displacements. A simple geometric relation (Fig. 9.4) will provide the two angles of the tower oscillation. For example, the inline oscillation angle at the universal joint is obtained as follows:

$$\theta(t) = \sin^{-1}\left[\frac{l_2}{l}\sin\{\delta' + \delta_1 + \delta(t)\}\right] - \delta' - \delta_1 - \delta(t) - \theta' - \theta_1 \qquad (9.11)$$

in which the subscripts 0 and 1 refer to the tower equilibrium values, and values under static load respectively. The quantity $\delta(t)$ is measured from static position at the gimbal during waves.

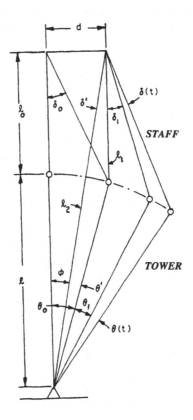

FIGURE 9.4
GEOMETRY OF A GIMBALLED STAFF MEASUREMENT

This method can minimize the effect of friction on the measurement. However, the mass of the staff will have some effect on the measured data. In particular, the vertical load is found to be affected by this rod. A measurement of the tower motion with the gimballed staff was made where the vertical pin load at the universal joint was also measured. In order to study the effect of the rod on the vertical load, tests were repeated with and without the rod for the motion measurement. The measured mean values of the transfer function for the vertical pin load in regular waves are shown in Fig. 9.5. It is clear that the gimbal transfers some of the vertical load in the load cell.

The noncontact method of measurement most commonly used now employs LEDs on the tower with a system similar to SELSPOT (Section 6.12) In this case, the errors appear only by the measurement system, but not due to setup in the testing.

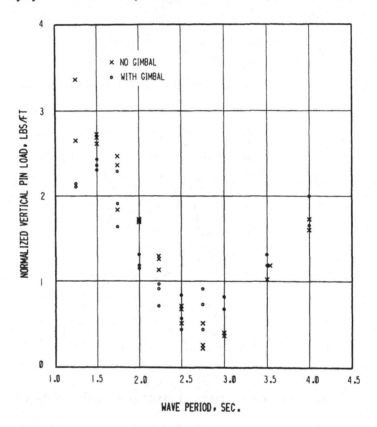

WAVE PERIOD, SEC.

FIGURE 9.5
VERTICAL PIN LOAD WITH AND WITHOUT GIMBALLED STAFF

In an analytical development, the oscillation of the tower about the pivot point in the plane of the wave propagation may be modeled as a forced-damped spring-mass system with linear and nonlinear damping and solution obtained in a closed form [Chakrabarti and Cotter (1979)].

Data from a model test on an articulated tower are compared here with this simple solution. The model resembles the tower at the Statfjord "B" location of the North Sea operated by Mobil. The test consisted of the model tower placed in the wave tank on a universal joint. Regular wave runs were made with no steady load on the tower. The comparison of the measured maximum responses (normalized with respect to the wave height) is made with the theoretical results. The dynamic tower oscillation is shown in Fig. 9.6. The comparison is good in spite of any out of plane motion of the tower neglected in the analysis. The natural period of tower is outside the test range near 3.5 sec.

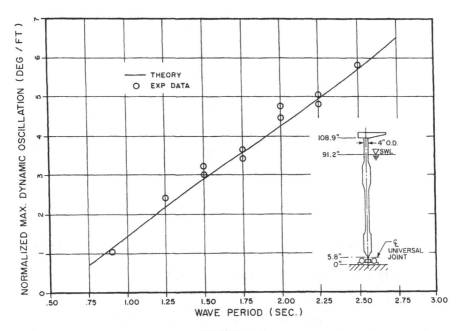

FIGURE 9.6
COMPARISON OF INLINE OSCILLATIONS OF AN SPM

The response of the tower oscillation due to irregular waves is shown in Fig. 9.7. The theoretical response spectrum is computed from the transfer function shown in Fig.

9.6. The RAO is assumed to be linear so that the response spectrum may be simply obtained as the product of the square of the RAO and the wave energy density spectrum (Chapter 10). The second peak in the measured spectrum is at the natural frequency of the tower.

When subjected to oscillatory waves, an articulated rigid tower is free to move in all planes about its pivot point near the ocean floor. Therefore, the transverse force due to separated flow introduces a motion of the tower normal to the inline motion. The magnitude of this motion depends on the magnitude of the transverse force, which in turn depends on the nature and number of eddies shed by the tower. An example of the transverse force on a submerged instrumented section of a fixed vertical tower versus the inline force as a function of time for about two cycles is given in Fig. 9.8.

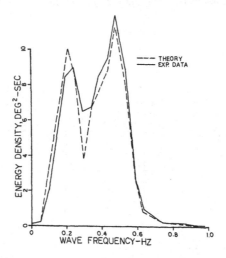

FIGURE 9.7
COMPARISON OF TOWER OSCILLATION IN IRREGULAR WAVES

The oscillation of waves past a fixed tower may be equivalently represented by an oscillating tower without the wave motion. The approach is to oscillate the tower harmonically in an otherwise still fluid. While the effect of free surface of the wave cannot be reproduced in this method, a submerged tower section experiences similar motion. This method of testing is sometimes preferred over the test in waves because the tests can be more easily controlled, uncertainty of water particle kinematics can be avoided and higher values of Reynolds number can be achieved compared to a wave test. Moreover, these results may be directly applied to cases where added mass and drag

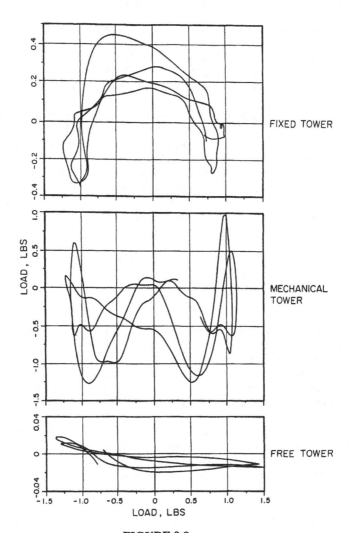

FIGURE 9.8
LOADS ON AN INSTRUMENTED SECTION OF A BUOYANT ARTICULATED TOWER

coefficients are   required for moving members of an offshore structure, e.g., an articulated mooring tower, riser, or tendon of a TLP.

In comparison to a fixed tower, the resultant force profile on the instrumented section of the tower harmonically forced-oscillated in still water shows (Fig. 9.8) that the transverse force is on the same order of magnitude as the inline force, but the tower experiences predominantly double frequency. The KC number in both cases is about eight while the Re number is on the order of 10^5.

In a mechanical oscillation test, noise in the measured data is commonly found. Although the cylinder angle is generally a smooth sinusoidal function of time, the loads in the inline direction show high frequency noise. This noise occurs because a servo-controlled device tends to overcompensate for errors between the reference and feedback signals and therefore oscillate about the mean reference value. This 'hunting' has little effect on the cylinder's angular position (the feedback to the drive system) but does cause cylinder acceleration that appears as spikes in the inline loads, especially where the reference signal changes most rapidly. A digital filter may be used after the data has been collected to remove any noise or unwanted data.

The desired result of the mechanical oscillation test is the estimation of the added mass coefficient (C_A), the drag coefficient (C_D) and the lift coefficient (C_L). These coefficients are formulated as functions of the Keulegan-Carpenter number and the Reynolds number.

Problems may be encountered in a test with a free tower in waves in which the purpose is to compute hydrodynamic coefficients from measured local loads on the submerged section of the tower. When the damping and the drag terms are relatively small, the tower begins to move in phase with the driving force of the waves.   At this point the tower velocity is in phase with the water particle velocity (this tends to make the relative velocity term quite small) and the tower acceleration and water particle acceleration are in phase which causes the load to be the residual of the two dominant loads. Although both of these loads may be estimated to within 10% accuracy (which may easily be done as evidenced by the first two phases of the test), the residual load may be over 100% in error. This obviously makes the use of load data from the free tower in the evaluation of the hydrodynamic coefficients questionable.

With the tower free to oscillate in the same waves as the fixed tower the corresponding two-dimensional force is shown in Fig. 9.8. The Keulegan-Carpenter number based on relative velocity between the tower and the wave is of the order of one. At this KC number, the transverse force on the tower is small and the transverse force has the same frequency as the inline force.

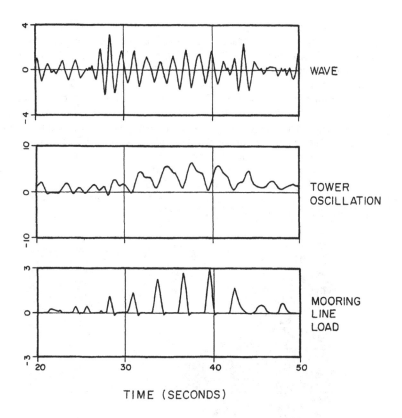

TIME (SECONDS)

FIGURE 9.9
MEASURED TOWER OSCILLATION AND MOORING LINE LOADS IN
IRREGULAR WAVES FOR AN SPM

9.4 TOWER-TANKER IN IRREGULAR WAVES

Articulated towers may be used to moor a shuttle tanker with a single point mooring hawser. Besides the extent of oscillation of the tower, the load experienced by the hawser is an important design criterion. Model tests are often conducted to study this interaction phenomenon. An example of the resonance of the tower near its natural period is evident in Fig. 9.9 in which the tower-tanker system is subjected to a simulated irregular wave. The low frequency of surge motion of the tanker coexists with the high frequency tower motion. However, the subharmonic oscillating tower motion occurred only for a certain range (e.g., 32-42 sec) of tanker position in surge as shown in Fig. 9.9. This arises from certain instability experienced by the nonlinear hawser e.g., in a catenary moored tanker (Fig. 9.10), which may become slack during the oscillation in waves if the steady loads are not large enough. The result of this nonlinearity has been found even in regular waves in model tests as an occasional increase in the hawser load. This is illustrated in Fig. 9.11. The model test results have been scaled up to the prototype values. The wave elevation, SPM and tanker motion have been presented in meter while the hawser load is in metric ton. Note that the wave is regular of a period of about 14 sec. The tanker motion shows the long period oscillation and high frequency wave motion superimposed on it. The hawser experiences slackness during motion and thus shows spikes. At an intermediate range of data, the tower motion becomes violent associated with a high load in the hawser. This instability was first discovered from model test results and may be explained mathematically as the Mathieu type instability generated from the Duffing's equation [Chakrabarti (1990)].

FIGURE 9.10
TEST SETUP FOR A CATENARY TANKER MOORING SYSTEM

9.5   TESTING OF A FLOATING VESSEL

A floating structure is one designed to accommodate the motions caused by wave action, rather than to resist motion. The testing of a floating vessel differs significantly from testing of a fixed structure. On a fixed structure, the overall exciting load and, possibly, local pressures are important. Sometimes, other wave effects such as run up may also be important. However, instrumenting, constructing, and modeling a floating structure are more complex. As previously described for articulated towers, the loads at critical points as well as the overall motions are needed in tests of a floating structure model. Unlike an articulated tower which has limited degrees of freedom, a rigid floating structure, in general, has all six degrees of freedom. A photograph showing the test setup and instrumentation of a floating storage vessel is shown in Fig. 9.12.

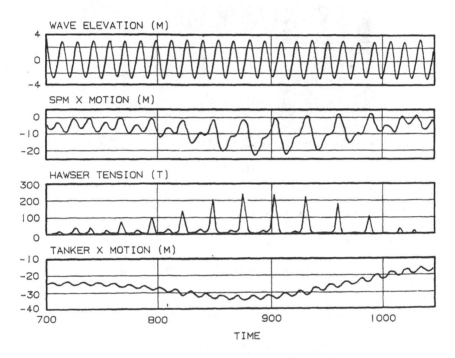

FIGURE 9.11
RESPONSES OF AN SPM IN REGULAR WAVES
[Chantrel, et al. (1987)]

The vessel in this case has two degrees of freedom of importance. The vessel is an open bottom buoyant cylindrical tank in which the draft and anchoring force are maintained automatically by the geometry of the buoyancy section. It has a single tensioned anchoring leg so that the tank appears as an inverted double pendulum. In this case, the displacement of the top of the storage tank in waves is an important design factor. This may be measured by a gimballed staff equipped with a potentiometer. An example of the comparison of this measured displacement with computed values for a two degrees of freedom motion analysis is shown in Fig. 9.13.

FIGURE 9.12
FLOATING STORAGE TANK MODEL IN WAVE TANK

9.6   TENSION LEG PLATFORMS

A tension leg platform is a floating production platform which is anchored to the ocean floor by vertical mooring lines. The mooring lines are called tendons (or tethers), because they are held in high tension by the excess buoyancy of the floating platform. Thus, the in-service draft of the platform is considerably higher than (about twice) that of the floating hull. The stiff mooring system permits limited motions of the platform in heave, pitch and roll when subjected to waves. The geometry of the hull and the tendon placement are almost invariably symmetric so that the pitch and roll periods are about the same. The period range of appreciable ocean waves generally is 6-20 seconds. The typical natural periods in heave, pitch and surge of a TLP are such that practically no wave energy exists in the sea at these periods. Thus, virtually no amplification in these responses are expected from the linear wave exciting forces arising from a wave spectrum.

However, the combination of multiple frequencies in a wave or the higher components of a nonlinear wave may produce exciting forces near the natural frequencies in these motions both in the low frequency (e.g., surge period) and in the

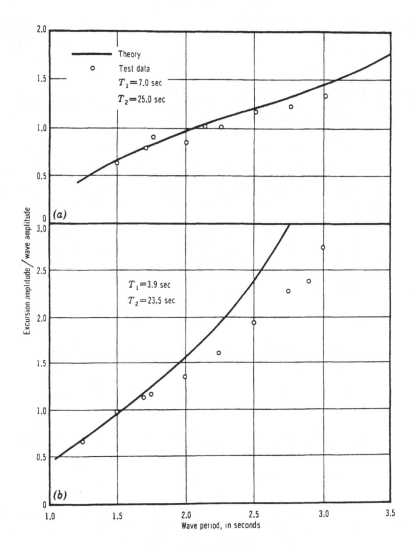

FIGURE 9.13
DISPLACEMENT OF TOP OF STORAGE TANK FOR TWO BALLAST
CONDITIONS IN REGULAR WAVES
(T_1, T_2 - Natural Periods)

high frequency (e.g., heave and pitch period) regions. The low frequency force is known as drift force while the high frequency force is commonly referred to as springing force. The springing force on a TLP arises from the heave force or the pitch moment. These exciting forces being of higher order are an order of magnitude lower compared to the first-order exciting forces at the wave frequencies. However, the damping in the TLP system both for the high frequency vertical motion and low frequency horizontal motion is very small. Thus, the dynamic amplification near the natural period is relatively large. The damping measurements of floating structures are discussed in Section 9.8.

Typically, the natural periods of a TLP in heave and pitch for the deep water applications (greater than 305 m or 1,000 ft) are between one and five seconds. On the other hand, the system is soft in the direction of surge because the restoring force in the vertical tendons in the surge direction is generally small. The natural period of the TLP in surge (or sway) is large, being of the order of 100 seconds or more. A TLP experiences two types of forces: a primary exciting force at the wave period and a secondary nonlinear exciting force having steady and oscillating components outside the range of typical wave periods. In particular, a high frequency springing force develops in the tendon in the pitch natural period.

9.6.1   Model Testing Program for a TLP

The objectives of a TLP test program are to determine:

- Overall dynamic behavior of the TLP in wave, wind and current environment

- Global response characteristics, such as six degrees of motion, and tendon loads

- Excitation and damping for the slow drift response of the TLP in random wave fields

- High frequency response and damping of the TLP motion

The hull construction (Chapter 3) consists of building the columns and pontoons made of steel or aluminum, generally rolled and welded. Access is provided in the hull for ballasting. It is preferable to ballast the model with dead weights anchored inside the hull rather than with water. Having a free water surface inside the hull can further complicate the dynamic behavior of the structure. The hull is checked by pressure testing to verify that it is watertight. The deck is not replicated in the model except that framing is needed for the rigidity of the model. A flat deck is provided for mounting the desired instruments (e.g., optical light sources, accelerometers, tendon tensioning device). Moreover, the deck allows to study the under-deck wave effects. The properties of the hull are established by swinging the model in air. Simple calculations provide the location of ballasts to model the center of gravity and mass moments of inertia for various angular directions.

The tendons are modeled to achieve the scaled weight and buoyancy characteristics. The scaled axial and bending stiffness of the tendon are desirable in the model tendon. However, in reality, both of them are difficult to achieve. For a TLP model, the axial stiffness is more important. The effective axial stiffness can be provided by an adjustable cantilever spring at the bottom of a tendon. The stiffness is established by loading tendon/spring combination in an axial direction. The bending stiffness may be achieved by the proper choice of the tendon material (see Section 3.4.3.2). The tendon loads are monitored by a three-component load cell mounted between the spring and the foundation.

The foundation consists of a mounting plate or frame on which the tendon load cells are mounted. The plate is either welded or bolted to the tank floor. Sometimes, weights are placed on the plates sufficient to withstand the shear load or the vertical load at the base, simply to avoid attachment to the wave tank floor.

9.6.2 Typical Measurements for a TLP

In a model test of a TLP, the following measurements are desired:

- Six degrees of motion
- Accelerations at extreme deck locations
- Wave profile in front of the model
- Wave profile in line with the model center for phases
- Tendon loads
- Wind and current loads, if simulated in tests
- Photographic and video recording.

These measurements allow the evaluation of the performance of the TLP design.

9.6.3 Wave Frequency Response of a TLP

An example of the motions of a TLP model at various wave frequencies is shown in Fig 9.14. The TLP system for the test case was designed to have natural frequencies in the wave period range. An external electro-magnetic damping system was introduced in each leg of the four-legged platform in order to study the effect of additional damping in the system. This excess damping was expected to be introduced in the prototype by an equivalent system. In this case the damping factor is duplicated in the model. The external damping, as may be expected, reduced the amplification near the natural period.

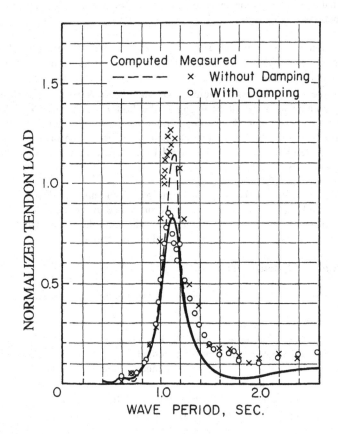

FIGURE 9.14
RESPONSE OF A TLP MODEL EQUIPPED WITH A MECHANICAL DAMPER
[Katayama, et al. (1982)]

9.6.4 Low and High Frequency Loads

The second order inertial load on structures may be important, particularly when the natural period in a particular degree of motion is such that the structure is excited at the higher (or lower) harmonic frequency of this load. This phenomenon may be investigated in the laboratory by measuring either the second-order load or the nonlinear responses due to this load. Generally, the second-order loads are small, but the corresponding responses are large. The difficulty in the direct measurement of the second order load on a structure was addressed in Chapter 7. This section describes the relative ease in measuring the corresponding responses.

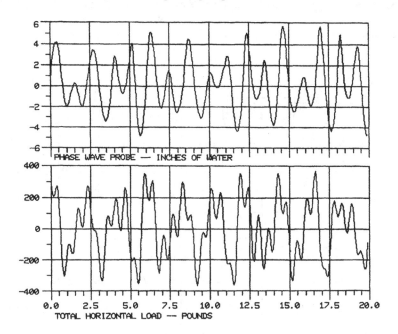

FIGURE 9.15
MEASURED FORCES ON A FIXED CYLINDER DUE TO A WAVE GROUP

An example of the forces measured on a fixed vertical cylinder due to a wave group is shown in Fig. 9.15. The wave profile corresponds to waves of frequencies 0.44 Hz and 0.88 Hz. The wave energy density spectra show peaks at the above frequencies. The measured forces show the presence of higher frequency components. The force spectrum have additional peaks present corresponding to second harmonics of the above frequencies, and their sum frequency component at 1.32 Hz. These higher order loads are generally small, being on the order of 3 to 5 percent of the first order loads.

The effect of these exciting forces on a moving structure due to dynamic amplification is illustrated by another example. The above cylinder was moored in a floating position with linear fore and aft springs. The springs are chosen such that the natural period of the system in surge was about 30 seconds. The springs were pretensioned to avoid slackness during testing. Regular wave groups having three frequency components were chosen. The frequencies were chosen such that the difference frequency between two adjacent frequencies was equal to the natural frequency of the system. Figure 9.16 shows the wave and the corresponding mooring line loads. It is clearly seen that the response is amplified at the natural frequency (due to low damping) such that its amplitude is larger than the amplitude at the wave frequency.

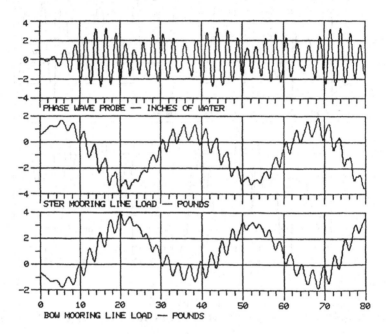

FIGURE 9.16
MOORING LINE LOADS DUE TO A WAVE GROUP ON A MOORED
FLOATING CYLINDER

An example of the response due to high frequency springing force on a four column TLP is taken from DeBoom, et al. (1984). Comparison of the high frequency tendon force in regular waves is made in Fig. 9.17 with the corresponding measured data from a wave tank test. The phase relationship among the forces and wave profiles suggested that the springing force is a result of the pitching force and motion of the TLP.

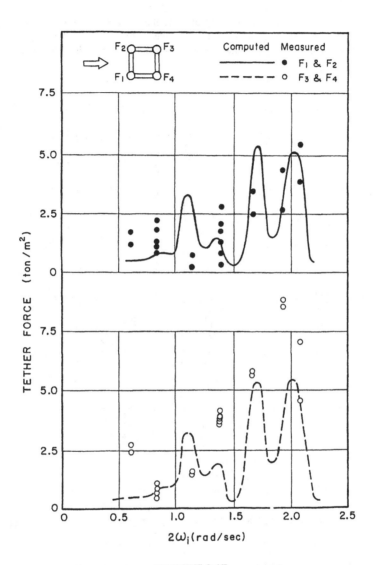

FIGURE 9.17
NORMALIZED TENDON LOADS IN REGULAR WAVES ON A TLP
[DeBoom, et al. (1984)]

A typical TLP test setup is shown in Fig. 9.18. The setup includes a false floor in the wave tank which allows generation of current as well as deep water waves. A deeper circular pocket in the tank allows a larger model for deep water depth typical of a TLP without scale distortion. The tendons are shown to go through the false floor.

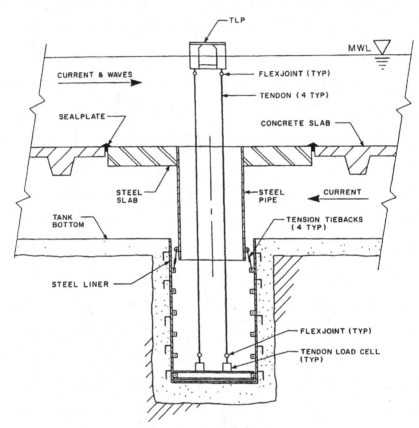

FIGURE 9.18
A TYPICAL TLP TEST SETUP IN A WAVE TANK EQUIPPED WITH A DEEP SECTION

The springing forces on a large scale (1:16) TLP model were measured in a test in a wave tank [Petrauskas and Liu (1987)] with a four-legged TLP hull connected to the sea floor with four vertical tendons. The tendon lengths were distorted due to large scale, but axial stiffness was modeled. The springing forces arised from the resonant pitch periods of about 3 seconds. Regular waves at twice the pitch period amplified the resonant springing force in the tendon. The amplification of the force at the tendon due to a random wave is clearly shown in Fig. 9.19. The tendon load at the wave frequencies is almost negligible compared to the resonant load at twice the wave frequencies.

9.7 DRIFT FORCE TESTING OF A MOORED FLOATING VESSEL

The measurement of slow drift oscillation is demonstrated in a tanker test set up. In general, a large floating structure moored at location with a soft mooring system will experience long period oscillation when subjected to irregular waves. In practice, such systems are found in exposed location tanker mooring systems. The following illustrates the setup and measurement technique using a simple system.

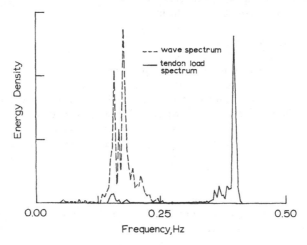

FIGURE 9.19
WAVE SPECTRUM AND CORRESPONDING TENDON LOAD SPECTRUM ON A TLP [Petrauskas and Liu (1987)]

9.7.1 Test Setup

The test setup is intended to produce results on the wave drift force in surge. Two identical spring sets are attached with aircraft cables on the fore and aft side of a tanker model, influencing its surge motion. Springs are chosen to be linear in the range of the anticipated loads. The points of attachment on the deck of the vessel restrict the pitch of the vessel to some extent. However, by making the lines long (of the order of 4.6-6.1 m or 15-20 ft), this error is minimized.

Loads in the mooring lines and the three-directional motions (surge heave and pitch) of the floating vessel are generally measured during these tests. The springs are pretensioned so that they never go slack during the tests.

9.7.2 Hydrodynamic Coefficients at Low Frequencies

It is important that the low frequency added mass and damping coefficients of a floating vessel be known in addition to the wave drift forces on them so that the mooring line loads may be accurately predicted. These coefficients are obtained in two ways: (1) the free oscillation of the moored vessel in surge is obtained by what is commonly known as the pluck test, or extinction test in still water; (2) the transient oscillation of the vessel at a low frequency caused by a few initial regular waves is filtered to obtain the long period of decaying oscillation. Thus, the variation of the added mass and damping coefficients, not only with the vessel geometry and spring systems, but also with the wave height and frequency can be studied.

9.7.2.1 Free Oscillation Tests

The model which is setup as a spring-mass system is oscillated in surge by displacing the model structure from equilibrium and then releasing it. The free oscillations of the model in surge are recorded by the load cells and LEDs. In still water the extinction test produces a low-frequency decaying oscillation in the surge motion or in the mooring line load which may be described by a differential equation. The method of analysis for the added mass and damping is discussed in Chapter 10.

9.7.2.2 Forced Oscillation Tests

In regular waves, the long period oscillating drift force component is absent. However, the initial waves introduce a transient low-frequency model oscillation which decays at a rate depending on the damping in the system. Thus, the natural period of the system and its damping factor at different exciting wave frequencies can be determined from these runs. These quantities can be used to determine the added mass and damping coefficients for various wave frequencies.

9.8   DAMPING COEFFICIENTS OF A MOORED FLOATING VESSEL

The extent of motion of the system at its natural frequency is controlled by the system damping. Therefore, knowledge of the damping of such a dynamic system is an important consideration in its design. The damping of a moored system arises from two natural sources, namely, material characteristics and hydrodynamic forces. The material damping of the mooring lines is small and is generally determined from model tests of the component material. The hydrodynamic damping is considered to be the limiting factor in the low frequency motion of the moored vessel. In a low-damped system at resonance, the motion amplitude is approximately proportional to the system damping.

The hydrodynamic damping may be characterized as originating from four sources:

1. radiation wave damping proportional to structure velocity,
2. linear viscous damping proportional to structure velocity,
3. nonlinear viscous damping proportional to square of the structure velocity, and
4. wave drift damping

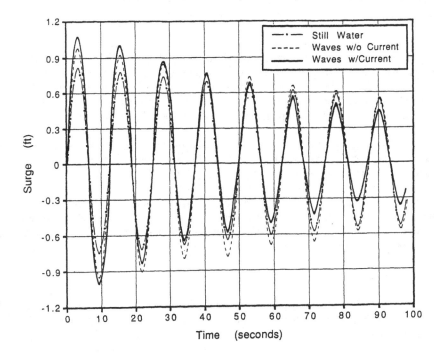

FIGURE 9.20
DAMPED OSCILLATION IN STILL WATER, WAVES AND WAVES WITH CURRENT

Model test results on various streamlined floating bodies e.g., tankers [Wichers (1988)] have shown that the nonlinear (drag type) viscous damping in slow drift oscillation in waves is negligible, and most of the damping is linear viscous. For other structures including vertical columns at the water surface, such as semi-submersible, the nonlinear viscous damping is important. For a catenery moored system, the drag forces on the mooring lines themselves and the friction between mooring lines and seabed also contribute to the slow drift damping [Huse and Matsumoto (1988,1989)].

Once nonlinear viscous damping has been eliminated from the total damping, the damping can be considered to be the linear superposition of damping sources. Thus, the damping coefficient, C, can be described by

$$C = C_R + C_V + C_W + C_C \qquad (9.11)$$

where C_R = radiation damping, C_V = viscous damping, C_W = damping in waves (wave drift damping), and C_C = damping in current.

An example of the damped oscillation of a moored tanker in still water, in waves, and in waves plus current is shown in Fig. 9.20. Note that for the same spring constant the natural frequencies of the moored vessel in still water are the same as those in regular waves and current. However, the damping coefficients in waves are, generally, higher than the coefficients in still water. When the current is additionally introduced, the amount of damping increased as evidenced by the faster decay.

Figure 9.21 presents the natural logarithm of the amplitude versus cycle for the same test runs as presented in Fig. 9.20. Even in the presence of current, the damping is found to be linear with the tanker surge velocity (Fig.9.21). Nonlinear viscous damping contributions will show up as deviations from a constant slope. The lines from each scenario are clearly linear. Structures consisting of submerged cylindrical members, however, show evidence of nonlinear damping. A simplified method of handling the nonlinear term in the second-order equation of motion [Faltinsen, et al. (1986)] is described in Chaper 10.

9.8.1 Tanker Model

A 67,000 dwt tanker model, representing the ESSO Houston, was ballasted to 40 percent capacity and moored fore and aft. The fore and aft mooring lines were instrumented with ring type load gauges. The horizontal lines were long enough such that the effect of pitch and heave on the line loads was negligible. The loads were converted to the displacement of the tanker by dividing by the spring constant.

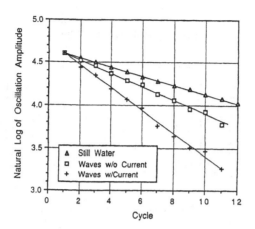

FIGURE 9.21
NATURAL LOG OF AMPLITUDE OF OSCILLATION VERSUS CYCLE

Extinction tests without waves were performed to determine the still water damping of the tanker at its surge natural frequency. A series of different spring sets were used to vary the resonant frequency.

In order to determine the surge damping of the tanker in waves, the tanker was displaced horizontally in head sea from its equilibrium position and held in place until waves generated by the wavemaker reached the tanker. Then the tanker was released and the decayed oscillation in waves were recorded. The high frequency response was filtered out and the low frequency oscillation was analyzed to determine the added mass coefficient and damping factor. An example of the surge calculated from the recorded mooring line load before and after digital filtering of the high (wave) frequency response is shown in Fig. 9.22.

9.8.2 Semi-submersible Model

A semi-submersible model consisting of two pontoons of rectangular cross-section and six columns with (roughly) elliptical cross-section was similarly tested in still water, waves and current. Table 9.1 presents the value of the spring constant and the resulting system natural period for the various spring sets tested with the semi model. The surge added mass coefficient of the semi-submersible in still water is found to be insensitive to the natural period. The results are presented in Fig. 9.23 and compared to the added mass coefficients previously obtained for the tanker. The added mass coefficient of the semi-submersible is seen to be a factor of 2 to 3 times larger than for the tanker. This is mainly due to the difference in geometry between the streamlined tanker hull and the presence of semi-submersible columns.

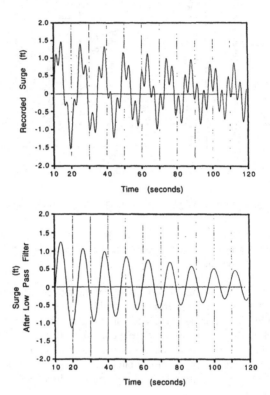

FIGURE 9.22
DAMPED OSCILLATION OF MOORED TANKER IN WAVES

The decaying oscillation curve was analyzed for the damping factor and drag coefficient using a simplified nonlinear theory (see Chapter 10). A typical X-Y relationship is presented in Fig. 9.24. With reference to this figure, the Y intercept is the damping factor, and the slope is a function of the nonlinear drag coefficient. The damping factor and the drag coefficient in still water for the semi-submersible and the tanker are compared in Fig. 9.25. This figure demonstrates that the damping factors for both floating vessels are similar in still water once any nonlinear contributions are removed. The system damping of the semi-submersible has a much stronger contribution from the nonlinear term, resulting in higher overall damping. As can be seen in Fig 9.25, the nonlinear term in still water tends to be relatively independent of the oscillation period. The damping factor for the semi-submersible in the presence of waves exhibited far greater damping than the tanker.

TABLE 9.1
SPRING SETS FOR STILL WATER SURGE OSCILLATION TEST

Spring Set	Spring Constant (lbs/ft)	Natural Period (sec)
A	1.05	23.27
B	1.32	20.75
C	1.75	18.01
D	2.63	14.69
E	3.76	12.31
F	5.26	10.40
G	8.77	8.03

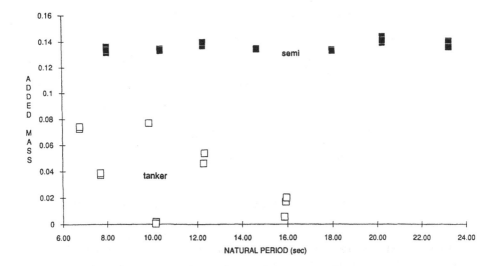

FIGURE 9.23
COMPARISON OF ADDED MASS COEFFICIENT IN STILL WATER

9.8.3 Heave Damping of a TLP Model

As with other floating vessels, the TLP system experiences damping from two natural sources, material properties and hydrodynamic effects. Sometimes, mechanical dampers [Katayama, et al. (1982)] or other active damper systems are introduced externally in order to reduce tendon loads. The material damping appears from the tendons and their attachments to the TLP as well as to the sea floor. The subsea template also provides some damping. The hydrodynamic damping appears in the form of the radiation damping as well as viscous damping as discussed previously. The rediation damping at the high frequency is generally quite small.

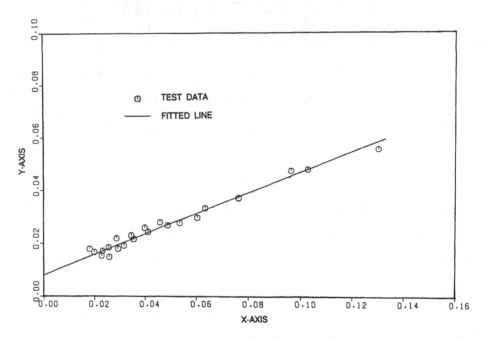

FIGURE 9.24
FITTING OF DATA ON DAMPED OSCILLATION OF SEMI

The first-order loads on a TLP do not induce resonant motions in the system, particularly, in the tethers. However, the tethers may still experience resonance at the second-order (sum frequency) components in heave and pitch. The amount of damping limits amplitude of tether loads due to springing. Due to inadequate theory in computing this damping, experiments are carried out to determine the damping coefficients in heave. Since the damping coefficient at the high frequency is extremely small, care must be taken in setting up the test so that accurate information may be obtained from the test.

In a test with a TLP column model [Huse (1990)], heave damping coefficients were obtained. The model represented a TLP column of 25m diameter and 37.5m draft at a scale of 1:20. The steel model had a sharp corner at the lower edge with a corner radius of less than 0.1mm. The model was suspended from the middle of two 6m long horizontal steel beams providing the springs of the oscillatory system in the vertical direction. The ends of the beam were welded to rigid structures to avoid additional damping introduced in the system from any fasteners. The spring constant of the suspension was high (1.414 x 10^6 N/m.).

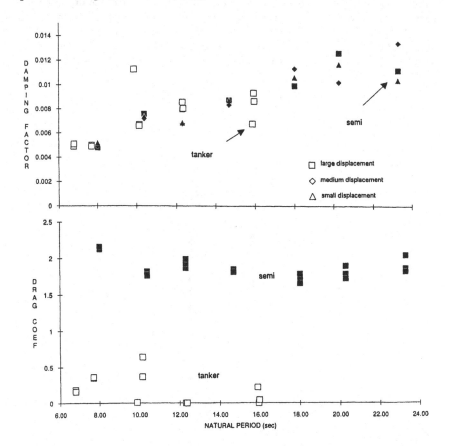

FIGURE 9.25
**DAMPING FACTORS AND DRAG COEFFICIENTS OF A TANKER VS. SEMI
IN STILL WATER**

The model was ballasted with steel and water. The steel ballast inside was secured to the hull. The mass of the model was 3900 kg. The vertical and angular motions were measured by accelerometers. The tests were carried out as heave decay tests. The heave amplitudes ranged from 0.1 to 2.0 mm. The angular motions were analyzed to ensure pure heave motions of the column. The frequency of vertical oscillation in water was 3.03 Hz. To start the test, the model was pulled upward from equilibrium by a crane and the steel wire connecting the model was cut. The subsequent heave oscillation was recorded. The initial displacement of the model was controlled by an electronic balance. Tests were also carried out in the dry to determine material damping of the setup so that the decay due to hydrodynamic damping alone could be established by subtraction. The hydrodynamic damping was found to be mainly linear, the quadratic damping being less than 10 percent. The linear damping coefficient was found to be $b_1 = 122.6$ N s/m or a normalized damping coefficient b_1/M = 0.03.

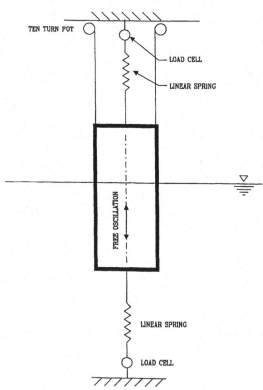

FIGURE 9.26
DAMPING TEST SETUP OF A TLP COLUMN

In another test with a floating vertical caisson [Chakrabarti and Hanna (1990, 1991)], the added mass and damping coefficients in heave were evaluated. The model consisted of a single vertical cylinder scaled (1:20) from a vertical column of a typical TLP. A 0.76 m (2-1/2 ft) diameter model was selected for this test. The draft of the model (about 1.22 m or 4 ft) also represents the typical draft of a TLP. The model surface was painted to a smooth finish. The bottom edges were sharp.

The ratio of the tank width to model radius was 26.4. A limiting ratio of 10 for the tank blockage effect was reported by Calisal and Sabuncu (1989). Tests were run in still water and in waves. The cylinder was held in place in the water by two vertical sets of springs (Fig. 9.26). The springs were linear and pretensioned and were arranged so that they could be easily changed during the test. Load cells were attached to the ends of springs to record the loads (and hence, motion). The cylinder was displaced vertically from its equilibrium position and released, and the decaying oscillation was recorded. The test was repeated twice for nine sets of springs.

In a first series of tests [Chakrabarti and Hanna (1990)], the motions of the caisson were recorded (to examine the effect of any pitching motion of the cylinder) with the help of two potentiometers placed at two diametrically opposite faces (Fig. 9.26). These potentiometers introduced substantial mechanical damping making the damping factor values erroneous. Subsequent tests were made without the potentiometers.

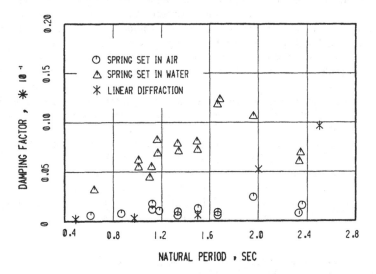

FIGURE 9.27
DAMPING OF A TLP LEG-SPRING SET IN WATER VS. AIR

In order to study hydrodynamic damping, it is important to extract the material damping in the spring system used in the tests. In this case, the spring sets were hung in air with an equivalent dead weight representing the pretension in the springs. The weight was displaced vertically an amount equivalent to the displacement of the caisson in water. The loads in the spring system were measured with a load cell at the fixed end of the setup. The results are shown in Fig. 9.27 for various spring sets. The material damping is generally quite small compared to the hydrodynamic damping, but not negligible.

An interesting observation of the test results is that the heave damping for the vertical cylinder is linearly dependent on the cylinder velocity in low amplitude heave. For larger amplitude of oscillation, there is some evidence of nonlinearity. However, for a TLP caisson, the heave is expected to be small. The damping which is composed of radiation and linear viscous components is mostly of the order of 1 percent of critical. However, it tends to decrease with higher frequencies. At a frequency near 2 Hz, the damping is about a third of 1 percent. Several runs made with different (but small) initial displacements show the consistency of the data. At a higher frequency of 3Hz [Huse (1990)] the damping factor was even lower.

9.9   MODELING AIR CUSHION VEHICLES

Froude scaling is used for conventional vessels in scaling the model test results on wave-induced motions and loads to their corresponding prototype values. In this case, the model and waves follow the usual geometric scaling. Speed of the vessel is based on Froude similitude between the model and the prototype. For this model, all the scaling laws for various quantities are known (Chapter 2). However, this scaling law is found to be inapplicable to the air cushion vehicles (ACV) such as surface effect ships and hovercraft. The ACV is a seagoing vessel that has its weight supported primarily by an increased static air pressure (compared to the atmosphere) within the confines of rigid sidewalls that penetrate the water surface. Because atmospheric pressure enters in the modeling of the prototype (see also Section 8.5), an interesting scaling problem arises in this case. The following discussion is essentially taken from Kaplan (1989).

For a Froude model, accelerations scale as 1:1. However, for an ACV, model tests produced unconservative full scale values for the heave acceleration. This is mainly due to the effect of nonscaling of the ambient atmospheric pressure. The pressure affects the natural frequency and damping of the coupled pressure-vertical motion mode.

A simple linear analysis in head seas with adiabatic laws for an air cushion shows [Kaplan (1989)] a natural frequency for the coupled heave-pressure mode of

$$\omega_y = \left[\frac{\gamma g (1 + p_a/p_0)}{h_b} \right]^{\frac{1}{2}} \qquad (9.13)$$

and the corresponding damping factor

$$\zeta_y = \frac{K_3}{2\omega_z K_1} \qquad (9.14)$$

where

$$K_1 = \frac{\rho_a A_b h_b}{\gamma \left(1 + p_a/p_o \right)} \qquad (9.15)$$

and

$$K_3 = \frac{1}{2} \rho_a Q_o - \rho_a \left(\frac{\partial Q}{\partial p} \right)_0 p_o \qquad (9.16)$$

The quantities in Eqs. 9.13 - 9.16 are ρ_a= density of air, A_b = cushion area, h_b = height of cushion, γ = ratio of specific heats for air, p_a = atmospheric pressure, p_o = equilibrium cushion pressure (gauge), Q_0 = volume of air delivered by the fan at a pressure p_0, and $(\partial Q/\partial p)_0$ = slope of the fan characteristic curve at p_0. The details of the derivation of these relationships may be found in Kaplan (1989). The purpose of presenting the results here is to discuss the scaling laws that may be used for an ACV. The damping of the motion is found to depend primarily on the slope of the p-Q fan curve which increases with a decrease in the slope.

In a Froude model, the frequency scales as $1/\sqrt{\lambda}$. Thus, a model which is 1/25 in size in linear dimension (i.e. $\lambda = 25$), will have a corresponding frequency 5 times that for a full-scale vehicle. For an ACV, the equilibrium cushion pressure as well as the cushion height in model scale varies as $1/\lambda$. Therefore, the natural frequency in model scale will vary approximately as λ due to the large value of the ratio p_a/p_o in Eq. 9.13. Similarly, the damping factor is proportional nearly to λ rather than being the same as the prototype value in a Froude model. From Eq. 9.13 the natural frequency ω_y decreases with the increasing size of the ACV. In this case a resonant effect may be expected for a full scale ACV. However, in a Froude model, this resonance may be missed since the model natural frequency (with same atmospheric pressure) will be too large and thus unconservative .

One approach to avoid this problem in model testing is to carry out the model testing in a tank facility where the ambient atmospheric pressure in the facility itself may be properly scaled by depressuring the entire tank facility. Such a facility exists at MARIN, The Netherlands. The facility, however, does not have a wavemaker. A mechanical oscillator may be able to simulate the vertical motion. This then can illustrate the effects of ambient atmospheric pressure on the scaling relations.

A more acceptable method of prediction of the ACV dynamic motion characteristics is to simultaneously develop a computer simulation model and a model testing program. The computer model is validated with the results from extensive series of model tests at smaller scales, by ensuring good correlation. The computer model is then applied directly to the full scale case using appropriate full scale parameter values to predict the design information on the dynamic characteristics of the ACV.

9.10   ELASTIC FLOATING VESSEL

In designing a ship hull, the stresses or strains in the structure of the ship due to the distributed loads from waves must be known. Usually the stresses are computed theoretically from the computed or measured distributed loads (such as pressures on the hull). Measurements of the strain in the hull are also made in a model test. For the measurement of bending moment at a few cross-sections, the model hull is distorted by segmenting it at these locations and back-bone beams of proper bending stiffness are mounted across these sections which are fitted with strain gauges to measure strains due to waves.

A more appropriate and complete modeling calls for an elastic ship model of suitable material [Lin, et al. (1991)] which follows the appropriate law of similitude. The material (such as foamed vinyl chloride) should be of stable characteristics and of small Young's modulus.

The model ship, in this case, must satisfy all the geometric, hydrodynamic (i.e. Froude) and structural (i.e. Cauchy) conditions of similitude. Considering the longitudinal bending for the structural similitude, the relative displacement or strain in the longitudinal direction between the model and the full size ship must be the same

$$\frac{M_B y}{EI} = \text{constant} \tag{9.17}$$

which gives the relationship for the bending stiffness (flexural rigidity)

$$(EI)_p = \lambda^5 (EI)_m \tag{9.18}$$

where M_B is the vertical bending moment and y is the vertical distance to the center line of the hull. The natural frequency corresponding to the vertical bending mode of the dry hull satisfies the condition

$$\omega_m = \sqrt{\lambda}\,\omega_p \qquad\qquad (9.19)$$

9.11 MODELING OF A LOADING HOSE

Another example of an interesting modeling problem is that of a very flexible object having both axial and bending stiffness of significance, such as a floating hose. A loading hose is generally used to transfer oil from an articulated loading tower (or buoy) to a shuttle tanker. The success of a single point mooring terminal greatly depends on the performance of the hose in waves and an effective life span of the hose string.

One of the higher maintenance items that both floating buoys and articulated towers have in common is the loading hose string. A floating hose string permits the mooring of side loading tankers as well as bow loading tankers. The single point mooring dynamics coupled with the prevailing sea subject the hose string to an extremely complex system of motions and loads. This together with the flexibility of the hose string leads to problems of (1) chafing from contact with the tower, ship, and the hose itself, (2) kinking due to high local bending stresses, and (3) entanglement with the tower or hawser. Since all of these problems may put the mooring system out of service temporarily, a reduction of the severity of any of these problems will lead to more time on-line, longer hose service life, and lower annual costs.

9.11.1 Hose Model

In modeling a hose system [Brady, et al. (1974)], the outside dimensions, mass and buoyancy of the hose are considered. In addition to scaling the geometry, the axial and bending stiffness are also modeled. When modeling underbuoy hoses, buoyancy tanks, beads and tie wires are also scaled with appropriate elasticity and net buoyancy. There are many techniques that may be used to model these parameters. However, satisfying all properties at the same time is not a simple task. Note that different parts of the hose section, such as floating hose, submerged hose, etc. have different properties. The validity of simple Froude scaling of the hose is subject to question, but considering that inertia plays a dominant role in the hose dynamics, it is a reasonable choice. One method of modeling described below [Young, et al. (1980)] illustrates some of these difficulties.

The hose models were made of helical springs coated with an elastomer. The spring essentially provides the axial stiffness while the required bending stiffness is provided in combination. The elastomer also gave the hose model its outer geometry.

Two types of hose modules in the hose strings were constructed from the same basic geometry--floating hose and submarine hose. The inertial properties of the hose were modeled as closely as possible. In particular, the reserve buoyancy of floating hose sections and the submerged weight of submarine hose with flotation collars in place were modeled to obtain the scaled mass. Likewise, the outside diameter was modeled to obtain reasonable values for displaced volume and virtual mass. The theoretical values of these properties were obtained from the manufacturer's catalogue.

To model the floating hose sections, it was necessary to increase the outside diameter of the existing models of submarine hose and to provide additional buoyancy. Washers and grommets were fabricated from 6.4 mm (1/4 in.) closed cell foam rubber. The outside diameter was cut to the scaled diameter while the inside diameter was such that the grommets would just slide over the submarine hose model. To prevent significant changes in the bending stiffness, the grommets were spaced 9.5 mm (3/8 in.) apart on center. However, insufficient buoyancy was obtained by the addition of grommets.

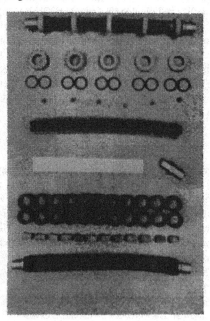

Buoyant hose string

Buoyancy modules

O'rings (retainers)
Ballast

Spring covered with latex

End connector

Grommets

Corks

Submarine hose string

FIGURE 9.28
TYPICAL FLOATING AND SUBMARINE HOSE MODEL SECTION WITH THEIR COMPONENTS

To compensate for the additional buoyancy required, short cork segments were inserted in the model hose bore. By placing the segments inside, the correct buoyancy could be obtained without affecting the added mass of the model. In the few instances where additional weight was required to ballast a model, it was added by distributing the required weight of lead shot along a piece of monofilament line and threading it through the model hose bore. Figure 9.28 shows typical completed floating and submarine segments with their components.

TABLE 9.2
LOADING HOSE STRING COMPONENTS
(Code Letter Identifies Components in Fig. 9.29)

		PROTOTYPE			MODEL	
CODE LETTER	DESCRIPTION	SIZE	BUOYANCY FULL OF OIL (kg)	APPROXIMATE OUTSIDE DIA. (in.)	OD (in.)	BUOYANCY (gm)
a	Full Length Reinforced Submarine Hose	12" IDx35'	-417	15.6	0.33	-3.8
b	Half Length Reinforced	12" IDx35'	-377	15.2	0.32	-3.4
c	Submarine Hose	12" IDx35'	-391	15.2	0.32	-3.5
d	Floating Hose	12" IDx35'	732	22.8	0.47	6.6
e	Reducing Floating Hose	12" IDx35'	604	19.7	0.41	5.5
f	Submarine Hose	10" IDx35'	266	12.9	0.27	-2.4
g	Extra Buoyant Tanker Rail Hose with Auxillary Equipment	10" IDx35'	1122	22.2	0.46	10.2
h	Bead Float for 10" End	—	81	30.1	0.63	0.7
i	Bead Float for 10" Body	—	59	26.4	0.55	0.5
j	Bead Float for 12" End	—	113	34.8	0.73	1.0
k	Bead Float for 12" Body	—	81	30.1	0.63	0.7
l	Supporting Buoy with Chain	—	570	—	—	5.2
m	Heavy Coflexip Type Hose	12" IDx35'	-1073	15.25	0.32	-9.7
n	45° Elbow	12" ID	—	—	—	-0.7
o	Submarine Hose	24" IDx35'	-4004	27.5	0.57	-16.4
p	Submarine Hose	24" IDx35'	-2746	27.3	0.57	-11.3
q	Floating Hose	24" IDx35'	4882	39.3	0.82	20.0
r	Lightweight Reducing Floating Hose	24" IDx35'	4146	35.2	0.73	17.0
s	Lightweight Floating Hose	20" IDx35'	2866	31.0	0.65	11.8
t	Bead Float for 24" End	—	261	47.3	0.98	2.4
u	Bead Float for 24" Body	—	236	43.5	0.91	2.1

1 in = 25.4 mm; 1 ft = 0.3 m

Table 9.2 lists all the prototype hose components used to assemble the hose configurations in this case along with their outside diameter and buoyancy. The eight hose configurations are shown in Fig. 9.29. A code letter is included in Table 9.2 to identify each hose component in the configuration.

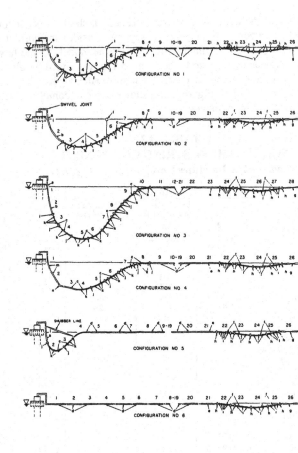

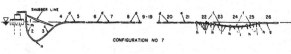

FIGURE 9.29
HOSE STRING CONFIGURATIONS 1-8

Starting at the hose to tower connection for configuration 1, the string consists of eight sections of 0.3 m (12 in.) diameter submarine hose with buoyancy collars attached to permit the hose to lie in a shallow U shape. The downstream end receives additional support from a buoy located at the junction of sections 5 and 6. Sections 9 through 20 are of 0.3 m (12 in.) dia. floating hose. Section 21 is a 0.3 m (12 in.) dia. to 0.25 m (10 in.) dia. reducing floating hose, and it is followed by four lengths of 0.25 m (10 in.) dia. submarine hose with flotation collars located to provide a very shallow submerged loop. The last section is an extra buoyant tanker rail hose with ancillary equipment attached. Configurations 2 through 8 are simple modifications of the basic hose string. The loading at the tower end of hose is a critical design consideration that should be measured in a model test. Load cells may be placed to measure bending and axial load at the hose to tower connection. Load cells are aligned such that bending is measured from a vertical plane radially away from the tower and at right angles to the radius, as shown on Fig. 9.30.

9.11.2 Hose Model Testing

In an SPM model testing, attached with hose, visual observations and video recordings are made to study the behavior of particular hose strings. Such things as erratic motion, snaking, flexing etc. are observed during the testing. Snaking is a phenomenon where the hose takes on a sine wave shape transverse to the direction of wave travel and in the plane of the water surface. Configurations that reduce the hose loads at the tower also reduce the amplitude of the transverse waves. Thus, hose designs that reduce tower loads may reduce axial loads in the tower end of the hose string as well.

Since the hose loop travels along its axis and since its flexing action is similar to a catenary, the downstream bending moments would be expected to be larger than the upstream moments. An obvious way to eliminate this moment is to put a swivel at the tower hose connection. This was done in Configuration 2. The result, as anticipated, reduced the bending moments measured in the load cell. This, however, allows the hose loop to come much closer to the tower when the tower makes a downstream excursion in waves.

Configuration 4 is the basic string with the first two sections made with hose having the weight and the geometry of a 0.3 m (12 in.) dia. coflexip hose (the stiffness was not modeled). The effect of this loop modification was to significantly reduce the dynamic loads in the tower to hose connection. Likewise, the snaking action of the hose was attenuated.

In Configuration 5, a snubber line was connected to the two ends of the tower hose loop. Results show a significant reduction in loads at the tower connection. These

FIGURE 9.30
LOAD CELL STRING AT TOWER CONNECTION

reduced loads compare on an equal basis with Configuration 4. The general motion appeared to be somewhat better because snaking in this hose string was of smaller amplitude than the others tested.

Configuration 6 is a hose string modeled after a monobuoy configuration. There is no submerged loop at the tower. The hose connects to the tower horizontally at the mean water line. This string showed the worst performance of the series. It was the only one in which jerking of the hose at the tower could clearly be observed. The loads were the highest and snaking was as severe as any of the configurations tested.

Configuration 7 was a combination of Configurations 4 and 5, whose purpose was to see if the two would work together to further reduce the dynamic loads at the tower, which was found to be the case.

The submerged hose loops have been shown to dampen oscillatory motions for different hose sizes and at different hose string locations. Drag appears to play a significant role in the action of submerged hose loops and may well be an important influence on the entire hose string. This casts serious doubt on the ability of the Froude model to accurately represent a prototype hose structure.

9.12 MOTIONS IN DIRECTIONAL SEAS

The discussion thus far has been generally directed to motions of floating bodies in two-dimensional waves. However, the method outlined earlier is equally applicable to three-dimensional waves as well. The 3-D waves are generally simulated in a small region of the test basin. Therefore it is important to plan the test setup carefully. The test and measurement system is otherwise similar to the 2-D wave tanks.

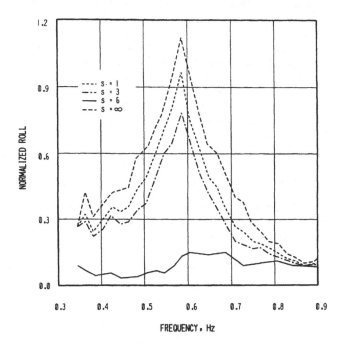

FIGURE 9.31
MEASURED ROLL RESPONSE AMPLITUDE OPERATOR IN MULTI-DIRECTIONAL WAVES [Nwogu (1989)]

The 3-D waves have a considerable influence on moored systems. Considering that the motion response is linear, the transfer function in a directional sea can be obtained in a similar fashion as in the case of a unidirectional sea. In this case, however, the additional dependency on the angle θ is attached to the transfer function, $H(\omega,\theta)$. Then the motion response spectrum in a particular mode of motion j of a floating vessel may be obtained from this transfer function H_j and the directional sea $S(\omega,\theta)$ as

$$S_{xj}(\omega) = \int_{-\pi}^{\pi} \left| H_j(\omega,\theta) \right|^2 S(\omega,\theta)d\theta \qquad (9.20)$$

The transfer function for a multi-directional sea state represents directionally averaged values since they depend only on the wave frequency. The effect of wave directionality may be clearly found in the response amplitude operator in the frequency domain. An example for the roll response of a barge in several multidirectional seas ($s=1,3,6,\infty$) is shown in Fig. 9.31 [Nwogu (1989)]. The roll is seen to increase in the multidirectional seas. The directionality of the waves is found to result in a decrease in the surge and pitch motion amplitudes and an overall increase in the response of sway, roll, and yaw motions. The effect on the heave of a floating moored system is mixed, but small. The low frequency motion, such as that found in surge is affected more by the directional sea. This has a direct influence on the mooring line loads. Tests on CALM systems should generally be performed in a 3-D tank for a more realistic assessment of the responses of the system.

9.13   REFERENCES

1.   Brady, I., Williams, S., and Golby, P., "A Study of the Forces Acting on Hoses at Monobuoy Conditions", Proceedings on the Sixth Offshore Technology Conference, Houston, Texas, OTC 2136, 1974, pp. 1057-1060.

2.   Calisal, S.M. and Sabuncu, T., "A Study of a Heaving Vertical Cylinder in a Towing Tank", Journal of Ship Research, Vol. 33, No. 2, June 1989, pp.107-114.

3.   Chakrabarti, S.K., and Hanna, S.Y., "Added Mass and Damping of a TLP Leg", Presented at the Twenty-second Annual Offshore Technology Conference, Houston, Texas, OTC 6406, May 1990, pp. 559-571.

4.   Chakrabarti, S.K., and Hanna, S.Y., "High Frequency Hydrodynamic Damping of a TLP Leg", Proceedings of the Offshore Mechanics and Arctic Engineering Symposium, Stavanger, Norway, Vol. 1, Part A, June 1991, pp. 147-152.

5.   Chakrabarti, S.K., "Moored Floating Structures and Hydrodynamic Coefficients", Proceedings on the Ocean Structural Dynamic Symposium, Corvallis, Oregon, Sept 1984, pp. 251-266.

6.   Chakrabarti, S.K., "Wave Interaction on a Triangular Barge", Proceedings on the Fifth International Offshore Mechanics and Arctic Engineering Symposium, Tokyo, Japan, ASME, 1986.

7.   Chakrabarti, S.K., and Cotter, D.C., "Analysis of a Tower-Tanker System", Proceedings on the Tenth Offshore Technology Conference, Houston, Texas, OTC 3202, 1978, pp. 1301-1310.

8.   Chakrabarti, S.K. Nonlinear Methods in Offshore Engineering, Elsevier Publishing Co., Netherlands, 1990.

9.   Chakrabarti, S.K., and Cotter, D.C., "Motion Analysis of an Articulated Tower", Journal of the Waterway, Port, Coastal and Ocean Division, ASCE, Vol. 105, August 1979.

10.   Chakrabarti, S.K., and Cotter, D.C., "Nonlinear Wave Interaction With a Moored Floating Cylinder", Proceedings on the Sixteenth Annual Offshore Technology Conference, Houston, Texas, OTC 4814, May 1984.

11.   Chakrabarti, S.K. and Cotter, D.C., "Interaction of Waves with a Moored Semi-submersible", Proceedings on the Third International Offshore Mechanics and Arctic Engineering Symposium, ASME, New Orleans, Louisiana, Feb., 1984, pp. 119-127.

12.   Chantrel, J. and Marol, P., "Subharmonic Response of Articulated Loading Platform", Proceedings of the Sixth International Offshore Mechanics and Arctic Engineering Symposium, ASME, Houston, Texas, 1987, pp. 35-43.

13.   Cotter, D.C. and Chakrabarti, S.K.., "Effect of Current and Waves on the Damping Coefficient of a Moored Tanker", Proceedings on Twenty-First Annual Offshore Technology Conference, Houston, Texas, OTC 6138, May 1989, pp. 149-159.

14.   DeBoom, W.C., Pinkster, J.A., and Tan, P.S.G., "Motion and Tether Force Prediction for a TLP," Journal of Waterway, Port, Coastal and Ocean Division, ASCE, Vol. 110, No. 4, November 1984, pp. 472-486.

15.   Faltinsen, O.M., Dahle, L.A. and Sortland, B., "Slowdrift Damping and Response of a Moored Ship in Irregular Waves," Proceedings on Fifth International Offshore Mechanics and Arctic Engineering Symposium, Tokyo, Japan, ASME, 1986.

16.   Goodrich, G.J., "Proposed Standards of Seakeeping Experiments in Head and Following Seas", Proceedings on Twelfth International Towing Tank Conference, 1969.

17.   Huse, E. and Matsumoto, K. "Practical Estimation of Mooring Line Damping", Proceedings on Twentieth Offshore Technology Conference, Houston, Texas, OTC 5676, 1988, pp. 543-552.

18.   Huse, E. and Matsumoto, K., "Mooring Line Damping Due to First and Second-Order Vessel Motion," Proceedings on Twenty-First Annual Offshore Technology Conference, Houston, Texas, OTC 6137, May 1989, pp. 135-148.

19.   Huse, E., "Effect of Mechanical Friction in Model Test Setup," MARINTEK Project Report No. 511151.00.03, Trondheim, Norway, 1989.

20.   Huse, E., "Resonant Heave Damping of Tension Leg Platforms", Proceedings on Twenty-Second Offshore Technology Conference, Houston, Texas, OTC 6317, 1990, pp. 431-436.

21.   Kaplan, P., "Scaling Problems of Dynamic Motions in Waves from Model Tests of Surface Effect Ships and Air Cushion Vehicles," Preprint 89-FE-1, Joint ASCE/ASME Mechanics, Fluids Eng., and Biomechanics Conference, San Diego, California, July 1989, 9 pages.

22.   Katayama, M., Unoki, K., and Miwa, E., "Response Analysis of Tension Leg Platform with Mechanical Damping System in Waves," Proceedings on Behavior of Offshore Structures, MIT, Boston, Vol. 2, 1982, pp. 497-522.

23.   Lin, J., Qui, Q., Li, Q. and Wu, Y.,"Experiment of an Elastic Ship Model and the Theoretical Predictions of its Hydroelastic Behavior," Proceedings on Very Large Floating Structures,Univ. of Honolulu, Hawaii, 1991, pp. 265-276.

24.   Nwogu, O., "Analysis of Fixed and Floating Structures in Random Multi-directional Waves," Ph.D. Thesis, Univ. of British Columbia, Vancouver, B.C., Canada, 1989.

25.   Petrauskas, C. and Liu, S.V., "Springing Force Response of a Tension Leg Platform", Proceedings on the Nineteenth Annual Offshore Technology Conference, Houston, Texas, OTC 5458, 1987, pp. 333-341.

26. Pinkster, J.A. and Remery, G.F.M., "The Role of Model Tests in the Design of Single Point Mooring Terminals," Proceedings of Seventh Annual Offshore Technology Conference, Houston, Texas, OTC 2212, 1975, pp.679-702.

27. Tan, S.G., and DeBoom, W.C., "The Wave Induced Motions of a Tension Leg Platform in Deep Water", Proceedings of the Thirteenth Annual Offshore Technology Conference, Houston, Texas, OTC 4074, 1981.

28. Wichers, J.E.W., "A Simulation Model for a Single Point Moored Tanker", Maritime Research Institute , Wageningen, The Netherlands, Ph.D., Delft U. of Technology/MARIN Publication No. 797, 1988, 243 pages.

29. Young, R.A., Brogren, E.E. and Chakrabarti, S.K., "Behavior of Loading Hose Models in Laboratory Waves and Currents," Proceedings of the Twelfth Annual Offshore Technology Conference, Houston, Texas, OTC 3842, May, 1980.

CHAPTER 10

DATA ANALYSIS TECHNIQUES

10.1 STANDARD DATA ANALYSIS

One of the most important aspects of a model testing program is the analysis of the data collected during test runs. Some model tests are only qualitative, requiring minimal instruments. These are quite helpful in evaluating a conceptual design or an offshore operation. However, the majority of the tests are designed to provide many useful data that are applied in verifying the design of the particular system or operational procedure. This chapter will deal with the most common data reduction procedures that are used in the analysis of the test results.

There are two types of data reduction routines. Standard routines are commonly used for all test data collected in a wave tank environment. These are used in any model testing work. Many of these programs are commercially available today, although most testing facilities find it advantageous to develop their own softwares. The second type of routines are specialized software written for specific types of tests performed. There could be numerous programs developed for model test data. Those that have been chosen for discussion here relate to the model testing covered in the earlier chapters. It is recognized that descriptions of many special data reduction routines have been left out in this chapter. Included is only a representative sampling from the testing described earlier in Chapters 5, 7, 8 and 9. Since it is important to adequately quantify the waves generated in the tank, a major part of this chapter deals with both the 2-D and 3-D waves (Chapter 5). Data reduction on the wave forces on fixed models (Chapter 7) includes computation of hydrodynamic coefficients which is discussed here. Evaluation of towing loads have been adequately described in an earlier chapter (Chapter 8) and hence is omitted here. Computations of hydrodynamic damping and associated quantities of floating moored models (Chapter 9) are also addressed in this chapter.

The recorded data contain three generic types of measurement errors: calibration errors, bias errors and random errors. Calibration errors result from the variation between the measuring device's actual input/output relationship and the calibration curve used with the device. This type of error will result in a deviation (systematic or scatter) about the recorded values when they are expressed in engineering units. Bias errors evolve from data reduction procedures such as the windowing operations associated with the calculation of power spectral densities, and they appear with constant magnitude and direction from one analysis to the next. Random errors are the result of averaging operations that must be performed over a finite number of sample records.

Time history records are the primary source of information from the wave tank testing. The time history data are first inspected for evidence of spurious signals. The following estimates of statistical and probabilistic parameters are generally obtained for further analysis:

- mean value;
- root mean square;
- standard deviation;
- random error;
- bias error;
- calibration error or uncertainty;
- probability and cumulative distribution functions plotted in normal
    and/or Rayleigh distribution format, as appropriate;
- Kurtosis and skewness of record;
- spectral (autospectral) density, including other spectral moments as
    appropriate for single peak spectra to estimate spectral bandwidth,
    peak frequency, etc.

Some of the standard data reduction routines that are available at any model testing facility and their function are as follows:

- Plotting package – ability to plot the recorded data in the time domain.

- Summing routine – ability to add two or more channels in the time domain from one or more data files. This enables the development of force and moment traces from direct load cell readings, for example.

- Maximum/Minimum routine – to compute the mean maximum and minimum amplitudes of a regular wave trace along with standard deviations of data from the mean values.

- Phase routine – ability to compute the phase difference between two channels in which phase angles between corresponding peaks in two traces are found.

- Normalization – ability to generate tables of transfer functions by normalizing amplitude of measured response time history by the amplitude of regular wave time history. For irregular waves, the corresponding spectra are normalized.

Some of the other data reduction routines that would be available at a model testing facility will be discussed subsequently in greater detail.

10.2   REGULAR WAVE ANALYSIS

Regular waves generated in a tank are generally two-dimensional in character and have a single frequency associated with them. While the input signal to the wavemaker in generating these waves is a sinusoid, the actual waves generated in the tank may not have a sinusoidal form due to the gravity, tank bottom and other effects. Moreover, the single frequency waves may produce responses that have multiple frequencies. This is particularly possible in responses for moving structures. An example of this may be found in Fig. 7.1 in which the measured forces have multiple frequencies due to a single-frequency wave. Theoretical analysis justifies the existence of these multiple frequencies in the response due to a regular wave (Section 10.6.1).

In a wave tank of limited size, the waves at the test site are often contaminated by the presence of standing waves and reflected waves. It is important that these effects are isolated from the measured data before further analysis is carried out.

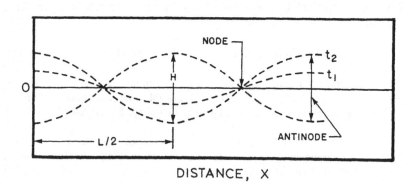

DISTANCE, X

FIGURE 10.1
STANDING WAVE PROFILE IN A TANK

10.2.1 Standing Wave

A standing wave is encountered in a closed or partially open basin when the confined water is disturbed. A standing wave produces a vertical oscillation with time t at a given location (Fig. 10.1). The amplitude of the standing wave is expressed as

$$\eta = \frac{H}{2}\cos kx \cos \omega t \tag{10.1}$$

where H is the height of the standing wave. According to this equation, a standing wave experiences the same amount of oscillation about the x axis from one cycle to the next. Also, where $\cos kx = 1$, it experiences the maximum vertical oscillation which is called the antinode. The point, where $\cos kx = 0$ (1/4 the wave length L away), experiences no vertical oscillation and is called the node. A standing wave in a basin may have one or more nodes and antinodes.

The oscillation period (sloshing) in the basin is called the natural period. There are several modes of natural period in the basin depending on its length (or width for cross oscillation). For a rectangular basin, the nth mode of oscillation has a wave length given by

$$L = \frac{2l}{n}, n = 1, 2,... \tag{10.2}$$

where l = length of basin. The sloshing period may be computed from the dispersion relationship:

$$T_n = \left[\frac{4\pi l}{ng \tanh kd}\right]^{\frac{1}{2}} \tag{10.3}$$

where d is the uniform water depth in the basin.

For a given test basin, the longitudinal and transverse sloshing periods should be computed from the above relationship. Testing at these natural periods may then be avoided. Tests at these periods have indicated that the wave heights may vary considerably from point to point in the longitudinal direction. If the sloshing period is discerned in the test data, its effect may be filtered out through digital filtering.

10.2.2 Reflected Wave

Another area of data contamination is the reflected wave from the walls of the wave tank. These waves have the same frequency as the incoming wave and, therefore, are indistinguishable from the incident wave. The methods of computing the wave reflection coefficient have been detailed in Chapter 4. These may be established for various wave frequencies in a given wave basin before any testing commences in the basin. Most test basins have this information available for their clients to review. Note that while it is important to know the amount of reflection in a basin, data measured in the test area include the reflected wave and its effect on responses. Sometimes it is desirable to reduce the effect of reflection. Methods to implement this have been discussed in Chapter 4.

10.2.3 Spurious Wave Data

The data collected in a wave tank environment suffer from several other shortcomings [Mansard and Funke (1988)]. Some of the spurious data are removed from the recorded data before any analysis is performed. The preprocessing of data is essential to reduce the error in subsequent analysis.

The first and foremost of preprocessing is the mean and trend removal. Most data have an offset from zero. The trends could be linear or second-order. The linear trend appears as a drift in the mean value. The second-order trend, while generally quite small, describes a parabolic drift of the mean value. These trends are removed from the wave data before temporal (zero-crossing) or spectral analysis is performed. However, a word of caution is warranted here. Some of the response data will produce a zero shift from the waves which is physical. Therefore, while the wave elevations may be assumed to have a zero mean for linear Gaussian waves, nonlinear waves or responses may have a non-zero mean which should not be removed before examining the physical phenomenon carefully.

In the determination of the amplitudes or heights of a record (e.g., the wave height), the maximum recorded value of the sample between zero-crossings is considered the crest (or minimum for troughs). This method works well for samples taken at a high sampling rate. However, if the sampling rate is low, it may introduce error in the height value. In this case, a parabolic fit to the three adjacent highest samples is made and the highest point is chosen as the crest value. The (wave) height, then, is obtained from the adjacent crest and trough thus chosen.

10.3 IRREGULAR WAVE ANALYSIS

Offshore structures are often tested in irregular or random waves. The generation of these waves in the tank has already been discussed in Chapter 4. Some of the ambiguities in the irregular wave data are similar to the regular wave data discussed in the earlier section. The analysis of the random wave data can be made both in the time domain and in the frequency domain. Obviously, the analysis time is lengthy in the time domain, but it has the advantage of maintaining any nonlinearity present in the data. The spectrum analysis is generally linear in nature and readily provides a transfer function between the waves and responses. However, nonlinear frequency domain analysis methods are currently being developed.

Some of the data analysis methods commonly used with the wave tank data analysis is described subsequently.

10.3.1 Fourier Series Analysis

Fourier series analysis is a useful tool in analyzing data with multiple frequencies. It allows the determination of amplitudes and phases of different frequencies present in a time history of recorded data. Thus, a one-to-one correspondence between the responses and wave at given frequencies is possible. If a small enough frequency increment is chosen, then it is possible to decipher all of the frequency components present in a system of recorded data.

The derivation of amplitudes and phases is as follows. Let the time history of the signal recorded be given by $f(t)$ and the length of the record be T_R. Then, the Fourier representation of the signal $f(t)$ is given by

$$f(t) = a_0 + \sum_{n=1}^{N} \left[a_n \cos \frac{2n\pi t}{T_R} + b_n \sin \frac{2n\pi t}{T_R} \right] \tag{10.4}$$

where N is the number of finite Fourier components of interest and a_0, a_n, and b_n are constants which are evaluated from the integrals

$$a_0 = \frac{1}{T_R} \int_0^{T_R} f(t) dt \tag{10.5}$$

$$a_n = \frac{2}{T_R} \int_0^{T_R} f(t) \cos \frac{2n\pi t}{L} dt \tag{10.6}$$

$$b_n = \frac{2}{T_R} \int_0^{T_R} f(t) \sin \frac{2n\pi t}{L} dt \tag{10.7}$$

The coefficient a_0 is often expected to be near zero for irregular waves. However, as stated before, it may be non-zero for the recorded responses, e.g., loads. A numerical integration routine is applied to the integrals in Eqs. 10.5 - 10.7. Once the Fourier coefficients are known, the amplitude and phase of the harmonic components are computed from

$$A_n = \sqrt{a_n^2 + b_n^2} \tag{10.8}$$

and

$$\varepsilon_n = \tan^{-1} \frac{b_n}{a_n}, \quad n = 1, ..., N \tag{10.9}$$

It is quite common to find higher harmonics of wave and response records. The above analysis allows one to study the harmonic components of the wave and the corresponding response.

10.3.2 Wave Spectrum Analysis

It is a common practice to represent a measured sea state in terms of its spectral density. Because of the short duration of the wave records, they suffer from spectral variability. One source of the variability is the algorithm used in the derivation of the spectral parameters, e.g., the peak frequency. Another variable is the choice of the upper and lower cut-off frequencies in order to avoid background noise. In order to overcome these and other difficulties and source of variability, Mansard and Funke (1988) advocated spectral fitting. In this case, a mathematical model, e.g., a JONSWAP spectrum, is fitted to the spectral estimate and parameters are derived from the best fit curve (e.g., by a least square technique). Once the mathematical description is known, all statistical parameters may be easily derived from this model without additional variability.

A bias correction factor is often applied to the parameter estimate to obtain an enhanced estimate of the parameter. For example, it is customary to use cut-off frequencies in the spectral estimates to derive moments of the spectrum. However, the imposition of upper and lower cut-off frequencies will underestimate the zeroth moment, m_0. The integration of a smooth spectral model between 0 and infinity will provide the theoretical value of m_0. Then, a bias correction factor for m_0 may be defined as

$$C_m^2 = \frac{\int_0^\infty S(f)df}{\int_{f_1}^{f_2} S(f)df} \tag{10.10}$$

where f_1 and f_2 are the lower and upper cut-off frequencies. The significant wave height may be modified by

$$H_s' = C_m H_s \tag{10.11}$$

For a JONSWAP spectrum (peakedness parameter, $\gamma = 1$ to 12), if the cut-off frequencies are chosen as $f_1 = 0.5f_0$, and $f_2 = 2.5f_0$, where f_0 is the peak frequency, then the correction factor is obtained [Mansard and Funke (1988)] from

$$C_m^2(\gamma) = 1.0015 + 1 / [19.9178(\gamma + 2.6937)] \tag{10.12}$$

The energy (or power) spectrum is computed using an FFT algorithm. The scale factor and sampling rate are chosen such that the number of data points is a power of 2. As an example, let us consider a prototype system which has a natural period of 90 seconds and we require 100 cycles of data for proper data reduction. Let the scale factor for the model test be chosen as 1:100 and a sampling rate of 9 Hz is considered sufficient. Then,

Test duration in tank = 90 x 100 /$\sqrt{100}$ = 900 sec.

Number of data points = 900 x 9 = 8100

In this case the total number of data points is chosen as 8,192 being the nearest power of 2. In order to obtain a smooth (averaged) estimate, the test record is divided into several intervals, such as 8 intervals of 1024 points each for the above example. A spectrum estimate S_k for the kth interval is made. Then the average (smoothed) estimate is obtained from

$$S(\omega) = \frac{1}{8} \sum_{k=1}^{8} S_k(\omega) \qquad (10.13)$$

The main characteristics of the spectrum is given in terms of the moments of the spectrum,

$$m_n = \int_{\omega_1}^{\omega_2} S(\omega) \omega^n d\omega \qquad (10.14)$$

where m_n is defined as the nth moment of the spectrum computed between the cut-off frequencies ω_1 and ω_2. Then, the significant wave height is obtained from the formula [Longuet – Higgins (1957)]

$$H_s = 4\sqrt{m_0} \qquad (10.15)$$

The mean and zero-crossing periods are defined as

$$T_m = 2\pi \frac{m_0}{m_1} \qquad (10.16)$$

and

$$T_z = 2\pi \sqrt{\frac{m_0}{m_2}} \qquad (10.17)$$

10.3.3 Wave Group Analysis

The motions of moored floating structures are very sensitive to low frequency second-order forces. The square of the wave envelope is important when considering this low frequency behavior since the slowly-varying second-order response of the vessel is significantly affected by grouping patterns of waves in the time domain. Therefore, particular attention must be paid in the generation of random waves so that the generated wave grouping is accurate when compared to the theoretical grouping. The groupiness function is defined as

$$G(\mu) = 8\int_0^\infty S(\omega)S(\omega+\mu)d\omega = \frac{16}{\pi}\int_0^\infty R^2(u)\cos(\mu u)du \qquad (10.18)$$

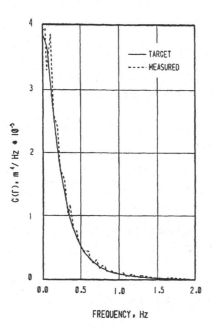

FIGURE 10.2
MEASURED VERSUS COMPUTED GROUPINESS FUNCTION FOR A
JONSWAP SPECTRUM

where R(u) is the autocorrelation function. The second equality avoids cumulative numerical errors. An example of the measured versus computed JONSWAP spectrum groupiness function is shown in Fig. 10.2. The groupiness function is plotted versus frequency. The trend and correlation shown are typical.

10.3.4 Statistical Analysis

For signals recorded during a test in irregular waves, a statistical analysis is performed on each measurable quantity and the following statistical parameters are sought:

- Mean value--This quantity is obtained as the mean of the data over the recorded length.

$$\bar{x} = \frac{1}{N} \sum_{i=1}^{N} x_i \tag{10.19}$$

where x_i are the values of the response at each sample point and N is the total number of sample points in the length of the record.

- Root mean square value--The rms value of the response is calculated from the definition

$$x_{rms} = \left[\frac{1}{N} \sum_{i=1}^{N} x_i^2 \right]^{\frac{1}{2}} \tag{10.20}$$

- Maximum positive value--This is the maximum recorded value, X_{max}, on the positive side of the zero axis.

- Maximum negative value--This is the minimum recorded value, X_{min}, on the negative side of the zero axis.

- Significant height value--This significant height $(2x)_{1/3}$ is defined (by Rayleigh distribution) as the average value of the highest one-third heights (i.e., peak-to-peak) in a record. It is also obtained from Eq. 10.15.

- Significant positive amplitude--This quantity is the average of the highest one-third positive amplitudes in a record.

- Frequency response functions--The transfer function for the response under consideration may be computed from a single irregular test. This technique will be described in a subsequent section.

- Peak frequency--The frequency at which the energy spectrum value is maximum.

Because of the erratic profile of the measured wave spectra, it is generally not sufficient to simply choose the frequency at which the spectrum peaks. A better alternative has been recommended by the IAHR (1986). In this method, a threshold of 80% of the maximum spectrum value is chosen and the two frequencies, low and high for the first and second threshold crossings are computed. The centroid of this portion between the two frequencies is taken as the peak frequency. The simple method of determining peak value shows a much larger variability (about 25% greater) than the centroid method [Mansard and Funke (1988)]. The centroid is given by

$$\omega_0 = \frac{\int_{\omega_1}^{\omega_2} \omega S(\omega) d\omega}{\int_{\omega_1}^{\omega_2} S(\omega) d\omega} \tag{10.21}$$

where ω_1 and ω_2 are the frequencies corresponding to 80% (60% has also been proposed as a modification) threshold value.

Peak frequency has also been defined as

$$\omega_0 = \frac{\int_0^\infty \omega S^n(\omega) d\omega}{\int_0^\infty S^n(\omega) d\omega} \tag{10.22}$$

where values of n have been proposed as 8 [Read(1986)] or 6 [Mansard & Funke (1988)]. Because of the asymmetry in the spectrum shape, a bias may be introduced in the computation for which a correction factor has been proposed (Eq. 10.10). For a JONSWAP spectrum with low γ value, this factor is about 1.02.

10.4   ANALYSIS OF DIRECTIONAL WAVES

For the analysis and description of multidirectional waves, the wave field requires simultaneous measurement of the water surface elevation at a number of

neighboring locations.   Typically, four or more wave probes are arranged near the
desired   location for the   wave measurement.   Alternatively, orthogonal horizontal
components of the water particle velocities along with the water surface elevation may
also be measured. Of the methods that can be used in the estimation of the directional
distribution of wave energy, the most common are the Direct Fourier Transform (DFT)
method,   parametric method,   the Maximum Likelihood Method   (MLM) and the
Maximum Entrophy Method (MEM). Most of these use a cross spectral method in the
computation of the wave energy as a function of wave frequency and direction of wave
propagation.

The direct transform method and the parametric method are the earlier methods
and have been described by Chakrabarti (1990). The first was proposed by Barber
(1963) while the second was applied by Longuet-Higgins, et al (1963). These techniques
are limited in the resolution of the directionality of waves. The MEM is considered more
useful in discerning the details of a directional sea and has become the most commonly
applied method for the estimation of multidirectional spectrum. The MEM is found to
resolve directional seas better than the MLM. However, it does not converge for very
narrow spreading functions (i.e., large values of s). The spectral derivations by these
methods have been detailed in the thesis by Nwogu (1989) and the reader is referred
to his work for details. These methods will be briefly discussed.

For the analysis of the directional waves, data is usually available in a series of
time history of water surface elevation (or slopes or velocities) at various locations of the
open water. These time series are transformed to the frequency domain by cross-spectral
density calculation.

The cross spectral calculation is similar to the autospectral estimation of wave
energy except that it works with two different time histories. Initially, a covariance
function of the profiles is obtained from

$$R_{ij}(\tau) = \int_0^{T_R} \eta_i(t)\eta_j(t+\tau)dt \tag{10.23}$$

where T_R is the record length and i and j refer to two separate probe readings. The
coincident spectra and quadrature spectra are computed   as   the cosine and sine
transforms of the covariance function

$$C_{ij}(\omega) = \int_0^{T_R} R_{ij}(t)\cos\omega t \, dt \qquad\qquad (10.24)$$

$$Q_{ij}(\omega) = \int_0^{T_R} R_{ij}(t)\sin\omega t \, dt \qquad\qquad (10.25)$$

The two-dimensional (ω,θ) energy spectrum is the Fourier transform of the co- and quad-spectra. Knowing the distance D_{ij} among probes i and j and the corresponding angles to the reference axis ϕ_{ij}, the directional spectra may be written as

$$S(\omega,\theta) = \sum_{ij} \left[C_{ij}(\omega)\cos\left\{kD_{ij}\cos\left(\phi_{ij}-\theta\right)\right\} + Q_{ij}(\omega)\sin\left\{kD_{ij}\cos\left(\phi_{ij}-\theta\right)\right\} \right] \qquad (10.26)$$

The summation is taken over a pair of probes among the total number of probes used. This is a simple estimate of the directional spectral form by the direct transform method. Other more commonly used methods are much more complex mathematically as summarized in the thesis of Nwogu (1989). Standard routines are available at facilities capable of generating 3-D waves. One important point to note is that these are estimates of true spectrum and are subject to errors. The more the number of recordings, the better is the estimate. Generally 4 to 5 wave probes suffice. They are arranged in several geometric patterns and research has been done to examine the optimum configuration [Chakrabarti and Snider (1972)].

Fewer time history recordings are needed if orthogonal water particle velocities are simultaneously measured along with a wave profile. Mathematical routines are available [Nwogu (1989)] to reduce this type of data.

10.5 FILTERING OF DATA

Filtering permits isolation of components of waves at one frequency band from those at a different frequency band. Filters are of two types: analog and digital. Analog filters are electronic in nature and have been described in Chapter 6. Digital filtering may be accomplished in the time domain or in the frequency domain. A digital filter is described as a transfer function which when applied to the original time-history function (or its Fourier transform) will provide the filtered data. Thus, mathematically

$$g(t) = f(t) * h(t) = \int_{-\infty}^{\infty} f(\tau) h(t - \tau) d\tau \qquad (10.27)$$

where f(t) is the input function such as the wave profile, h(t) is the weight function and g(t) is the resulting response function. The asterisk denotes convolution given by the integral on the following line. Alternatively, in the frequency domain Eq. 10.27 may be written as

$$G(\omega) = F(\omega) H(\omega) \qquad (10.28)$$

where H(ω) is the transfer function corresponding to its time history representation. They are related to each other by a Fourier cosine transform

$$h(t) = \frac{1}{\pi} \int_0^{\infty} |H(\omega)| \cos[\omega t - \Phi(\omega)] d\omega \qquad (10.29)$$

Digital filtering of the recorded data is performed for a number of different purposes:

- removing noise from data
- preparing data for spectral analysis
- removing transient data
- separating low and high frequency data
- comparing input-output relationship between two channels

There are three types of digital filters: low pass, high pass and band pass. Low pass filters eliminate high frequency components and may be used for smoothing data by removing   short   period wave components riding on the wave frequencies,   for example.   High pass filters may be used to eliminate low frequency transients such as the decaying natural period oscillation at the start of a test run. Band pass filters retain the middle frequency band of interest. There is also a band reject filter which rejects a frequency band from the middle of the spectrum.

If a record f(t) consists partly of signal s(t) and partly of noise n(t), then

$$f(t) = s(t) + n(t) \qquad (10.30)$$

In the frequency domain this becomes:

$$F(\omega) = S(\omega) + N(\omega) \tag{10.31}$$

An ideal filter to remove noise will have the form

$$H(\omega) = \frac{S(\omega)}{S(\omega) + N(\omega)} \tag{10.32}$$

However, in practice numerical leakage occurs with digital filters and distortion in the amplitude of data is evident due to this leakage. As explained earlier, filtering may be accomplished in the time domain or in the frequency domain. In the time domain a convolution integral is performed with the function h(t) (Eq.10.27). There are many mathematical expressions of h(t). One description of the weight function is

$$H(\omega) = W(0) + 2\sum_{p=1}^{P} W(p)\cos(\omega p \Delta t) \tag{10.33}$$

where

$$W(0) = 2\Delta t(\omega_2 - \omega_1) / \pi; W(p) = [\sin(\omega_2 p \Delta t) - \sin(\omega_1 p \Delta t)] / \pi p \tag{10.34}$$

and where P is the number of weights. The weight functions are generally modified with a "window" to reduce spurious data in the filter. The larger the value of P the sharper is the cut-off frequency of the filter and the leakage in the filtered data is correspondingly less. This is illustrated in Fig. 10.3 in which a band pass filter has been constructed using P=30 and 50. The filter with P=50 appears to be sharper, but both filters deviate from the ideal band pass filter.

It should also be kept in mind that this procedure loses original recorded data at its both ends through numerical integration (convolution). The phase information in the data may be restored with the knowledge of the number of weights and the time increment. For example, data lost in the beginning is equivalent to P(Δt)/2. Shifting data by this amount will restore the original phase. It is generally customary to filter all channels in a test run using the same filter parameters. This allows the phase relationship to be maintained among them.

In the frequency domain analysis, the time domain data are transformed into frequency domain using a Fast Fourier Transformation (FFT) routine. Then, the energy between the prescribed frequency limits (depending on high, low or band pass filter) is removed. The remaining modified frequency domain representation of the data is transformed back into the time domain by an inverse FFT routine. The entire process can be accomplished by one FFT routine with a positive (plus) and a negative (minus) exponent. There are many FFT routines commercially available [e.g., Brigham (1974)].

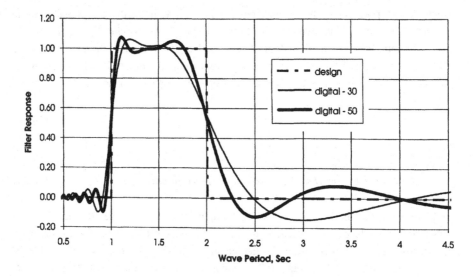

FIGURE 10.3
IDEAL AND DIGITAL BAND PASS FILTER

10.6 RESPONSE ANALYSIS

In analyzing the responses of a model in waves, it is customary to obtain the transfer function between the response and the wave. The transfer function relates the response amplitude to the wave frequency and amplitude. In regular waves, it is straightforward to compute the transfer function. In random waves, there are several methods to compute the transfer function.

10.6.1 Frequency Domain Analysis

The frequency-domain analysis is suitable for linear systems. They are also carried out for a weak nonlinear system. Sometimes, the responses are separated into linear and nonlinear signals and the transformation is carried out separately and then results added to obtain the total effect.

An example (Fig. 10.4) is given here for such an analysis for a tension-leg platform model subjected to a steady current and random waves [Botelho, et al (1984)]. In this case, the total displacement of the TLP is composed of four parts: (1) steady offset due to current, (2) primary response due to the wave frequency, (3) steady drift offset, and (4) slowly varying response at the low frequency. A flow chart to analyze the extreme response due to these four components is shown in Fig. 10.5.

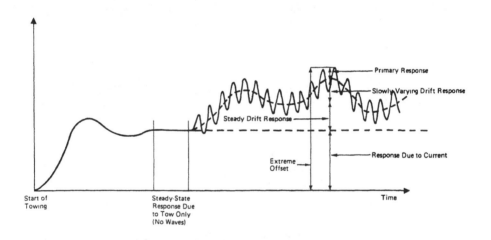

FIGURE 10.4

COMPONENTS OF HYPOTHETICAL RESPONSE OF A TLP MODEL TO RANDOM WAVES AND CURRENT [Botelho, et al. (1984)]

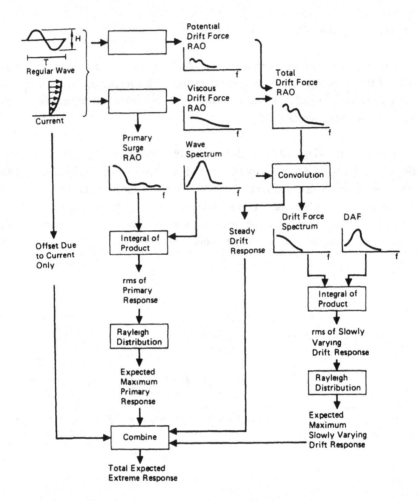

FIGURE 10.5

FLOW CHART FOR A FREQUENCY DOMAIN PROCEDURE

[Botelho, et al. (1984)]

The open boxes in the frequency domain represent appropriate model tests to derive the required RAOs. It is also possible to obtain this information through computer programs or a combination of the two methods. The wave spectrum may similarly be a measured wave spectrum or a theoretical model. The subsequent analysis is carried out mainly in the frequency domain. The first-order and second-order responses are treated individually and then combined to obtain the total extreme response. An appropriate distribution function should be selected, e.g. Rayleigh. Linear combination of extremes is simple, but may be conservative.

The required input to this procedure are the wave spectrum, current profile and transfer functions for the surge displacement at the first-order and second-order respectively. The response due to current is obtained separately, disregarding interaction. The transfer functions are obtained from the tests in regular waves. In this case, the response amplitude at the wave frequency is normalized by the corresponding wave amplitude. This provides the first-order Response Amplitude Operator:

$$\hat{X}_a^{(1)}(\omega) = \frac{X_a(\omega)}{a}$$ (10.35)

where X_a is the first-order response and a is the regular wave amplitude. The RAO is corrected to account for the Doppler frequency shift due to the current. The correction consists of shift in the RAO from the encounter frequency (frequency in the presence of current) to the wave board frequency (i.e., at zero current).

The second-order RAO is obtained from the regular wave steady drift force measured in the mooring system and the dynamic amplification factor (DAF). The steady drift force is calculated from the test data as the mean over a complete number of cycles. The steady drift force is normalized by the square of the wave amplitude and divided by the spring constant to give

$$\hat{X}^{(2)}(\omega) = \frac{\overline{F}(\omega)}{Ka^2}$$ (10.36)

where $\overline{F}$ = steady drift force in regular waves and K = spring constant of the system. Botelho (1984) obtained the second-order steady force separately from current and waves as shown in Fig. 10.5. The current load is simulated by towing the model in tank before and during waves. Note that while the two steady components are obtained separately in one run, any interaction effect is included in the second component because of the method of testing used here.

The spectrum of the primary response is computed from the product of the wave spectrum and the square of the first-order RAO. The root-mean square (rms) of the primary response is the square root of the area under the computed response spectrum.

$$\sigma^{(1)} = \left[\int_0^\infty [\hat{X}_a^{(1)}(\omega)]^2 S(\omega) d\omega \right]^{1/2}$$

(10.37)

where $\sigma^{(1)}$ is the rms of the primary response and $S(\omega)$ is the wave spectrum.

The steady drift force in random waves is computed from

$$F_s = 2 \int_0^\infty \bar{F}(\omega) S(\omega) d\omega$$

(10.38)

The spectrum of the slowly-varying displacement is computed by first computing the second-order displacement RAO from

$$\hat{X}_a^{(2)}(\omega) = \bar{X}^{(2)}(\omega)(DAF(\omega))$$

(10.39)

where $\bar{X}^{(2)} = \bar{F}/K$ is the steady second-order displacement. The slowly-varying displacement spectrum is computed numerically from

$$S_x(\omega) = 8 \int_0^\infty \left[\hat{X}_a^{(2)} \left(\frac{\omega}{2} + \mu \right) \right]^4 S(\omega) S(\omega + \mu) d\mu$$

(10.40)

Then the rms of the slowly-varying drift, $\sigma^{(2)}$ is given by

$$\sigma^{(2)} = \left[\int_0^\infty S_x(\omega) d\omega \right]^{1/2}$$

(10.41)

10.6.2 Linear System

One of the goals of most model testing is to develop the Response Amplitude Operator. The RAO is the magnitude of the linear transfer function between the forcing function and the response function.   A computer program determines this RAO given two measurements recorded as a function of time. A simple calculation of the RAO from the spectral density is given below. A more detailed theoretical derivation is discussed in the following section.

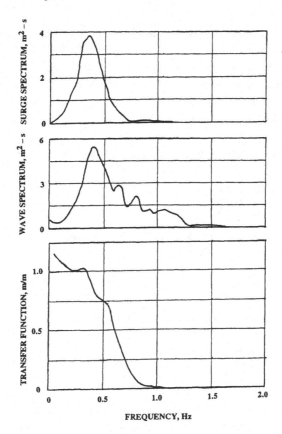

FIGURE 10.6

RAO COMPUTED FROM FORCE AND WAVE SPECTRUM

The time domain records of the measured wave and structure responses are first converted to the frequency domain. Consider that $S_{xx}(\omega)$ [i.e.. $S(\omega)$ rewritten for clarity here] is the autospectral energy density spectrum of the wave and $S_{yy}(\omega)$ is the corresponding energy density spectrum of a measured response. The transfer function is then computed from

$$H(\omega) = \left[\frac{S_{yy}(\omega)}{S_{xx}(\omega)}\right]^{\frac{1}{2}}$$ (10.42)

An example of this method is illustrated in Fig. 10.6. The measured response and wave spectra are given by the top two curves from which the RAO (bottom curve) has been derived. This method may produce uncorrelated signal noise especially at the tail ends of the spectrum (removed in Fig. 10.6). Additionally, the information on the phase relationship between the response amplitude and wave amplitude is lost through this method. In order to avoid these problems, a cross spectral density analysis method may be employed. To apply this method, the cross spectral density function $S_{xy}(\omega)$ between the wave and the response under consideration is computed. Then,

$$C(\omega) + iQ(\omega) = \frac{S_{xy}(\omega)}{S_{xx}(\omega)}$$ (10.43)

where C and Q are the coincident and quadrature spectra as functions of frequency, ω, and i is the imaginary quantity. The transfer function is computed as the amplitude

$$H(\omega) = \left[C^2(\omega) + Q^2(\omega)\right]^{\frac{1}{2}}$$ (10.44)

and the corresponding phase angle is obtained from

$$\varepsilon(\omega) = \tan^{-1}\left(\frac{Q(\omega)}{C(\omega)}\right)$$ (10.45)

A coherence function that states the degree of correlation between the input and output signal may be calculated from the spectral estimate . The coherence function is obtained from

$$\gamma^2(\omega) = \frac{S_{xy}(\omega)}{\left[S_{xx}(\omega)S_{yy}(\omega)\right]^{\frac{1}{2}}}$$

(10.46)

The correlation between the input and output signals is good as the value of γ approaches unity, while a poor correlation is indicated as γ approaches zero.

10.6.3 Theory of Cross Spectral Analysis

As shown here, the transfer function under the cross spectral theory is a complex quantity and retains the amplitude and phase relationship between the forcing function and the response function. The following presentation documents the theoretical background, possible calculation errors and one of the example problems used to verify the computation method.

Consider the equation of motion in time for any linear dynamic single degree of freedom system (SDOFS)

$$m\ddot{y}(t) + C\dot{y}(t) + Ky(t) = x(t)$$

(10.47)

where y = response function and x = forcing function. The forcing function $x(t)$ can be any deterministic or random function.

Equation 10.47 can also be expressed by the following convolution (or Duhamel) integral [Boas (1966)]

$$y(t) = \int_{-\infty}^{\infty} x(\tau)h(t-\tau)d\tau$$

(10.48)

where the function "h" is known as the "weighting function" for the system. By taking Fourier transform of both sides of Eq. 10.48 and making an appropriate substitution, the following frequency domain relation results:

$$Y(\omega) = H(\omega)X(\omega)$$

(10.49)

where H(ω) = Fourier transform of h(t), Y(ω) = Fourier transform of y(t), X(ω) = Fourier transform of x(t).   Keep in mind that "H", "Y", and "X" are all complex quantities.   Equation 10.49 is the frequency domain SDOFS equation of motion, and it directly results from applying the convolution theorem to Equation 10.48 [Brigham (1974), pg. 58].

The quantity H(ω) is known as the transfer function for a linear system. It may be expressed in terms of the system properties. By assuming a complex forcing function of unit strength and solving the equation of motion for H(ω)   [Clough and Penzien (1975)], the system property description of H(ω)   (i.e. the exact transfer function) becomes:

$$H(\omega) = \frac{1}{-\omega^2 m + i\omega C + K} = \frac{(K - \omega^2 m) - i\omega C}{(K - \omega^2 m)^2 + (\omega C)^2} = A(K - \omega^2 m) - iA(\omega C) \qquad (10.50)$$

where A = $[(K - \omega^2 m)^2 + (\omega C)^2]^{-1}$. The magnitude of the exact transfer function (i.e. the RAO) is:

$$RAO = |H(\omega)| = \sqrt{A^2 (K - \omega^2 m)^2 + A^2 (\omega C)^2} \qquad (10.51)$$

Thus if the system properties (i.e. "C", "K", & "M") are known one can find the RAO from Eq. 10.51.

If the system properties are not known, but the forcing time function and the response time function can be measured, the RAO may be calculated from the following spectral relations [Bendat and Piersol (1980)]:

$$S_{yy}(f) = |H(f)|^2 S_{xx}(f) \qquad (10.52)$$

$$S_{xy}(f) = H(f) S_{xx}(f) \qquad (10.53)$$

from which equations 10.42 and 10.43 are derived. Note that $\omega = 2\pi f$ here. The forcing autospectrum is written as

$$S_{xx}(f) = \lim_{T \to \infty} \frac{1}{T} E[X(f,T)X*(f,T)] \qquad (10.54)$$

while the response autospectrum becomes

$$S_{yy}(f) = \lim_{T \to \infty} \frac{1}{T} E[Y(f,T)Y*(f,T)] \qquad (10.55)$$

The cross-spectrum between force and response function is given by

$$S_{xy}(f) = \lim_{T \to \infty} \frac{1}{T} E[X(f,T)Y*(f,T)] \qquad (10.56)$$

The variables "X" and "Y" are defined in Equation 10.49. The notation E [ ] denotes the average value of the quantity inside the brackets. The superscript asterisk implies the complex conjugate. By substituting Eqs. 10.54 and 10.56 into Eq. 10.53, Eq. 10.49 will result. The substitution of Eqs. 10.54 and 10.55 into Eq. 10.52 yields:

$$|Y(f)|^2 = |H(f)|^2 |X(f)|^2 \qquad (10.57)$$

Eq. 10.57 shows that only the moduli of the Fourier transforms are used in Eq. 10.52, hence all phase information is lost, whereas Eq. 10.53 retains the phase information.

The ideal although unrealistic situation for applying Eqs. 10.52 and 10.53 is to have time records of infinite length from which to take the Fourier transforms. In reality, the time record length T_R must be finite, and for this reason, all of the Fourier transforms are a function of T_R as well as f [Bendat and Piersol (1980)]. The "real life" situation also dictates that the Fourier transforms shown in Eqs. 10.54 - 10.56 be the average transform which is obtained by averaging the transforms of many subrecords. These subrecords are obtained by dividing one long time record into a finite number of sections (or subrecords). If the long time record depicts a stationary and ergodic process, increased averaging will insure that the measured Fourier transform will approach the true transform (i.e. increased averaging means less random error, thus less Fourier Transform dependence on T_R). The variables S_{xx} and S_{yy} are the (average) autospectrums obtained from the average Fourier transforms, and "S_{xy}" is the (average) cross-spectrum obtained from the average Fourier transforms.

The spectrums shown in Eqs. 10.52 and 10.53 are referred to as "double sided spectrums". This terminology arises from the fact that a Fourier transform of any time function (or record) will result in an even frequency function i.e., the frequency function is symmetric about zero [Brigham (1974), pg.132]. For many applications, the area under the frequency function is very important. By multiplying the positive frequency portion of the frequency function by two and zeroing out the negative frequency portion, the area is preserved and one is left with a "single sided spectrum". This manipulation is acceptable since the primary quantities of interest are the area and the frequency distribution, neither of which will be affected. The loss of the negative frequencies is inconsequential since they have no physical meaning. In equation form, the single sided spectrum (G) and the double sided spectrum (S) are related as follows:

$$G = 2S, f > 0 \tag{10.58}$$

Most FFT programs calculate only the positive frequency portion of the Fourier transform. They do not calculate the negative portion. The resulting spectra are single sided spectra i.e., twice the positive frequency portion of a double sided spectrum.

10.6.4 Error Analysis

Since discrete frequency increments are used for the Fourier transform instead of the ideal continuous case, negative bias errors will result. This means that all of the measured density spectra will be slightly smaller than the actual spectrum. This problem will always occur for "density" functions, such as those used here [Bendat and Piersol (1980), pg. 41]. The bias error normalized with respect to the quantity being measured associated with any density spectrum is as follows:

$$\varepsilon_b = -\frac{1}{3}\left(\frac{\Delta f}{BW}\right)^2 * 100 \tag{10.59}$$

where ε_b = % maximum normalized bias error, Δf = frequency increment, BW = spectrum half-power bandwidth. The quantity "ε_b" is the "maximum" normalized bias error. This maximum is measured by locating the largest peak in the spectrum and calculating the bias error associated with that peak (i.e. by dividing Δf by the bandwidth). Bias errors less than 2% are considered negligible.

The inability to take infinite time records introduces random errors. The normalized random error formulas are shown in Table 10.1. The coherence function (γ) indicates the amount of nonlinearities, noise and spectral bias errors present in the measurements. A completely noise-free linear system (containing no bias errors in its spectral measurements) will have a coherence of one for all frequencies. A system whose output is completely unrelated to its input will have a coherence of zero for all frequencies. A coherence of 0.6 or above is considered acceptable. It can also be shown that the coherence represents the ratio of the cross-spectral RAO to the autospectral RAO; thus for any system exhibiting a coherence less than one, the autospectral RAO will always be larger than the cross-spectral RAO.

TABLE 10.1
RANDOM ERRORS IN SINGLE INPUT/OUTPUT PROBLEMS

QUANTITY	RANDOM ERROR	QUANTITY DESCRIPTION
$G_{xx}(f), G_{yy}(f)$	$\dfrac{1}{\sqrt{n_d}}$	autospectrum
$\left\|G_{xy}(f)\right\|$	$\dfrac{1}{\gamma_{xy}(f)\sqrt{n_d}}$	cross-spectrum modulus
$\gamma_{xy}(f)$	$\dfrac{\left[1-\gamma^2_{xy}(f)\right]}{\gamma_{xy}(f)\sqrt{2n_d}}$	coherence function
$\|H_c(f)\|$	$\dfrac{\left[1-\gamma^2_{xy}(f)\right]^{\frac{1}{2}}}{\gamma_{xy}(f)\sqrt{2n_d}}$	cross-spectral RAO
$\|H_a(f)\|$	$\dfrac{\left[1-\gamma^2_{xy}(f)\right]^{\frac{1}{2}}}{\gamma^2_{xy}(f)\sqrt{2n_d}}$	autospectral RAO
n_d= the number of averages (i.e., subrecords or sections)		

Assuming that the coherence will be some value less than one, the table shows that the autospectral RAO will always possess more random error than the cross-spectral RAO. The RAO random error formulas also show that as long as the coherence is nonzero and the bias errors are not excessive, the RAO may be estimated to any degree of accuracy one desires simply by increasing the number of averages.

Assuming that the time record being analyzed is of a given duration, increasing the number of averages means decreasing the subrecord length. Whenever the subrecord length is decreased, the frequency resolution becomes coarser (Δf equals the reciprocal of the subrecord length) and this implies an increase in the bias error. It is evident that there are conflicting requirements between bias error and random error considerations when working with a time limited set of data. If it is impossible to increase the duration of the run, one must find the appropriate number of averages that will give a reasonable random error and the appropriate subrecord length to give an acceptable bias error. If the duration of the run is not time limited, the user must first determine the proper subrecord length yielding an acceptable bias error and then multiply that subrecord length by the required number of averages (to reduce the random error) to obtain the necessary run duration. As previously mentioned, the subrecord length equals the reciprocal of the frequency resolution (Δf). To determine the necessary subrecord length, one must estimate the half-power bandwidth of the data being analyzed and then solve the bias error formula for Δf.

10.6.5 Example Problem

If the forcing function is known as a function of time, Equation 10.47 may be numerically solved in time by using a piecewise linear acceleration scheme [Clough and Penzien (1975)]. In the example, it is assumed that a measured irregular wave time trace is the forcing function. The linear acceleration scheme was used to solve for the response, then the response time trace was created. The values used for the example are:

$$K = 0.80 \text{ lb/ft}, M = 0.075 \text{ slugs}, C = 0.0583 \text{ slugs/sec}, f_N = .52 \text{ Hz}$$

The exact transfer function is as follows:

$$H(\omega) = A\left[0.80 - \omega^2(0.075)\right] - Ai[\omega(0.0583)] \tag{10.60}$$

$$A = \frac{1}{\left[0.80 - \omega^2(.075)\right]^2 + \left[\omega(.0583)\right]^2} \tag{10.61}$$

The wave was measured in the wave tank using a 5 hertz low-pass Butterworth filter. A spectrum of the measured wave form (Fig. 10.7) shows the peak at approximately 0.5 Hz. One will note that there is relatively little energy between 2 and 5 HZ. Since the Butterworth filter removes frequencies above 5 Hz, it is safe to say that absolutely no energy exists in the recorded wave data above 7 Hz. Any energy above 7 Hz that is present in the spectrum can only be due to numerical leakage [Brigham (1974), pg.140]. Figures 10.8 and 10.9 show the exact RAO and phase functions for this system. These exact functions are plotted as solid lines on the corresponding function estimate plots. Figure 10.10 present the response function energy density spectrum generated by Eq. 10.61.

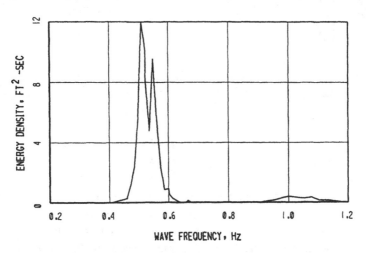

FIGURE 10.7
MEASURED WAVE SPECTRUM

For the case of Hanning smoothing, Figure 10.8 shows that the spectral analysis estimates of the RAO are very close to the true RAO function. Figure 10.9 shows an excellent estimation of the phase function below 5.5 Hz. Figure 10.11 shows that the coherence function has a value very close to unity at frequencies below 5.5 Hz. Above 5.5 Hz the coherence falls far below unity. The coherence function for any system can fall below the value of one due to any of (or combination of) the following reasons:

1. Extraneous noise is present in the measurements.
2. Resolution bias errors are present in the spectral estimates.
3. The system relating y(t) to x(t) is not linear.
4. The output y(t) is due to other inputs besides x(t).

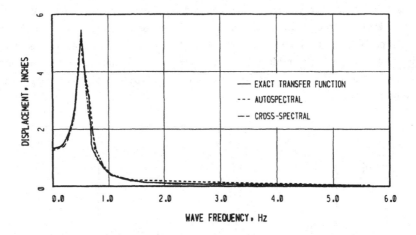

FIGURE 10.8
SYSTEM RAO

Since the output data being analyzed is contrived data, causes 1,3, and 4 will not pose a problem. This leaves cause 2, resolution bias errors, as the only possible reason that could cause the coherence to drop. Large bias errors will occur in spectral estimates at frequencies where the frequency resolution is not fine enough to pick up all of the peaks and troughs of the spectrum. This can typically occur at frequencies where the spectrum is extremely peaked, or jagged. Although the magnitude of the calculated spectra is very small above 5.5 Hz, an expanded scale view of the spectra reveals a very jagged spectra above 5.5 Hz, hence large bias errors are very likely to be present.

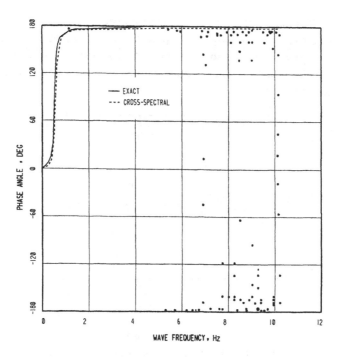

FIGURE 10.9
SYSTEM PHASE ANGLE

Estimates of the frequency response function gain $|H(f)|$ and phase (ε)   are also analyzed in order to determine the existence of random and bias errors utilizing the following guidelines:

- If $\gamma_{xy}(f)$ falls broadly over a frequency range where $|H(f)|$ is relatively
  small, this might indicate measurement noise in output and/or contributions
  of other uncorrelated inputs.
- If $\gamma_{xy}(f)$ falls broadly over a frequency range where $|H(f)|$ is not near a
  minimum value and $G_{xx}(f)$ is relatively small, then measurement noise at the
  input should be suspected.

- If |H(f)| peaks sharply at system resonance frequencies and γ_{xy} does not, then system nonlinearities might be suspected, as well as resolution bias errors in the spectral estimates. To distinguish between resolution problems and system nonlinearities, the spectral estimates may be repeated utilizing a different sampling window.

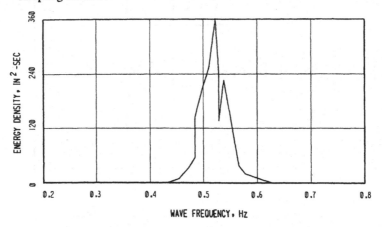

FIGURE 10.10

RESPONSE ENERGY DENSITY FUNCTION

The above example discusses linear responses and methods of computing them for a given wave input. An example of motion response was shown but the method is similar for other linear responses. If a higher order response is expected from a wave input then special care should be taken to account for these higher order responses. The case of second-order response is discussed below.

10.6.6 Nonlinear System

When the higher order responses are expected, such as the low frequency response of a moored floating system, then the second-order spectrum of the wave profile is also matched in the random wave generation. The second-order spectrum is obtained as

$$S^{(2)}(\omega) = 8 \int_0^\infty S_{xx}(\mu) S_{xx}(\omega - \mu) d\mu \tag{10.71}$$

One of the significant problems associated with the determination of the second order transfer function is the length of the record required to obtain a stable estimate of the second-order transfer function. Tests of this nature are typically half an hour long or longer [Pinkster and Wichers (1987)]. For statistical analysis the time history should be considerably longer (> 1 hr). Assuming $H_2(\omega)$ as the quadratic transfer function, the total response spectrum is computed from

$$S_{yy}(\omega) = |H_1(\omega)|^2 S_{xx}(\omega) + 8\int_0^\infty |H_2(\mu)|^2 |H_2(\omega-\mu)|^2 S_{xx}(\mu) S_{xx}(\omega-\mu) d\mu \qquad (10.72)$$

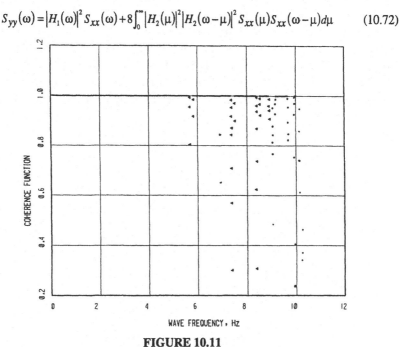

FIGURE 10.11
COHERENCE FUNCTION

10.7 ANALYSIS OF WAVE FORCE COEFFICIENTS

Wave forces on small tubular members of a jacket type structure (as well as other cylindrical structural members) are determined experimentally. The method of testing and some of the results from these tests have been given in Chapter 7.

Typically, the wave profile at the test cylinder and the wave force on a small section of the cylinder are recorded as functions of time. The velocity and acceleration time histories are computed from the wave profile using a particular wave theory (e.g., linear theory or stream function theory). Sometimes, the particle velocities at the test section are directly measured in addition to the wave profile. The acceleration profile is obtained by numerical differentiation of the velocity profile. An example of these measurements is shown in Fig. 10.12. The wave represents a rather long period wave in shallow water. The horizontal and vertical components of the particle velocity are measured at a submerged point at the same location as the wave profile measurement. Note that the horizontal velocity profile is in phase and the vertical velocity profile is out of phase with the wave profile. The profiles contain multiple frequencies. The differentiation and integration of a single frequency and a double frequency profile are shown in Figs. 10.13 and 10.14. The latter is obtained by a Fourier series technique.

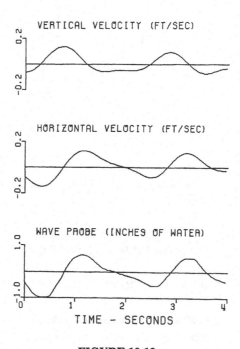

FIGURE 10.12
MEASURED WAVE PROFILE AND WATER PARTICLE VELOCITIES IN TANK

The purpose of measuring forces on structural members in waves is to compute the hydrodynamic coefficients, C_M and C_D, for different wave conditions. The equation for the forces of the type of Eq. 7.1 is applied in this analysis. Notice that for a vertical cylinder, the maximum inertia and drag forces are 90° out of phase of each other. However, calculating C_M when drag is zero and C_D when inertia is zero is generally not adequate. It is more customary to compute the average values of these coefficients over a wave cycle. Thus, it is assumed that the coefficients C_M and C_D are constant throughout a given wave cycle. There are two acceptable methods used in the computation of hydrodynamic coefficients.

10.7.1 Fourier Averaging Method

The Fourier averaging analysis is applied to cases where the wave kinematics are sinusoidal. In cases where the forces are obtained through a simple harmonic motion (whether of the fluid or the cylinder) this method is applicable. When wave profiles are applied on a fixed cylinder and linear theory is used to derive the wave kinematics, the Fourier averaging technique may also be used.

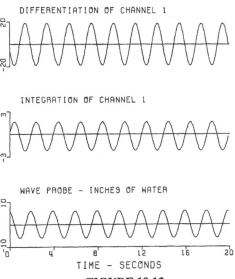

FIGURE 10.13

DIFFERENTIATION AND INTEGRATION OF A WAVE PROFILE

In this method, the force profile is fitted to a Fourier series. The first two terms of the series are fitted to the inertia and drag components of the Morison equation. The coefficients of these terms provide average values of C_M and C_D. The higher order terms in the Fourier series are grouped together to form the error term or the remainder term, ΔR. The necessity for the term ΔR is associated with the fact that the point values of C_M and C_D in a wave cycle deviate from their average values over the cycle. In this respect, the method of analysis applied here is similar to the one adopted by Keulegan and Carpenter (1958) in analyzing data from a cylinder test.

On the assumption that the kinematics are represented by harmonic function, the force time history on a unit length of a cylinder is represented by the Morison equation as follows:

$$f(\theta) = -C_M A_I \dot{u}_0 \sin\theta + C_D A_D u^2_0 |\cos\theta|\cos\theta + \Delta R(\theta) \qquad (10.73)$$

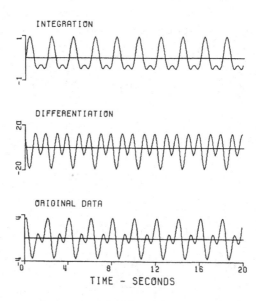

FIGURE 10.14

DIFFERENTIATION AND INTEGRATION BY FOURIER TECHNIQUE

where $\theta = \omega t$ and the kinematics u_0 and $\dot{u}_0$ are the particle velocity and acceleration measured at the center of the instrumented section. The coefficients C_M and C_D are evaluated over one wave cycle from

$$C_M = -\frac{1}{\pi A_I \dot{u}_0} \int_0^{2\pi} f(\theta) \sin\theta \, d\theta \qquad (10.74)$$

and

$$C_D = \frac{3}{8 A_D u_0^2} \int_0^{2\pi} f(\theta) \cos\theta \, d\theta \qquad (10.75)$$

The remainder term is the difference between the measured force and the predicted force based on C_M and C_D and should be expected to be small. Also, by the nature of the computation, ΔR should contain higher harmonics of the force. An example of the comparison between the two forces and the resulting error is shown in Fig. 10.15. The computation has been carried out on the measured forces on a 0.3m (1 ft) submerged section of a vertical cylinder. The remainder term is principally second harmonic, which is the next higher term for a Fourier series for $f(\theta)$.

10.7.2 Least Square Technique

The Fourier series method is not applicable to a case where the kinematics are nonlinear (i.e., non-sinusoidal). This will occur when the nonlinear wave profiles are involved (Fig. 10.12). In this case, a least square method is applicable. This method minimizes the difference between the measured and predicted forces over one wave cycle in a least square sense. Thus, using ΔR as the difference between the two traces having N samples over a wave cycle, we write

$$(\Delta R)^2 = \sum_{n=1}^{N} \left[f_n - C_M T_n^I - C_D T_n^D \right]^2 \qquad (10.76)$$

where T_n^I and T_n^D are quantities in the inertia and drag parts of the Morison equation (minus the coefficient) at every time increment in a cycle ($n = 1, \ldots N$). Note that average values of C_M and C_D over one wave cycle which minimize the left hand side of the equation are sought. Also, if Δt is the sampling rate, then $T = N \Delta t$ where T is the wave period. Since ΔR is minimum for the chosen values of C_M and C_D, we

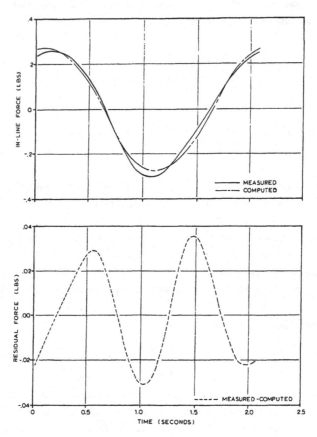

FIGURE 10.15

MEASURED FORCE VERSUS COMPUTED FORCE; RESIDUAL

UNACCOUNTED FORCE

set to   zero the derivatives of ΔR with respect to C_M and C_D respectively. This provides two equations in the two unknowns C_M and C_D. Solving for C_M and C_D from these equations, the following expressions are derived (the summations of the products in these expressions run from 1 to N, i.e. use one cycle of wave data).

$$C_M = \frac{\sum f_n T_n' \sum (T_n^D)^2 - \sum f_n T_n^D \sum T_n' T_n^D}{\sum (T_n')^2 \sum (T_n^D)^2 - \sum (T_n' T_n^D)^2} \tag{10.77}$$

$$C_D = \frac{\sum f_n T_n^D \sum (T_n^D)^2 - \sum f_n T_n' \sum T_n' T_n^D}{\sum (T_n')^2 \sum (T_n^D)^2 - \sum (T_n' T_n^D)^2} \tag{10.78}$$

For a sinusoidal description of the wave kinematics, either of the two methods can be used and should provide similar results.

10.8 FREE VIBRATION TESTS

Any test setup in the wave tank, whether for load tests or motion tests, may be treated as a spring mass system. Therefore, valuable information may be obtained from the free vibration of the system. For fixed structures, the vibration frequency determines if problem will be encountered from the dynamic amplification of the system (Section 7.5.7). For floating structures, information regarding the system natural period and damping may be determined from the vibration analysis. The data analysis is similar in both cases and is described below.

10.8.1 Low Frequency Hydrodynamic Coefficients

The magnitude of damping determines the extent of motions and corresponding mooring loads in a moored floating platform near its natural frequency (refer to Section 9.8). The free oscillation of the platform takes place at the natural frequency. In an experimental setup, this oscillation may be easily measured when the platform is disturbed from its equilibrium position. The platform returns to its equilibrium position and the duration of oscillation depends strictly on the damping of the system.

10.8.1.1 Linear System

The low-frequency hydrodynamic coefficients of the platform in still water can be determined from the recorded extinction curve. The equation of motion is described by a second-order differential equation having a single degree of freedom:

$$(M_0 + M_a)\ddot{x} + C\dot{x} + Kx = 0 \tag{10.79}$$

where x is the surge amplitude and dots represent first and second derivatives, and M_0
and K are the structure displacement (mass) and linear spring constant of the spring
set, respectively.   These quantities are measured directly. The quantities M_a and C are
the added mass and linear damping coefficients, respectively. They are considered to be
functions of the frequency of oscillation, ω_d.   Note that in this case, ω_d is the
damped natural frequency of the system. Values for M_a and C are determined in the
following manner. By assuming a solution to Eq. 10.79 of the form $x = e^{st}$ and defining M
$= M_0 + M_a$, the equation can be rewritten in the form

$$\left(s^2 + \frac{C}{M}s + \frac{K}{M}\right)e^{st} = 0 \tag{10.80}$$

and thus

$$s_{1,2} = -\frac{C}{2M} \pm \sqrt{\left(\frac{C}{2M}\right)^2 - \frac{K}{M}} \tag{10.81}$$

The damping factor, ζ, is defined as the ratio of the amount of damping C
present in the system to that amount of damping, C_c, which will cause the part of the
equation under the radical to go to zero. Therefore,

$$\frac{C_c}{2M} = \sqrt{\frac{K}{M}} = \omega_N \tag{10.82}$$

and

$$\frac{C}{2M} = \zeta\omega_N \tag{10.83}$$

In case of light damping, the radical is imaginary and Eq. 10.81 can be written as

$$s_{1,2} = \left[-\zeta \pm i\sqrt{1-\zeta^2}\right]\omega_N \tag{10.84}$$

The general solution to Eq. 10.79 is

$$x = x_0 \exp(-\zeta \omega_N t) \sin\left[\sqrt{1-\zeta^2}\,\omega_N t + \varepsilon\right] \qquad (10.85)$$

in which x_0 is the magnitude of oscillation at t=0 and ε is its phase angle.

The solution of the equation of motion given by Eq. 10.85 represents harmonic oscillation values of subsequent amplitudes of oscillation in which the amplitudes decay exponentially. If two consecutive absolute values are given by $|x_k|$ and $|x_{k-1}|$, then the logarithmic decrement is defined as

$$\delta = \ln|x_k| - \ln|x_{k-1}| \qquad (10.86)$$

which gives

$$\zeta = \frac{\delta}{\sqrt{\pi^2 + \delta^2}} \qquad (10.87)$$

The logarithmic decrement may be quite accurately related to the damping factor simply by

$$\delta = \pi\zeta \qquad (10.88)$$

For small values of ζ the error is small (for example, for $\zeta = 0.1$, the error is about 0.5 percent).

The term $x_0\exp(-\zeta\omega_N t)$ represents the curve that can be drawn through the succeeding peaks of the damped oscillation. Strictly speaking, the curve does not pass exactly through the peaks, but a small difference is usually neglected. If the natural logarithm of these peaks is taken, the quantity $\zeta\omega_N$ represents the slope, m, of the line that can be drawn through the converted values. The frequency of the damped motion, ω_d, is also obtained from Eq. 10.85, and thus we obtain two equations and two unknowns:

$$m = -\zeta\omega_N \qquad (10.89)$$

$$\omega_d = \omega_N \sqrt{1-\zeta^2} \qquad\qquad (10.90)$$

The terms on the left hand side of Eqs. 10.89 and 10.90 are obtained by fitting exponential curves to the decayed oscillation data (Fig. 10.16). Once the values of ω_N and ζ are known from the above equations, the added mass and damping coefficients are computed:

$$M_a = M - M_0 = \frac{K}{\omega_N^2} - M_0 \qquad\qquad (10.91)$$

and

$$C = 2 M \zeta \omega_N \qquad\qquad (10.92)$$

Therefore, knowing the extinction curve for a moored floating structure, the damping of the system may be established by a simple analysis. This is illustrated by an example based on Fig. 10.16.

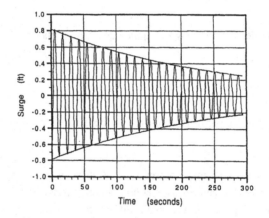

FIGURE 10.16

EXTINCTION TEST OF A TANKER IN SURGE

The extinction curve represents the free oscillation of a moored tanker. The displacement of the tanker is $M_0 = 372.3$ kg (25.5 slugs) and the spring constant K = 102 N/m (7.0 lbs/ft). The least square analysis described above gives an added mass coefficient and a damping factor of 0.049 and 0.008 respectively. The natural period between the positive peaks no. 2 and 3 in Fig. 10.16 is measured as 12.3 sec. Then,

$$M_a = 17.1\text{kg}(1.17\text{slug}) \tag{10.93}$$

and

$$C_a = \frac{M_a}{M_0} = 0.046 \tag{10.94}$$

Also, the amplitudes of peaks 2 and 3 are 0.235 and 0.22 m (0.77 and 0.72 ft) respectively. Therefore,

$$\delta = \frac{1}{2}(\ln 0.235 - \ln 0.22) = 0.03 \tag{10.95}$$

so that

$$\zeta = \frac{0.03}{\sqrt{\pi^2 + 0.03^2}} = 0.01 \tag{10.96}$$

This example shows that two peak values in the extinction curve determine the unknowns. In general, a least square fit of all data peaks in the extinction curve is desirable.

10.8.1.2 Nonlinear System

When nonlinear damping is present, the equation of motion for the damped free oscillation of a moored floating vessel, e.g., a semisubmersible in surge, is given by

$$M\ddot{x} + C\dot{x} + b_2|\dot{x}|\dot{x} + Kx = 0 \tag{10.97}$$

where M = total mass of the vessel in water, and b_2 = nonlinear damping coefficient. Since this equation is nonlinear, it is difficult to solve in a closed form. Therefore, the following simplification is made. On the assumption that each half cycle of the decayed oscillation is reasonably sinusoidal, the nonlinear term is linearized by a

Fourier series expansion as

$$|\dot{x}|\dot{x} = \frac{8}{3\pi}\omega_N x_k \dot{x} \tag{10.98}$$

where ω_N = frequency of oscillation corresponding to the natural frequency of the system, and x_k corresponds to the amplitude of the kth oscillation cycle. Upon substitution of Eq. 10.98 in Eq. 10.97, a linearized equation (with respect to time) is obtained

$$M\ddot{x} + C\dot{x} + \frac{8b_2}{3\pi}\omega_N x_k \dot{x} + Kx = 0 \tag{10.99}$$

Writing

$$C' = C + \frac{8b_2}{3\pi}\omega_N x_k \tag{10.100}$$

Eq. 10.99 becomes the familiar form as in Eq. 10.79 whose solution may be written in the form similar to Eq. 10.85 with ζ replaced by ζ', where ζ' is the damping factor including the linearized term,

$$\zeta' = \frac{C'}{2M\omega_N} \tag{10.101}$$

Then, using Eqs. 10.100, 10.101 and 10.88

$$\ln\frac{x_{k-1}}{x_{k+1}} = \frac{2\pi}{2M\omega_N}\left[C + \frac{8b_2}{3\pi}\omega_N x_k\right] = \frac{T_N}{2}\left[\frac{C}{M} + \frac{b_2}{M}\frac{16}{3T_N}x_k\right] \tag{10.102}$$

where T_N is the natural period of oscillation [Chakrabarti and Cotter (1990)]. In terms of the traditional damping factor, ζ, a more convenient nondimensional form may be written as

$$\frac{1}{2\pi}\ln\frac{x_{k-1}}{x_{k+1}} = \zeta + \frac{4}{3\pi}x_k\frac{b_2}{M} \tag{10.103}$$

Assuming that the nonlinear damping term may be represented by the Morison equation drag term,

$$b_2 = \frac{1}{2} \rho A C_D \tag{10.104}$$

where A = the projected area of the vessel in the direction of flow, Eq. 10.103 reduces to

$$\frac{1}{2\pi} \ln \frac{x_{k-1}}{x_{k+1}} = \zeta + \frac{2}{3\pi} \left(\frac{\rho A C_D}{M} \right) x_k \tag{10.105}$$

which is the equation of a straight line with the left-hand side representing the Y-axis and x_k the X-axis. Thus, knowing the peak values of the oscillation, the points (X,Y) from the measured data may be fitted to a straight line by the least square method (Fig. 9.24). Then, the quantities C_D and ζ may be obtained from the slope and intercept of the fitted line. It should be noted that for sufficient accuracy in these estimates, a reasonable number of peaks are required. However, for a highly damped system, the amplitude reduces to a small value rather quickly and the estimates in these cases are rough. Because waves introduce further damping in the system, the resulting traces may be difficult to analyze by the above method.

The added mass coefficient is computed from the measured natural period, T_N, the spring constant, K, and the displaced mass, M_0, using the formula

$$C_a = \left(\frac{K T_N^2}{4\pi^2} - M_0 \right) / M_0 \tag{10.106}$$

10.8.2 Mechanical Oscillation

Consider the case of a floating structure model which is attached to a mechanical system similar to the ones described in Chapter 4. Consider also that the structure is forced to oscillate sinusoidally in a prescribed direction at a given amplitude and frequency, and the resulting forces are recorded. Assuming linear damping, the equation of motion due to sinusoidal oscillation has the form

$$M\ddot{x} + C\dot{x} + Kx = F_0 \sin \omega t \tag{10.107}$$

where x = prescribed oscillation, and F_0 = exciting force amplitude. On the assumption that the structure oscillation is described by $x = x_0 \sin(\omega t - \varepsilon)$, where ε = phase angle by which the oscillation lags the force, the solution of Eq.10.107 may be written after eliminating time, t, as

$$(K - M\omega^2)\cos\varepsilon + C\omega\sin\varepsilon = \frac{F_0}{x_0} \qquad (10.108)$$

$$(K - M\omega^2)\sin\varepsilon - C\omega\cos\varepsilon = 0 \qquad (10.109)$$

which reduces to

$$(K - M\omega^2)^2 + (C\omega)^2 = \left[\frac{F_0}{x_0}\right]^2 \qquad (10.110)$$

and

$$C\omega = (K - m\omega^2)\tan\varepsilon \qquad (10.111)$$

Equation 10.110 may be equivalently written as:

$$\left(\omega_N^2/\omega^2 - 1\right)^2 + (C/M\omega)^2 = \left(F_0/M\omega^2 x_0\right)^2 \qquad (10.112)$$

Note that x_0, and F_0 are known from the test. From Eqs. 10.108 and 10.109,

$$K - M\omega^2 = \frac{F_0}{x_0}\cos\varepsilon \qquad (10.113)$$

and

$$C\omega = \frac{F_0}{x_0}\sin\varepsilon \qquad (10.114)$$

The added mass coefficient and damping factor may be derived from the above expressions as

$$C_a = \frac{M - M_0}{M_0} \qquad (10.115)$$

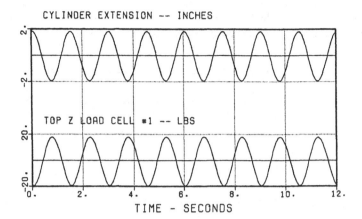

FIGURE 10.17
FILTERED MEASURED DATA ON FORCED OSCILLATION TEST

where

$$M = \frac{K}{\omega^2} - \frac{F_0 \cos\varepsilon}{\omega^2 x_0}$$ (10.116)

and

$$\zeta = \frac{C}{C_c} = \frac{F_0 \sin\varepsilon}{2\omega x_0 \sqrt{MK}}$$ (10.117)

in which M_0 = displaced mass of the structure.

The quantities F_0, x_0 and ε are determined from the test data by a Fourier series analysis. Note that some noise may be present in the data from the vibration of the hydraulic cylinder. These are digitally filtered out by choosing only the first harmonic data through the use of the Fourier series analysis.

While similitude does not strictly exist between forced oscillation and response to waves, it is often a common practice to simulate wave tests with forced oscillation. A

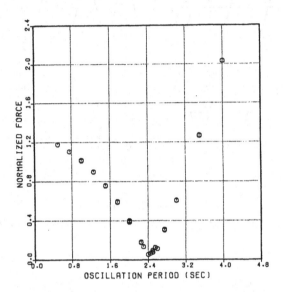

FIGURE 10.18

TRANSFER FUNCTION FOR FORCE

vertical cylinder was tested mechanically in the heave mode. Both the oscillation and force traces were passed through a digital filter in order to maintain the phase and amplitude relationship. A sample trace of measured displacement and load after filtering is shown in Fig. 10.17. The normalized force and the phase angle are plotted in Figs. 10.18 and 10.19 versus the oscillation frequency. The force is normalized by $M_0\omega^2x_0$ (see Eq. 10.112). Once the quantities on the right-hand side of Eqs.10.116 and 10.117 are computed, the added mass coefficient and damping factor are known. These are plotted versus the oscillation frequency in Figs. 10.20 and 10.21. The area of frequency (0.4 to 0.5 Hz), where the phase angle is significantly different from 0 or 180 degrees, is expected to produce the most reliable results in this method. Near 0 and 180, the stiffness or inertia force predominates and the accuracy in the damping coefficient diminishes.

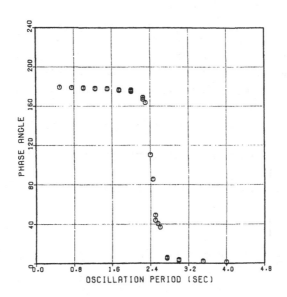

FIGURE 10.19

PHASE ANGLE BETWEEN MEASURED OSCILLATION AND FORCE

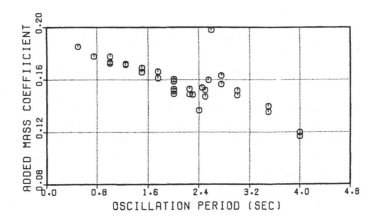

FIGURE 10.20

ADDED MASS COEFFICIENT FROM FORCED OSCILLATION TESTS

10.8.3 Random Decrement Technique

Since amplitudes and frequencies in a random wave are variable, low-frequency hydrodynamic coefficients are also expected to be time dependent. However, for engineering calculation purposes, we can assume the hydrodynamic coefficients to be time invariant and take on their average values. The average values will, however, be wave spectrum dependent.

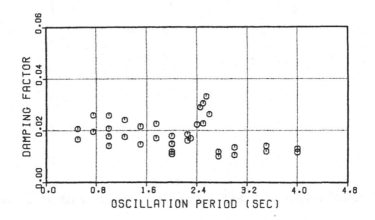

FIGURE 10.21
DAMPING FACTOR FROM FORCED OSCILLATION TESTS

Since the low frequency response is irregular, the standard technique of free oscillation may not be used to determine hydrodynamic coefficients. In this case, they may be determined by the random decrement technique [Yang, et al. (1985)]. The random decrement technique helps us to obtain curves of free extinction from the surge motion time history in irregular waves. Once the curve of free extinction is obtained, the hydrodynamic coefficients can be determined by the same method outlined in the case of regular waves.

The basic concept of the random decrement method is based on the fact that the random motion response of the platform due to irregular waves may be viewed to consist of two parts: 1) the deterministic part (impulse and/or step function) and 2) the random part (assumed to have a stationary average). By averaging enough samples of the same random response, the random part of the response will be averaged out, leaving the deterministic part of the response alone. The deterministic part is the free-decay response from which the added mass and damping may be computed as before.

To obtain this decayed oscillation, the oscillation time history is divided into an ensemble of segments of equal lengths. Each segment begins at the same chosen level and half the segments have positive initial slope while the other half have negative slope. These segments are then ensemble averaged, giving a signature starting at the chosen level. The initial slope is zero because an equal number of positive and negative slopes have been chosen in the ensemble. The constant level is arbitrary and can be chosen as a fraction of the zero-mean rms value of the random time history. What remains on averaging is a decayed oscillation of the signature. The accuracy of the signature depends on the total record length and the number of averages performed.

10.9 REFERENCES

1.   Arhan, M. and Ezraty, R., "Statistical Relations Between Successive Wave Heights," Oceanologica Acta, 1978, Vol. 1, No. 2, pp 151-158.

2.   Barber, N.F., "The Directional Resolving Power of an Array of Wave Detectors," Ocean Wave Spectra, Prentice-Hall, Inc., New Jersey, 1963, pp. 137-150.

3.   Bath, M., Spectral Analysis in Geophysics, Elsevier Scientific Publishing Co., Amsterdam, Holland, 1974.

4.   Bendat, J.S., and Piersol, A.G., Engineering Applications of Correlation and Spectral Analysis, John Wiley and Sons, Inc., New York, New York,1980, pp.41-42, 54-55, 78-81, 264-283.

5.   Bloomfield, P., Fourier Analysis of Time Series: An Introduction, John Wiley and Sons, New York, New York, 1976.

6.   Boas, M.L., Mathematical Methods in the Physical Sciences, John Wiley and Sons, Inc., New York, New York, 1966, pp. 607-611.

7.   Botelho, D.L.R., Finnigan, T.D. and Petrauskas, C., "Model Test Evaluation of a Frequency-Domain Procedure for Extreme Surge Response Prediction of Tension Leg Platforms," Proceedings of Sixteenth Annual Offshore Technology Conference, Houston, Texas, OTC 4658, May 1984, pp. 105-112.

8.   Brigham, E.O., The Fast Fourier Transform, Prentice - Hall, Inc., New Jersey, 1974, pp.58-61, 132-146, 148-169.

9.   Chakrabarti, S.K., and Cotter, D.C., "Damping Coefficient of a Moored Semisubmersible in Waves and Current," Proceedings of the Offshore Mechanics and Arctic Engineering Symposium, Houston, Texas., Vol. 1, Part A, February, 1990, pp. 145-152.

10.   Chakrabarti, S.K., and Snider, R.H., "Design of Wave Staff Arrays for Directional Wave Energy Distribution," Underwater Technology, Vol. 5, October., 1972.

11.   Clough, R.W., and Penzien, J., Dynamics of Structures, McGraw-Hill, New York, New York, 1975, pp. 83-85, 118-128, 482-484.

12.   Cotter, D.C., and Chakrabarti, S.K., "Effect of Current and Waves on the Damping Coefficient of a Moored Tanker," Proceedings on Twenty-First Annual Offshore Technology Conference, Houston, Texas., OTC 6138, May 1989, pp. 149-159.

13.   Dean, R.G., "Methodology for Evaluating Suitability of Wave and Force Data for Determining Drag and Inertia Forces," Proceedings on Behavior of Offshore Structures, Vol. 2, Trondheim, Norway, 1976, pp. 40-64.

14.   IAHR, "List of Sea State Parameters," Joint Publication by the IAHR Section on Maritime Hydraulics and PIANC, Brussels, Belgium, 1986.

15.   Keulegan, G.H., and Carpenter, L.H. "Forces on Cylinders and Plates in an Oscillating Fluid," Journal of the National Bureau of Standards, Vol. 60, No. 5, May, 1958, pp. 423-440.

20. Longuet-Higgins, M.S., "The Statistical Analysis of a Random Moving Surface," Philosophical Trans. Royal Society, London, Vol. 249, Ser. A, 1957, pp 321-387.

21. Longuet-Higgins, M.S., Cartwright, D.E., and Smith, N.D., "Observations of the Directional Spectrum of Sea Waves Using the Motions of a Floating Body," Ocean Wave Spectra, Prentice-Hall Inc., New Jersey, 1963, pp. 111-132.

22. Mansard, E.P.D. and Funke, E.R., "On the Statistical Variability of Wave Parameters," National Research Council, Canada, Technical Report TR-HY-015,1986.

23. Mansard, E.P.D. and Funke, E.R., "On the Fitting of Parametric Models to Measured Wave Spectra," Proceedings of the Second International Symposium on Wave Research and Coastal Engineering, University of Hanover, Mass., 1988.

24. Nwogu, O., "Analysis of Fixed and Floating Structures in Random Multi-directional Waves," Ph.D. Thesis, University of British Columbia, Vancouver, B.C., Canada, 1989.

25. Otnes, R.K., and Enochson, L., Digital Time Series Analysis, John Wiley and Sons, New York, New York, 1972.

26. Otnes, R.K., and Enochson L., Applied Time Series Analysis. Vol. 1 Basic Techniques, John Wiley and Sons, New York, New York, 1978.

27. Pinkster, J.A. and Wichers, J.E.W., "The Statistical Properties of Low-Frequency Motions of Nonlinearly Moored Tankers", Proceedings on the Nineteenth Offshore Technology Conference, Houston, Texas, OTC 5457, 1987, pp.317-331.

28. Read, W.W., "Time Series Analysis of Wave Records and the Search of Wave Groups," Ph.D. Thesis, James Cook University of North Queensland, Australia, 1986.

29.   Remery, G.F.M., "Model Testing for the Design of Offshore Structures," Proceedings of Symposium on Offshore Hydrodynamics, Publication No. 325, N.S.M.B., Wageningen, 1971.

30.   Yang, J.C.S., et al., "Determination of Fluid Damping Using Random Excitation," Journal of Energy Resources Technology, ASME, Vol. 107, June 1985, pp. 220-225.

LIST OF SYMBOLS

a	cylinder radius
a	wave amplitude
a_i	incident wave amplitude
a_r	reflected wave amplitude
A	cross-sectional or surface area
b_0, b_1, b_2	damping coefficent
B	beam of model
c	celerity or wave speed
C	damping coefficient or Chezy coefficient
C_D	drag coefficient
C_f	frictional coefficient
C_L	lift coefficient
C_M	inertia coefficient
C_r	residual coefficient or reflection coefficient at beach
C_R	reflection coefficient at wavemaker
C_t	total resistance coefficient
Cy	Cauchy number
d	water depth
d_{50}	mean material grain size
D	structure diameter
$D(\omega,\theta)$	directional spreading function
E	modulus of elasticity
Eu	Euler number
f	force per unit length or frequency
f_D	drag force for $C_D = 1$
f_f	friction factor
f_I	inertia force for $C_M = 1$
f_L	lift force
F	total force
F_D	drag force
F_e	elastic force
F_G	gravity force
F_H	horizontal component of force
F_I	inertia force
Fr	Froude number
F_v	viscous force or vertical component of force
g	gravitational acceleration
H	wave height
H_0, H_1, H_2	hydrostatic heads
H_s	significant wave height

Iv	Iverson number
k	wave number
ka	diffraction parameter
K	spring constant
l , ℓ	structure dimensional length
L	wave length
M	mass of structure
M_B	bending moment
N	number of frequency components or data points
N_R	grain size Reynolds number
N_S	sediment number
P, p	fluid pressure
P_a, P_o	atmospheric pressure
p (.)	probability function
r, θ	cylindrical polar coordinates
R	hydraulic radius
Re	Reynolds number
R_f	frictional resistance
R_r	residual resistance
R_s	soil resistance force
R_t	total resistance
s	specific gravity or wave spreading index
S ($\bullet$)	spectral energy density
t	time
T	wave period
T_R	length of time series
T_z	zero-crossing period
u	horizontal water particle velocity
u_F	free fall velocity
u_T	turbulent settling velocity
u_*	shear velocity
$\dot{u}$	horizontal water particle acceleration
U	current velocity
U_W	wind velocity
v	fluid velocity
V	volume of structure
w	normal component of water particle velocity
W	(model) weight
W($\bullet$)	filter weight function
x	horizontal (longitudinal) coordinate
y	vertical coordinate (measured from seafloor or SWL as specified)
z	lateral coordinate

β	vertical scale
γ	specific weight or specific heat
Δf	frequency increment
Δt	time increment
Δ_x or Δ_y	water particle displacement in x or y direction
ε	wave component phase angle or pipe submergence ratio
η	wave profile
θ	direction of wave or polar coordinate
λ	scale factor
μ	friction coefficient or dynamic viscosity
ν	kinematic viscosity
ξ	wavemaker displacement or entrainment function
π	dimensionless quantity (or = 3.1416)
ρ	mass density of water
ρ_a	mass density of air
ρ_s	mass density of steel
σ	surface tension
σ_y	standard deviation of y
τ	shear stress or scale factor for time
ϕ	velocity potential or angle of friction
ω	circular wave frequency ($= 2\pi f$)
ω_e	encounter wave frequency in current

Superscripts

$\wedge$	normalized quantity
$-$	nondimensional quantities
$\bullet$	first derivative with respect to time
$\bullet\bullet$	second derivative with respect to time
$*$	quantity in the presence of current

Subscripts

e	equivalent quantity
m	related to model
o,0	refers to amplitude (e.g., u_o, f_o) or deep water (e.g., d_0, L_0)
p	related to prototype
w	related to wind

LIST OF ACRONYMS

CBI	Chicago Bridge & Iron Co. (Plainfield)
CERC	Coastal Engineering Research Center
COV	Coefficient of Variation
DCDT	Direct Current Displacement Transducer
DHI	Danish Hydraulic Institute (Denmark)
DHL	Danish Hydraulic Laboratory (Denmark)
DTRC	David Taylor Research Center
HRS	Hydraulic Research Station (U.K.)
IMD	Institute of Marine Dynamics (Newfoundland)
ITTC	International Towing Tank Conference
KRISO	Korean Research Institute of Ship (Korea)
LED	Light-Emitting Diodes
LVDT	Linear Variable Differential Transformer
MARIN	Maritime Research Institute, Netherlands
MARINTEK	Norwegian Wave Basin Facility, Trondheim
MASK	Maneuvering and Seakeeping Facilities (DTRC)
NEL	National Engineering Laboratory (U.K.)
NIST	National Institute of Standard Testing
NRCC	National Research Council of Canada (Ottawa)
OTEC	Ocean Thermal Energy Conversion
RVDT	Rotary Variable Differential Transformer
SPM	Single Point Mooring
SWL	Still Water Level
TLP	Tension Leg Platform

AUTHOR INDEX

Aage, C., 135
Aas, B., 177, 178, 185
Abel-Aziz, H.S., 205-207, 230
Abramson, H.N., 283, 301
Abuelnaga, A., 205-207, 230
Aguilar, J., 188
Allender, J.H., 239, 301
Anastasiou, K., 186
Anderson, C.H., 166, 184
Arhan, M., 457
Barber, N.F., 418, 457
Bath, M., 457
Battjes, J.A., 152, 184
Bazergui, A., 217, 230
Beach Erosion Board
Bendat, J.S., 430-432, 457
Bergman, J., 170, 185
Berkley, W.B., 29, 37
Bhattacharyya, R., 63, 74
Bhattacharyya, S.K., 333, 336-337, 353
Biesel, F., 75, 78, 135, 136
Bishop, J.P., 199, 230
Bloomfield, P., 457
Boas, M.L., 429, 457
Borgman, L.E., 140, 149, 184, 185
Bothelho, D.L.R., 423-425, 458
Bowers, C.E., 138
Brabrook, M.G., 137
Brady, I., 395, 402
Brevik, I., 177, 178, 185
Brewer, A.J., 136
Bridgeman, P.W., 38
Brigham, E.O., 432, 435, 458
Brogren, E.E., 260, 301, 303, 405
Brown, D., 225-226, 231
Buckingham, E., 38
Bullock, G.N., 77, 136
Burns, G.E., 342, 346, 353
Burrows, R., 186
Calisal, S.M., 130, 136, 391, 402
Carpenter, L.H., 442, 458
Carstens, M.R., 291, 295, 301
Carter, D.J.T., 189

Chakrabarti,S.K., 31, 38, 141, 144, 150-
    152, 172, 185, 223, 232, 233,
    238, 240, 250, 259, 260, 262-
    264, 266, 277, 301-303, 365,
    370, 391, 402-403, 405, 418-
    419, 450, 458
Challenor, P.G., 189
Chamberlin, R.S., 342, 353
Chang, P.A., 196, 211, 231
Chantrel, J., 371, 403
Cartwright, D.E., 459
Chen, D.T., 186
Chen, M-C., 136, 186
Chen, Y., 136
Clauss,G., 118, 136, 170, 185, 268, 302,
    347-50, 352-353
Clifford, W.R.H., 336, 353
Clough, R.W., 430, 434, 458
Cornett, A., 161, 163, 164, 166, 185
Cotter, D.C., 233, 238, 302, 365, 403,
    450, 458
Dahle, L.A., 403
Datta, I., 115, 138
Davis, M.C., 167, 185
Dawson, T.H., 277, 279, 280, 282, 302
Dean, R.G., 138, 234, 239, 302-303, 458
DeBoom, W.C., 378-379, 403-405
Diez, J.J., 140, 188
Dunlop, W.A., 303
Eatock Taylor, R., 277, 303
Eggestad, I., 135, 136
Elgar, S., 152-154, 185
Enochson, L., 459
Eryzlu, N.E., 230
Ezraty, R. 457
Faltinsen, O.M., 403
Feldhausen, P.H., 185
Finnigan, T.D., 458
Forristall, G.Z., 163, 185
Frederiksen, E., 186
Funke, E.R., 76, 123, 126, 137-139,
    141, 143-146, 186-188, 410,
    412, 459
Gabriel, D., 186

SUBJECT INDEX